MICROBIAL INOCULANTS

Developments in Applied Microbiology and Biotechnology

MICROBIAL INOCULANTS

Recent Progress and Applications

Edited by

VIJAY KUMAR SHARMA
Agriculture Research Organization, Volcani Center, Rishon LeTsiyon, Israel

AJAY KUMAR
Centre for Advanced Study in Botany, Institute of Science, Banaras Hindu University, Varanasi, India

MICHEL RODRIGO ZAMBRANO PASSARINI
Laboratory of Environmental Biotechnology, Federal University of Latin American Integration, Latin American Institute of Life Sciences and Nature, Foz do Iguaçu, Brazil

SHOBHIKA PARMAR
Department of Environmental Science, College of Basic Sciences and Humanities, G. B. Pant University of Agriculture and Technology, Pantnagar, India

VIPIN KUMAR SINGH
Assistant Professor, Department of Botany, K.S. Saket P.G. College, Ayodhya, Uttar Pradesh, India

ELSEVIER

ACADEMIC PRESS
An imprint of Elsevier

Academic Press is an imprint of Elsevier
125 London Wall, London EC2Y 5AS, United Kingdom
525 B Street, Suite 1650, San Diego, CA 92101, United States
50 Hampshire Street, 5th Floor, Cambridge, MA 02139, United States
The Boulevard, Langford Lane, Kidlington, Oxford OX5 1GB, United Kingdom

ISBN 978-0-323-99043-1

For information on all Academic Press publications
visit our website at https://www.elsevier.com/books-and-journals

Publisher: Stacy Masucci
Acquisitions Editor: Kattie Washington
Editorial Project Manager: Tim Eslava
Production Project Manager: Selvaraj Raviraj
Cover Designer: Matthew Limbert

Typeset by STRAIVE, India

Contents

Contributors

Deepali Ashok
Department of Botany, Government Degree College Drang at Narla, Mandi, Himachal Pradesh, India

Subhi Avasthi
School of Studies in Botany, Jiwaji University, Gwalior, Madhya Pradesh, India

Ambreen Bano
IIRC-3, Plant-Microbe Interaction and Molecular Immunology Laboratory, Department of Biosciences, Faculty of Sciences, Integral University, Lucknow, UP, India

Maria Gabriela Azevedo Barros
Laboratory of Environmental Biotechnology, Federal University of Latin American Integration, Latin American Institute of Life Sciences and Nature, Foz do Iguaçu, Brazil

Ashwini Benjarwad
Department of Rasashastra and Bhaishajya Kalpana, SDM College of Ayurveda and Hospital, Hassan, Karnataka, India

Suzan Prado Bernandes Bernal
Laboratory of Environmental Biotechnology, Federal University of Latin American Integration, Latin American Institute of Life Sciences and Nature, Foz do Iguaçu, Brazil

Nikunj Bhardwaj
Department of Zoology, Maa Shakumbhari University, Saharanpur, India

Yogita Bohra
Department of Plant Pathology, Punjab Agricultural University, Ludhiana, Punjab, India

Debolina Chakroborty
Guru Nanak Institute of Pharmaceutical Science and Technology, Maulana Abul Kalam Azad University of Technology, Kolkata, West Bengal, India

Sourav Chattaraj
Department of Microbiology, Raiganj University, Raiganj, West Bengal, India

Cintia Caroline Gouveia da Silva
Federal Rural University of Pernambuco, Recife, Pernambuco, Brazil

Pradeep K. Das Mohapatra
Department of Microbiology, Raiganj University, Raiganj, West Bengal, India

Victor Lucas Vieira Prudêncio de Araújo
Luiz de Queiroz College of Agriculture, University of São Paulo, Piracicaba, São Paulo, Brazil

Emerson Rodrigues de Camargo
Interdisciplinary Laboratory of Electrochemistry and Ceramics, Department of Chemistry, Federal University of São Carlos, São Carlos, Brazil

Sergio de los Santos-Villalobos
Instituto Tecnológico de Sonora, Ciudad Obregón, Sonora, México

Amanda Carolina Prado de Moraes
Laboratory of Microbiology and Biomolecules—LaMiB, Department of Morphology and Pathology, Center for Biological and Health Sciences; Biotechnology Graduation Program, Federal University of São Carlos, São Carlos, SP, Brazil

Edivan Rodrigues de Souza
Federal Rural University of Pernambuco, Recife, Pernambuco, Brazil

Suzan Pantaroto de Vasconcellos
Laboratory of Health and Environment—LABMSMA, Department of Pharmaceutical Sciences, Federal University of São Paulo (UNIFESP), Diadema, SP, Brazil

Maddalena Del Gallo
Department of Life, Health and Environmental Sciences, University of L'Aquila, L'Aquila, Italy

Bruna Martins Dellagnezze
Division of Microbial Resources, Research Center for Chemistry, Biology and Agriculture (CPQBA), Campinas University, Campinas, Brazil

Rihab Djebaili
Department of Life, Health and Environmental Sciences, University of L'Aquila, L'Aquila, Italy

Rosane dos Santos Grignet
Laboratory of Environmental Biotechnology, Federal University of Latin American Integration, Latin American Institute of Life Sciences and Nature, Foz do Iguaçu, Brazil

Lidiane Maria dos Santos Lima
Laboratory of Health and Environment—LABMSMA, Department of Pharmaceutical Sciences, Federal University of São Paulo (UNIFESP), Diadema, SP, Brazil

Ravikant Dubey
Department of Humanistic Studies, Indian Institute of Technology (Banaras Hindu University), Varanasi, Uttar Pradesh, India

Amrita Dwivedi
Department of Humanistic Studies, Indian Institute of Technology (Banaras Hindu University), Varanasi, Uttar Pradesh, India

Felipe José Cury Fracetto
Federal Rural University of Pernambuco, Recife, Pernambuco, Brazil

Giselle Gomes Monteiro Fracetto
Federal Rural University of Pernambuco, Recife, Pernambuco, Brazil

Ajay Kumar Gautam
Patanjali Herbal Research Department, Patanjali Research Institute, Haridwar, Uttarakhand, India

Caroline Da Costa Silva Gonçalves
Laboratory of Environmental Biotechnology, Federal University of Latin American Integration, Latin American Institute of Life Sciences and Nature, Foz do Iguaçu, Brazil

B.E. Guerra Sierra
Universidad de Santander, Facultad de Ciencias Naturales, Grupo Microbiota, Bucaramanga, Colombia

Mohammad Haris
Plant Pathology and Nematology Section, Department of Botany, Aligarh Muslim University, Aligarh, India

Touseef Hussain
Plant Pathology and Nematology Section, Department of Botany, Aligarh Muslim University, Aligarh; Division of Plant Pathology, ICAR-Indian Agricultural Research Institute, New Delhi, India

C. Jayabaskaran
Department of Biochemistry, Indian Institute of Science, Bengaluru, Karnataka, India

N. Jennifer Michellin Kiruba
Faculty of Chemistry, Department of Engineering and Technology of Chemical Processes, Wroclaw University of Science and Technology, Wrocław, Poland

Prabhjot Kaur
Department of Nephrology, Postgraduate Institute of Medical Education and Research (PGIMER), Chandigarh, Punjab, India

Abrar Ahmad Khan
Plant Pathology and Nematology Section, Department of Botany, Aligarh Muslim University, Aligarh, India

Amir Khan
Plant Pathology and Nematology Section, Department of Botany, Aligarh Muslim University, Aligarh, India

Bahman Khoshru
Department of Soil Science, Faculty of Agriculture, University of Tabriz, Tabriz, Iran

Sachin Kumar
Department of Botany and Microbiology, Gurukula Kangri (Deemed to be University), Haridwar, Uttarakhand, India

Sandeep Kumar
Department of Botany and Microbiology, Gurukula Kangri (Deemed to be University), Haridwar, Uttarakhand, India

Paulo Teixeira Lacava
Laboratory of Microbiology and Biomolecules—LaMiB, Department of Morphology and Pathology, Center for Biological and Health Sciences; Biotechnology Graduation Program, Federal University of São Carlos, São Carlos, SP, Brazil

Midouin Lidelias
Federal Rural University of Pernambuco, Recife, Pernambuco, Brazil

Paulo Henrique Gomes Lisboa
Laboratory of Microbiology and Biomolecules—LaMiB, Department of Morphology and Pathology, Center for Biological and Health Sciences; Biotechnology Graduation Program, Federal University of São Carlos, São Carlos, SP, Brazil

H.N. Madhushree
Department of Biochemistry, Indian Institute of Science, Bengaluru, Karnataka, India

Manju
Department of Life-Sciences, RIMT University, Mandi Gobindgarh, Punjab, India

José Petrônio Mendes Júnior
Federal Rural University of Pernambuco, Recife, Pernambuco, Brazil

Debasis Mitra
Department of Microbiology, Raiganj University, Raiganj, West Bengal, India

Rittick Mondal
Department of Sericulture, Raiganj University, Raiganj, West Bengal, India

Navneet
Department of Botany and Microbiology, Gurukula Kangri (Deemed to be University), Haridwar, Uttarakhand, India

Shilpa Nayuni
SPMVV-Women Biotech Incubation Facility, Sri Padmavati Mahila Visvavidyalayam, Tirupati, Andra Pradesh, India

Adeyemi Nurudeen Olatunbosun
Department of Plant Physiology and Crop Production, Federal University of Agriculture, Abeokuta, Nigeria

Júlia Ronzella Ottoni
Laboratory of Environmental Biotechnology, Federal University of Latin American Integration, Latin American Institute of Life Sciences and Nature, Foz do Iguaçu, Brazil

Akanksha Pandey
Department of Botany and Microbiology, Gurukula Kangri (Deemed to be University), Haridwar, Uttarakhand, India

Chitranshu Pandey
Department of Biotechnology, MRD Life Science, Lucknow, India

Periyasamy Panneerselvam
Crop Production Division, ICAR-National Rice Research Institute, Cuttack, Odisha, India

R. Parthasarathy
Department of Biochemistry, Indian Institute of Science, Bengaluru, Karnataka, India

Michel Rodrigo Zambrano Passarini
Laboratory of Environmental Biotechnology, Federal University of Latin American Integration, Latin American Institute of Life Sciences and Nature, Foz do Iguaçu, Brazil

Marika Pellegrini
Department of Life, Health and Environmental Sciences, University of L'Aquila, L'Aquila, Italy

Ankita Priyadarshini
Crop Production Division, ICAR-National Rice Research Institute, Cuttack, Odisha, India

Minakshi Rajput
Department of Biotechnology, School of Applied and Life Sciences, Uttaranchal University, Dehradun, India

Deepak Rathore
Department of Humanistic Studies, Indian Institute of Technology (Banaras Hindu University), Varanasi, Uttar Pradesh, India

Lucas da Silva Ribeiro
Interdisciplinary Laboratory of Electrochemistry and Ceramics, Department of Chemistry, Federal University of São Carlos, São Carlos, Brazil

Mariana Rocha Roswell
Laboratory of Health and Environment—LABMSMA, Department of Pharmaceutical Sciences, Federal University of São Paulo (UNIFESP), Diadema, SP, Brazil

Agnieszka Saeid
Faculty of Chemistry, Department of Engineering and Technology of Chemical Processes, Wroclaw University of Science and Technology, Wrocław, Poland

Rafaella Costa Bonugli Santos
Federal University of Latin American Integration (UNILA), Institute Latin American of Nature and Life Sciences (ILACNV), Interdisciplinary Center of Life Sciences (CICV), Foz do Iguaçu, PR, Brazil

Ansuman Senapati
Crop Production Division, ICAR-National Rice Research Institute, Cuttack, Odisha, India

Amita Sharma
Department of Botany, Kurukshetra University, Thanesar, Haryana, India

Pooja Sharma
NUS Environmental Research Institute, National University of Singapore, Singapore

Salman Siddeeqh
Department of Maxillofacial Surgery and Diagnostic Sciences, College of Dentistry, King Saud Bin Abdulaziz University for Health Sciences, Riyadh, Kingdom of Saudi Arabia

Isabel Natalia Sierra-Garcia
Department of Biology and CESAM, University of Aveiro, Aveiro, Portugal

Monika Singh
Department of Biotechnology, School of Applied and Life Sciences, Uttaranchal University, Dehradun, India

Surendra Pratap Singh
Plant Molecular Biology Laboratory, Department of Botany, Dayanand Anglo-Vedic (PG) College, Chhatrapati Shahu Ji Maharaj University, Kanpur, India

Adriana Heiss Siqueira
Federal University of Latin American Integration (UNILA), Institute Latin American of Nature and Life Sciences (ILACNV), Interdisciplinary Center of Life Sciences (CICV), Foz do Iguaçu, PR, Brazil

D. Sruthi
Department of Biochemistry, Indian Institute of Science, Bengaluru, Karnataka, India

Sushma
Department of Biosciences, Chandigarh University, Gharuan, Punjab, India

Atirah Tauseef
Plant Pathology and Nematology Section, Department of Botany, Aligarh Muslim University, Aligarh, India

Yen Wah Tong
NUS Environmental Research Institute; Department of Chemical and Biomolecular Engineering, National University of Singapore, Singapore

Rajnish Kumar Verma
Department of Plant Pathology, Punjab Agricultural University, Ludhiana, Punjab, India

H.C. Yashavantha Rao
Department of Biochemistry, Indian Institute of Science, Bengaluru, Karnataka, India

Yasheshwar
Department of Botany, Acharya Narendra Dev College, Govindpuri, Kalka ji, University of Delhi, New Delhi, India

Preface

In the recent past, beneficial microorganisms were sustainably used in agriculture as a safe, economical, and better alternative to chemical fertilizers or pesticides. These beneficial microbes, including bacteria, actinomycetes, and yeast, were efficiently applied in the soil, seeds, fruits, or plants as inoculants to achieve the optimum agricultural yield. Application of microbial inoculum improves the overall performance of the plants by maintaining the bioavailability of nutrients in the soil, endogenous secretion of growth hormones, checking various infections related to pests and/or phytopathogens during pre- or postharvest periods, as well as mitigating abiotic stress conditions such as salinity, drought, and heavy metals. Besides crop improvement, microbial inoculants have also played an effective role in fermentation technology and tissue culture.

However, an efficient delivery method or enhanced shelf life of microbial inoculants in the soil or the seed is still a lagging area. The response of local genetic or ecological factors after microbial applications is still less known and less studied. Therefore, there is a need to explore and learn about an efficient delivery mechanism, the selection of microbial strains as inoculants, and related technological advances for the efficient and productive use of microbial inoculants. Moreover, factors like methods of formulation, the interaction between host plant and microbe, the impact of inoculation on the metabolomics of plants, the effect of microbial inoculants on soil dynamics, the proteomics approach of plant-microbes interactions as well as registration and the regulation process of bioinoculants for commercial production also need to be explored for the better suitability and effectivity of inoculants.

Microbial Inoculants: Recent Progress and Applications is an attempt to offer in-depth information of latest approaches and advancements in the field of microbial inoculants. The book comprises 16 quality chapters on the different and latest aspects of microbial inoculants from leading academicians and researchers. Chapter 1 discusses recent progress in formulations and application methods of microbial inoculants. Chapter 2 presents a detailed account of the application of microbial inoculants as an alternative to chemical products for the decomposition of organic wastes. The chapter also provides insights into the benefits of microbe-mediated compost, agriculture waste biodecomposition, applications in long-term crop production, and different forms of biofertilizers like N-fixing biofertilizers, mycorrhizal fungi. Chapter 3 describes the prospects of the microbial inoculants in salinity management. Chapter 4 discusses the potential of metal-tolerant microbial inoculants to improve phytoextraction. Chapter 5 presents insights into seed priming with microbial inoculants for enhanced crop yield. Chapter 6 provides a comprehensive account of organic waste decomposition using microbial inoculants, which can be an effective tool for environmental management. Chapter 7 describes the use of microbial inoculants for the management of herbicide toxicity in plants. Chapter 8 explains the prospect of immobilizing microbial inoculants for improving soil nutrient bioavailability. Chapter 9 discusses the plant growth-promoting microbes and nanoparticles and their biotechnological potential in agrobiological systems. Chapter 10 describes the potential of microbial inoculants for the rejuvenation of agricultural soils contaminated by recalcitrant compounds. Chapter 11 discusses tropical biomes as microbial sources for efficient biocatalysts for environmental purposes. Chapter 12 presents the current perspectives of microbial biofilms and applications in bioremediation, such as the decontamination of polycyclic aromatic hydrocarbons, petroleum, chlorinated aromatic compounds, heavy metals, and radionuclides. Chapter 13 provides insights into microbial enzymes and their role in bioremediation. Chapter 14 describes applications of microbial formulations in the pharmaceutical industry. Chapter 15 presents a brief account of an evaluation of the lacunae in current techniques using microbial inoculants for enhanced bioremediation and nutrient recovery. Chapter 16 explains the potential of fungal endophytes as a source for novel secondary metabolites and their beneficial aspects.

This book is the collaborative work of several individuals. The views expressed by the authors are their own. We appreciate and thank all the experts for readily contributing to various aspects of microbial inoculants presented in the book. We are hopeful that this book will be useful to academics, researchers, postgraduate students, and practitioners in the area of microbial technology.

Vijay Kumar Sharma
Ajay Kumar
Michel Rodrigo Zambrano Passarini
Shobhika Parmar
Vipin Kumar Singh

CHAPTER 1

Microbial inoculants: Recent progress in formulations and methods of application

Pooja Sharma[a,*], Ambreen Bano[b,*], Surendra Pratap Singh[c], and Yen Wah Tong[a,d]
[a]NUS Environmental Research Institute, National University of Singapore, Singapore
[b]IIRC-3, Plant-Microbe Interaction and Molecular Immunology Laboratory, Department of Biosciences, Faculty of Sciences, Integral University, Lucknow, UP, India
[c]Plant Molecular Biology Laboratory, Department of Botany, Dayanand Anglo-Vedic (PG) College, Chhatrapati Shahu Ji Maharaj University, Kanpur, India
[d]Department of Chemical and Biomolecular Engineering, National University of Singapore, Singapore

1.1 Introduction

Plant development is influenced by a wide range of microorganisms in the soil, including mycorrhizal fungi, plant growth-promoting rhizobacteria (PGPR), rhizobia, root endophytic, and phosphate solubilizers, through plant-mediated and direct processes, including during stressful times (Shilev et al., 2019; Berg, 2009; Swarnalakshmi et al., 2020). Specific beneficial plant microorganisms use either a separate or a multifunctional microbial consortium, which is essential for improving crop yield and health (Maron et al., 2018; Ahmad et al., 2018). The reviewed literature is replete with investigations on isolation and the characterization of beneficial plant microbes, but not many have made it to a commercial scale. Various bioinoculants at the commercial level do not operate as well in the field as they do in a greenhouse or laboratory studies (Malusà et al., 2016), attributed to insufficient or low-quality formulations, having lower carrier stability and affinity (Bashan et al., 2016; Baez-Rogelio et al., 2017; Bhattacharyya and Jha, 2012; Stamenković et al., 2018).

Key functions of inoculant formulations are:

- In order to avoid a sudden decline in cell viability throughout the depository, better, more stable microenvironments must be provided to inoculated microbial strains.

* Equal contribution.

Microbial Inoculants
https://doi.org/10.1016/B978-0-323-99043-1.00017-7

- To maintain strain competitiveness with the better-adapted microflora of soil.
- The purpose is to reduce the loss caused by macrofauna attack following soil introduction.

These activities are designed to provide a consistent flow of live cells for interaction with soil and plant microbiota (Bashan et al., 2014; Herrmann and Lesueur, 2013). Microbial cultures that have been treated with oil, water, or additives (i.e., polymers) to increase dispersion capacity stability and cell and suspension viscosity are known as liquid inoculants (Bashan et al., 2016; Catroux et al., 2001; Malusà et al., 2016). Issues with these products include whether the metabolic activity and microbial population drop rapidly following the cell suspension's introduction into the soil, especially if it does not contain appropriate additives.

In recent years, there has been a focus on cell-free formulations like fermentation broth filtrate (Kumar et al., 2012; Bashan et al., 2016; Vassilev et al., 2017; Vassileva et al., 2021). Because certain beneficial plant bacteria perform numerous functions, their culture extracts have a variety of metabolic products, including siderophores, lytic enzymes, toxins, and antibiotics, as well as phosphate solubilizers (Mendes et al., 2015; Vassilev et al., 2017). This has a favorable impact on plant growth. These types of products and related strategies can be called postbiotic. The solid formulations are based on organic or inorganic carriers, are granular, powdery, or solid, and are categorized depending on the size of particles or application method (Malusá et al., 2012; Adholeya and Das, 2012; Stamenković et al., 2018). Peat, agroindustrial wastes, compost, perlite, vermiculite, calcium sulfate, polysaccharides, and rock phosphate are the most common carriers used in solid formulations (Sahu and Brahmaprakash, 2016). Polysaccharide-immobilized inoculants have received increased interest in the solid formulation technologies field in recent years (Malusà et al., 2016), as well as inoculants created utilizing agroindustrial wastes in solid-state fermentation (SSF) conditions (Vassilev and de Oliveira Mendes, 2018). SSF procedures provide several benefits, such as induction of biocontrol activity, enrichment with soluble phosphate, co-cultivation of two microorganisms (Mendes et al., 2015), and solid substrates used separately or in combination, and dampened with liquid waste (Vassilev et al., 2009; Vassilev and de Oliveira Mendes, 2018).

Moreover, the gel-cell immobilized technique is a technical option that can ensure the formulated inoculum's uniformity and purity. This chapter summarizes the immobilized-cell approach and some precise qualities of the carrier formulation and structure, such as the role of supplementary compounds implemented into the cell gel structures and the impact of gel-forming polysaccharides and their derivates on plant health as well as growth. Table 1 shows various PGPR formulations (liquid and solid wet).

1.2 Inoculant characterization

A project must consider both the manufacturer's and the growers' interests, which are complementary, in order to design a proper inoculant. In actuality, and above all, farmers are always looking for the highest possible output. The key practical characteristics of inoculants that growers expect are consistent with ordinary field activities, for example, an infection of seed and the pesticide's widespread use. Secondarily, significant inoculant qualities are:

(1) During seeding, compatibility with seeding equipment is essential. Compatibility with the seeding equipment during seeding.

(2) Convenience of use.

(3) Capacity to assist the extended survival of injected bacteria for the period required by the plant.

(4) Adaptability to diverse field conditions and soil types.

(5) Tolerance of abuse during storage.

Manufacturers' additional needs include:

(6) Shelf-life that extends beyond a season,

(7) Field repeatable findings, and

(8) Removing hazardous materials used for the safety of plant, animal, and human. The major parameters to consider for a good formulation are shown in Fig. 1.

1.2.1 Physical and chemical characteristics

An inoculant's carrier should: be contaminant-free and virtually or inexpensively sterilized; be as homogeneous physically and chemically as is feasible; exhibit constant quality of batch; exhibit, for wet carriers, a high water-holding capacity; and be appropriate for as many bacterial strains and species of plant growth-promoting bacteria (PGPB) as is feasible. For all carriers, raw material consistency is a fundamental need. This is because carriers are an important component for inoculant synthesis. If such inoculants are modified, the established quality control procedure may not be adjusted for each batch of raw materials during manufacturing of industrial inoculants (Stephens and Rask, 2000).

Table 1 Plant growth-promoting rhizobacteria (PGPR) formulations (liquid and solid wet).

Microorganism	Type of formulation	Carrier, additive, or protector	Shelf life	Viable cells initial number	Crop assayed	References
Pseudomonas fluorescens AMB-8	Liquid formulations	Glycerol or polyvinyl pyrrolidone and coconut water	At room temperature for 6 months (10^7 CFU/mL)	10^9 CFU/mL	*Capsicum* and *Solanum lycopersicum*	Anith et al. (2017)
Bradyrhizobium sp. J-81		Arabic gum	At room temperature for 6 months (10^8 CFU/mL)	10^{10} CFU/mL	*Arachis hypogaea*	Valetti et al. (2016)
Rhodopseudomonas palustris PS3		Horticultural oil	At 4°C for 1 month (10^8 CFU/mL)	10^9 CFU/mL	*Brassica rapa* Pekinensis	Lee et al. (2016)
Bacillus siamensis SCFB3-1		Carrageenan	–	10^9 CFU/mL	*Capsicum annuum*	Pastor-Bueis et al. (2017)
Azospirillum brasilense Ab-V5		Biopolymers that are stress-resistant (polyhydroxybutirate and exopolysaccharides) formed by *A. brasilense* Ab-V5	At 20°C for 1 month	10^9 CFU/mL	*Zea mays*	Oliveira et al. (2017)
Kosakonia radicincitans DSM16656T		AviVital product carried out by ABiTEP GmbH, Berlin, Germany	at 4°C for 6 months	Approx 10^{11} CFU/g	*Zea mays*	Berger et al. (2018)
Paraburkholderia tropica MTo-293		Gelatin or xanthan gum, Polyvinyl alcohol	unspecified storage temperature for 12 months	10^{10} CFU/mL	*Triticum aestivum*	Bernabeu et al. (2018)
K. pseudosacchari TL13		Tetrasodium pyrophosphate (16% w/v)	–	10^6 bacterial cells/g of vermiculite	*Zea mays*	Romano et al. (2020)

Herbaspirillum seropedicae BR11417 (ZAE94)		Peat-based, PVP (polyvinylpyrrolidone)–PV, Xanthan gum–XI, Xanthan gum and PVP–XP	10^7 CFU/mL	N/A	*Zea mays*	Scheidt et al. (2020)
Bacillus licheniformis BDS31 and *Pseudomonas azotoformans* C2-114		Granule+M3: 10% starch, 10% PVP, 59% mannitol, 6% sodium alginate, 10% water, and 5% bacterial cells	1 month	10^{10} CFU/mL	*Dipterocarpus alatus*	Riddech et al. (2021)
A. brasilense AbV5/Biofilm formation inside the pores of the support	Solid wet formulations	Sugarcane bagasse, cassava starch, powdered skim rock phosphate, glycerol, crystal sugar milk, yeast extract, and phosphate buffered saline	At room temperature for 4 months	10^6 CFU/g	*Zea mays*	Marcelino et al. (2016)
Rhizobia meliloti		Ammonium molybdate with peat	–	10^9 CFU/g	*Medicago sativa*	Zhou et al. (2017)
Rhizophagus irregularis MUCL 41833 in alginate beads and *pseudomonas plecoglossicida* R67094		Alginate	–	–	*Solanum lycopersicum*	Loján et al. (2017)
Entrapment in alginate beads *Pseudomonas putida* A (ATCC 12633)		Perlite, alginate, paraffin	At 4°C or room temperature for 5 months 10^8 CFU/g)	10^8 CFU/g	*Arabidopsis*	Liffourrena and Lucchesi (2018)

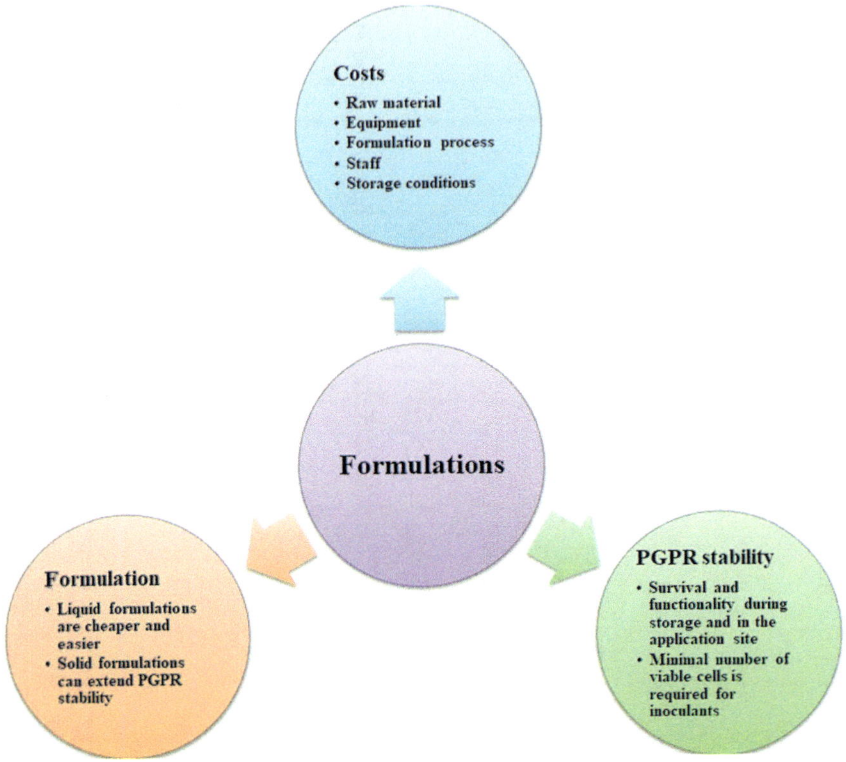

Fig. 1 Parameters to consider while creating an inoculant formulation.

1.2.2 Farm handling

Inoculants that are effective are easy to handle (which is important for growers), enable quick and regulated bacteria release in the soil, and can be applied with regular seeding tools. This is critical since farm practices seldom alter to assist a technology that produces a high-quality inoculant using a special apparatus, particularly in traditional agricultural communities (Date, 2001).

1.2.3 Manufacturing

The microbial fermentation industries must be able to quickly make and combine inoculants. They should be able to accept nutritional additions, have a readily adjusted pH, and should use low-cost raw material with sufficient quantity available and accessible (Catroux et al., 2001).

1.2.4 Environmental friendliness

Concerning the current environmental use of compounds that alter the properties of soil, the inoculants should be biodegradable, nonpolluting, nontoxic, and have no carbon footprint. The application should limit environmental concerns, such as PGPB cell dispersal into groundwater or the atmosphere.

1.2.5 Long shelf life

These inoculants should have a long enough storage capacity. In developed countries, storage of 2 years approximately at room temperature is frequently required for proper amalgamation into the agricultural distribution chain (Deaker et al., 2004; Bharti and Suryavanshi, 2021; Dey, 2021).

1.3 Inoculation techniques

1.3.1 Culture of microbes

A culture medium is available for every PGPB/PGPR/rhizobia. The majority of these are designed for the microorganism's laboratory handling. They do not produce enough mass to be used in inoculant manufacturing. Numerous media include refined chemicals, the costs of which are too great for microorganism mass production (Fig. 2). However, other media formulations, including rhizobia (Singleton et al., 2002) and *Azospirillum* (Bashan, 2011), were created expressly for large generation of an inoculum. Many approaches for low-cost mass growth of microorganisms include the utilization of food processing by-products, including maize steep liquor, cheese whey, malt bagasse, and sprouts, which may serve as adequate carbon sources. However, the quality of raw materials used in ordinary industrial manufacturing is doubtful and warrants more investigation (Stephens and Rask, 2000). The fermentation process is the most straightforward aspect of microbial culture. A specific protocol for each strain's manufacture must be devised; the fermentation industries can handle any specialized strains that the inoculant industries may require.

1.3.2 Life-cycle assessment

Liquid inoculant generated by fermenters, in the field as well as applied directly, is scarce, with just a handful available for turfgrass for golf courses as well as limited hydroponic culture. Inoculants consisting of peats or other inorganic and organic ingredients are often required for common agricultural usage, as is a storage interval between preparation and use. With any

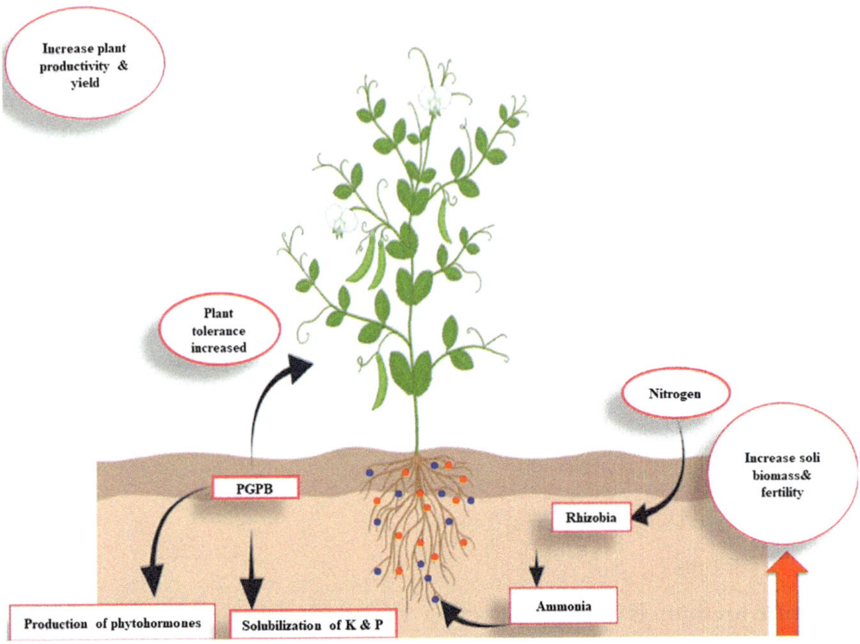

Fig. 2 The main advantages of PGPB inoculation and rhizobia. The following processes are represented: biological nitrogen fixation, solubilization of potassium (K) and phosphate (P), synthesis of phytohormone, along with induced plant systemic tolerance to biotic and abiotic stresses. Benefits include increased biomass yield, output, and soil fertility enhancement.

formulation, significant difficulty is encountered in extending the inoculant's shelf life while keeping its biological properties. Currently, one of the major solutions for this basic issue of increasing survival time is to decrease the formulation moisture content in the air-dried product, by drying in a lyophilized (freeze-dried) formulation, or fluidized bed, or storing at cold temperature. Bacteria remain dormant in fully dry formulations. Their metabolism is sluggish or even stopped. They are resistant to environmental challenges, not sensitive to contamination, and highly compatible with fertilizer application.

When using a wet formulation, for example, peat, a 40%–50% moisture level was shown to be ideal for the development and viability of the rhizobia variety (Deaker et al., 2004). As a result, a realistic approach, for example, to enhance rhizobia seed survival is to have a short 15-day curing period at 25°C or a prolonged period up to 120 days, whichever is best. This treatment promotes rhizobia adaptation to the carrier as well as increases drying tolerance (Albareda et al., 2008).

1.3.3 Sterilization

Carrier material sterilization is required to maintain a high number of desirable microorganisms in the final formulation throughout lengthy storage periods. Contaminants are well known to have a negative impact on the inoculant's shelf life. Despite this, various carriers, particularly in underdeveloped nations, are not sterile due to financial constraints. The less expensive nonsterile inoculants have a significantly higher market share and may have better sales potential in less industrialized areas. *Bradyrhizobium* spp., a slow-growing PGPB, requires sterilization to grow and survive. As a result, these PGPB will be able to outcompete faster-developing pollutants (Deaker et al., 2004). The present tendency is to create inoculants free from contaminants, and quite a number (mostly for rhizobia) are available on the market. Because the sterilization procedure has essentially little effect on the chemical and physical characteristics of the substance, irradiation is the best approach to sterilize any carrier. In practice, the material of the carrier is packed in thin polyethylene bags before being irradiated. After irradiation, various inoculants were examined and found to be free of contamination (Albareda et al., 2008). The primary drawbacks of nuclear sterilization are its high cost as well as delay. Aside from these process costs, other drawbacks involve the need to find a sterilization unit with sufficient capacity and fast control, and skilled personnel for inoculating the beneficial PGPB into the carrier under sterile conditions, which adds to the costs. Autoclaving is a viable option for carrier sterilization since autoclaves of various sizes and shapes are widely available and the operation may be performed by inexperienced persons. Carrier material is autoclaved for 60 min at 121°C in partially open, thin-walled polypropylene bags. A better alternative, although costlier, is tyndallization. In this type of sterilization, three independent sterilization procedures are carried out, each daily incubating at optimal bacterial growth conditions (25–30°C) for 24 h in between. This allows contaminated microbe spores to develop, and the vegetative cells were killed during the autoclaving process. While autoclaving works well for liquid inoculants (mostly chemicals and by-products of the other industries described earlier), the drawbacks of autoclaving organic materials are significant. Because of the enormous volume, the method is arduous, there is significant time and energy cost, and, most critically, some organic materials alter their chemical characteristics and may create substances poisonous to some bacteria during autoclaving. Nonetheless, autoclaving is advantageous for many rhizobia strains, which survive significantly better in sterile peat inoculants than in nonsterile circumstances (Temprano et al., 2002).

1.4 Formulation technologies

Microbial products, like ordinary agrochemicals, are available as liquids, solids, or slurries. Powders and granules are two types of solid formulations based on particle size. They are commonly used as soil supplements or seed coverings (Bashan et al., 2014). Several alternative choices have been studied in addition to peats as a conventional carrier for dry formulations (Fig. 3) (Bashan et al., 2014; Malusá et al., 2012). Organic carriers (soy/oat bran/wheat, sawdust, sewage sludge, vermicompost, grape bagasse, cork compost, animal manure), soil-derived carriers (turf, clays, charcoal), and inert materials (vermiculite, perlite, kaolin, silicates, bentonite, polymers, talc) are examples of these. Inoculant cell encapsulation in polymers has been suggested as a strategy to assure controlled release in the soil in the latter case (Bashan, 1986). Encapsulation technologies have improved significantly in recent years and are used to create microbial inoculants with varied shapes

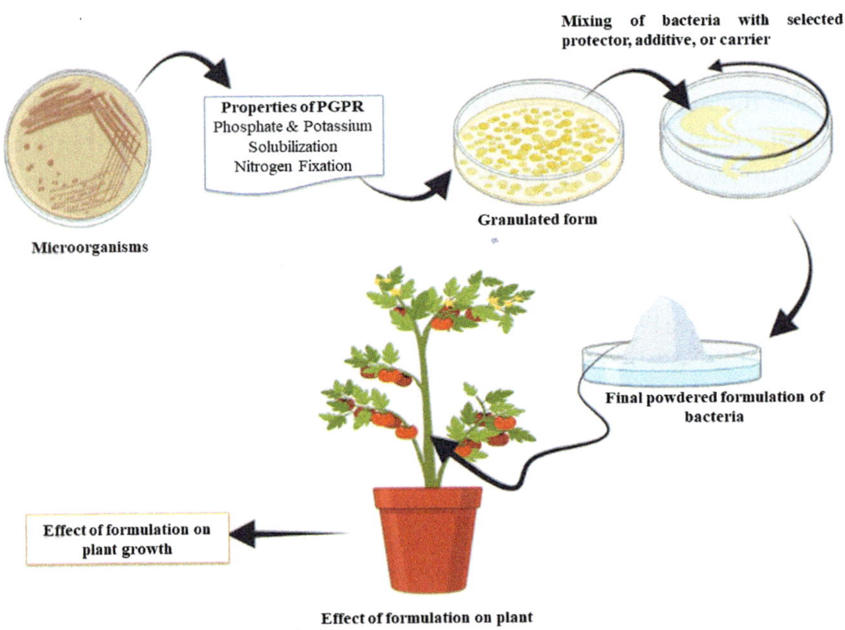

Fig. 3 The production of formulation from rhizobacteria in the granular form and its mixing with selected protector, additive or with the carriers for final powdered formulation that help in plant growth as well as development under various stress (biotic and abiotic).

as well as content (John et al., 2011; Schoebitz et al., 2013). Eventually, lyophilized pure culture can be utilized directly or in conjunction with a solid carrier (Malusá et al., 2012).

Furthermore, liquid formulations using water or oil-based suspension of cell concentrate, slurries, or emulsions having solid particles have been created (Malusá et al., 2012; Bashan et al., 2014). Liquids can be used in a variety of application technologies. These are similar to solid formulations, and can be applied directly to the seed (Bashan et al., 2014). Liquid can be applied during sowing in soil or subsequently using fertigation systems (Malusá et al., 2012). The latter method is especially useful for inoculating perennial crops. Thus favorable microbes must be inculcated into previously established plantations or orchards (Malusá et al., 2012). Moreover, liquid allows for above-ground plant treatment, such as foliar sprays (Jambhulkar et al., 2016). This may be beneficial if the active agents colonize plant aerial parts, as already reported with *Paraburkholderia phytofirmans* endophytic bacteria (Mitter et al., 2017). This colonization strategy of phytopathogens, as well as the location of biocontrol strain actions, may necessitate their treatment above ground, such as when suppressing fire blight in the Rosaceae plant (Pusey, 2002). Thus the formulation technology chosen is determined by the available equipment, farmer convenience, technique of application, the absence or presence of additional treatments, site of action, plant inherent characteristics cost, plant developmental stage, and inoculant colonization pathways (Deaker et al., 2004; Malusá et al., 2012; Bashan et al., 2014).

1.4.1 Freeze-drying

By exposing samples to higher vacuum conditions, freeze-drying, or lyophilization, involves two steps: prefreezing and water sublimation. Sublimation defines the transition of samples from a solid to a vaporous form, which is affected by the surrounding vacuum and temperature. Below a particular threshold that depends on the sample composition, a phase transition from solid directly into a vaporous state happens without going through a liquid phase. The lack of melting makes the procedure gentler and aids in the preservation of product properties. The prefreezing pressure, the temperature in the drying chamber, end-point of drying, amount of sample, input temperature, and instrument qualities all have an effect on the ultimate result of the sublimation process. This means that, while tailoring a lyophilization strategy for specific samples, there are unlimited permutations of process parameters to consider (Morgan et al., 2006). Nonetheless, lyophilization is a commonly used procedure in laboratory bacteria formulations. It has been tested for the conservation of several *Pseudomonas* spp. strains (Stephan et al., 2016), *Pseudomonas fluorescens* (Cabrefiga et al., 2014; Bisutti et al., 2015), and *Pantoea agglomerans* (Costa et al., 2000, 2002b; Soto-Muñoz et al., 2015). When no preventive measures were adopted, the viability of *Pseudomonas* spp. exposed to lyophilization was often decreased by one to two orders of magnitude (Stephan et al., 2016).

1.4.2 Spray-drying

Spray-drying involves atomizing a liquid matrix followed by hot airflow, resulting in rapid evaporation of water from the sample unless dry particles are formed (Morgan et al., 2006). About 20% of manufacturing expenses are projected to be due to lyophilization costs, making this approach highly economically viable (Santivarangkna et al., 2007). Unfortunately, because of rapid moisture elimination and the high temperature, the viability of cells is a frequent severe concern (John et al., 2011). When drying microorganisms, an inlet temperature of 100–200°C and outlet temperature of 60–85°C are generally used (Fu and Chen, 2011). One of the most important parameters for cell survival appears to be the outlet temperature. However, it cannot be directly modified and is reliant on other factors such as inlet temperature, airflow rate, solids content, and liquid feed rate (Fu and Chen, 2011). Because of this interdependence, intensive spray-drying process fine-tuning is frequently required. Spray-drying has been shown to be effective for strong, spore-forming microorganisms, for example, *Bacillus subtilis* (Yánez-Mendizábal et al., 2012). Spray-drying of sensitive bacteria, for example, *Pseudomonas agglomerans*, on the other hand, results in a drop in cell viability ranging from two to five orders of magnitude at 90°C outlet temperature, based on the carrier (Costa et al., 2002a,b). Lowering the exit temperature can boost the survival rate, but doing so may result in undesirable higher content of moisture and product clumping. Researchers recommend that bigger outlets allow for lower outlet temperatures, improving bacterial survivability without compromising physical product quality. After optimizing this technique for *Sinorhizobium meliloti* cells, a final number of cells of roughly 5×10^9 CFU/g was obtained with 42°C outlet temperature, yielding 11% final moisture content (Rouissi et al., 2013).

1.4.3 Fluidized bed-drying

This is another promising technology that runs at around 40°C temperatures, making it efficiently gentler than spray-drying (Krishnaiah et al., 2014). This is often used as a secondary drying process that reduces the remaining solid granules or particle moisture content (Broeckx et al., 2016). In an upward blowing stream of hot or warm air, the particles are suspended for this purpose, imparting fluid-like behavior to the majority of granules. The bacterial agent can be dispersed over the moving mass of the carrier, integrated into carriers before drying, while during alginate encapsulation, or delivered as a pure dry mass that is subsequently coated with a protecting shell in a fluidized bed (Broeckx et al., 2016). Despite the fact that it is a relatively low-cost drying technology, research on this is restricted, probably because of the infeasibility of having slurry-like or liquid original matrices. A *Pseudomonas fluorescens* fluidized bed-dried formulation was produced using a combination of vermiculite. When three drying procedures were compared with regard to the appropriateness of preserving *Pseudomonas agglomerans* cell viability, spray-drying had the biggest reduction in CFU by four orders of magnitude, fluidized bed-drying had two orders of magnitude, and freeze-drying had one order of magnitude (Soto-Muñoz et al., 2015).

1.4.4 Air-drying

Every drying procedure is connected with a distinct stress regime and may thus be somewhat suited for a specific strain of bacteria. As a result, desiccation occurs in the field after an inoculant is applied under uncontrolled conditions. Evaluating the viability of formulations during air-drying is thus extremely important. It has been noted that protectants that work well for one drying process may not work well for another, as demonstrated when designing a dry formulation for *Pseudomonas phytofirmans* (Berninger et al., 2017). In this scenario, yeast extract as well as gum arabic had higher survival rates after lyophilization but not during air-drying, whereas mannitol had the reverse effect. It appears to be extremely useful to explore drying under simulated circumstances that are as near to the ones that occur in the field as is feasible, as has been done in the *Pseudomonas fluorescens* strain case. When sprayed over rosaceous plant flowers, its survivability was explored as an antagonist to reduce fire blight (Bonaterra et al., 2007; Cabrefiga et al., 2011). There are few studies on employing this for inoculum generation, despite the fact that it best correlates to the drying regime that occurs under natural conditions. Schisler et al. (2016) produced this formulation for several *Pseudomonas fluorescens* strains and found extremely high viability rates in osmoprotectant, although solid carrier drying resulted in a three-orders-of-magnitude or more decrease of cell viability. Berninger et al. (2017) observed related results in *Pseudomonas phytofirmans* air-drying in a zeolite matrix, where survival was less than with osmoprotectants alone. As a result, rather than focusing on single parameters, the interaction between drying and matrix technique and the subsequent influence on ultimate moisture content, drying kinetics, and water activity should be considered. Finally, the drying process used is determined by the matrix composition, strain sensitivity, desired output form, and practical concerns, for example, the minimum size of the sample and the potential of the gnotobiotic operation.

1.5 Formulations of inoculants

The performance of an inoculant is frequently the most prevalent hurdle to be an inoculant. A microbial strain may work perfectly in a research facility under precise laboratory conditions and the supervision of skilled personnel. However, transforming this microorganism into inexpensive products for use by microbiologically untrained farmers, while related results are expected under field conditions, is difficult (Stephens and Rask, 2000). Compared to agricultural chemicals, formulations must be stable during manufacture, storage, distribution, and delivery to farmers. An inoculant must operate adequately in soil types wherever it is applied.

1.5.1 Organic inoculants

Without question, peat is the preferred transporter of rhizobia across Australia, Europe, and South and North America, and it is the primary element in commercially available inoculants. It is compatible with the majority of PGPRs. Peat, on the other hand, is scarce in some regions. All different PGPB-rhizobia carriers proposed are compared to the typical peat carrier. The limitations and performance and the specifics of the manufacture of numerous versions have been rigorously and continually scrutinized. Presently, technical aspects of fundamental peat-based inoculant manufacture, including pH, grain size, appropriate moisture, various additives, quality control standards, inoculant quality, and occupational safety and health, are widely known (Catroux et al., 2001; Date, 2001; Stephens and Rask, 2000; Deaker et al., 2004; Xavier et al., 2004). The peat formulation's physical nature might affect inoculation success. Population dynamics of

Rhizobium leguminosarum bv. viciae in rhizospheres of peas in a temperate zone infected by rhizobia revealed a similar pattern among solid formulations: an initial reduction enduring 2 to 5 days, followed by an increase lasting more than 2 months. In contrast, during spring, when soil moisture was high due to heavy rains, the soil population of similar strains originating in liquid formulations decreased to almost zero (Hynes et al., 2001). Granular inoculants applied to soil enhanced seed output as compared to seed-applied inoculants (Clayton et al., 2004b). In comparison to seed-applied inoculation, inoculants applied to soil improved nitrogen nutrition in field pea under field conditions. The rankings, from first to last, of the effects on nodule number, nitrogen build-up, and N_2 fixation are: peat powder, granular peat, liquid, and uninoculated (Clayton et al., 2004a). In Australia, however, comprehensive field testing revealed that granules composed of either peat or clay were not that much more efficient than peat slurry inoculants, but when application of inoculants on seeds is difficult, such as when insecticides and fungicides must be administered, the granules have a potential niche.

Lignite, composts, charcoal, and coir dust of different compositions and origins, filtered sugarcane mud bagasse, soil combined with other vermiculites, and organic amendment were popular replacements for peat inoculants in the 1980s and 1990s. Most were thought to be inferior to peat as a carrier (Bashan, 1998; Singleton et al., 2002). However, numerous organic inoculants created from various waste products have been tried with remarkable success in recent years, mostly in impoverished countries. A few were tested on larger scales, whereas others reported inoculant production but not in situ evaluation. Sawdust was first composted by inoculating it with the *Cephalosporium* sp. cellulose-decomposing fungus and diazotroph *Azospirillum brasilense*. An increase in yield was seen with groundnut, soybean, lucerne, and a grass mixture of ryegrass and bird's foot trefoil after crop inoculation with the carrier containing the *Bradyrhizobium-Rhizobium-Azospirillum* mixture, when compared to the culture medium without any additional formulations and uninoculated plants. Coir dust-coco peats have a high moisture-holding capacity and show considerable promise as PGPB carriers (Prabhu and Thomas, 2002). Where common peat is less available due to geographical and economic factors, coco peats may be part of the solution. Coco peats that are nonsterile can be utilized as low-cost farm-level mass multiplication and for field application of nitrogen-fixing cyanobacteria, due to its high water-holding capacity. Composts created from cork waste outperformed peat in terms of preserving the survival of various PGPB and rhizobia. This inoculant exhibited soybean yields that were comparable to those produced by peat-based inoculants (Albareda et al., 2008). As inoculants for *Pseudomonas fluorescens* and *R. leguminosarum*, five unsterile carriers, namely charcoal, alginate beads, sugarcane bagasse, sawdust, and sand, were compared. When employed alone and in combination, sawdust turned out to be the most effective carrier for bacterial population maintenance. The co-inoculant outperformed the control in terms of increasing seedling nodule and biomass number in *Trifolium repense* (Arora et al., 2008). *Azotobacter vinelandii* was cultivated in solid-state fermentation on lignin generated from kraft pulping operations. The impacts of lignin-to-corn-straw ratio, starting water content, and material bed depth substantially impacted bacteria development. Plants were not used in the testing of these two mixtures. Plant growth-promoting yeasts were cultivated on damp soybean brans and combined with soil before being seeded with disinfected seeds of sugar beet to exhibit growth enhancement (El-Tarabily, 2004). Similarly, to strongly control the fungus *Sclerotinia minor* in beans, actinomycetes and chitinolytic bacteria were cultured using wheat bran (substrate) that also worked as a carrier by combining with soil. Industrial as well as municipal wastewater sludge might be utilized as the only raw material to support rhizobia development. Rhizobia, both quick as well as slow growing, thrived in sterile sludge. The sludge content influenced the cell yield, generation time, and nodulation index. Alkaline, acid, and oxidative pretreatments enhanced the biodegradability of primary sludge, As a result, the cell counts of *S. meliloti* were increased. Barley fiber evaluation fractions on growing medium and transport material for natural bacterial fungicide culture *Streptomyces grisoviridis* demonstrated that brewer's waste grain was the optimum choice when price, storage period, nutritional composition, and simplicity of processing of the final and crude product were all considered. A large quantity of water-insoluble fiber affects the application and appearance of the product favorably (Tuomi et al., 2001). Torrefied grass fibers were tested as a carrier, with *Coniochaeta ligniaria* fungus inoculated with different inoculants of PGPB for tomato plant growth. The PGPB colonization of this substrate together with fungi created a disease-suppressing substrate that replaced up to 50% of the peat in potting soil as well as boosted growth of tomato plants (Trifonova et al., 2009). A comparison was made of the use of banana waste with chicken manure as an inoculant consortium carrier for *Azotobacter*, phosphate solubilizing bacteria, and *Azospirillum* strains to increase the performance of banana plants, and the microbiological and physical properties were measured. Despite the fact that both inoculants boosted soil qualities and plant growth, banana waste had a greater impact.

1.5.2 Inorganic inoculants

Natural inorganic minerals, synthetic materials, and natural polymers can all be used to make inorganic inoculants. Aside from polymeric inoculants, which are still experimental, inorganic inoculants are the most traditional type of inoculant (Bashan, 1998). The majority of these inoculants are utilized for crop production on a limited scale.

1.5.3 Soil inoculants

There have been few scientific publications on soil-made inoculants in recent years. Large-scale cyanobacterial inoculants were grown in open air utilizing soils from their habitat and nonhabitat as carriers. Cyanobacteria (nitrogen fixation) complement commercial nitrogen fertilizers for rice growth in Bangladesh and India on a sporadic basis. Field investigations revealed that cyanobacterial inoculants resolve issues in saline and acid soils, increase fertility, and supply nitrogen (25%–35%) required for rice growth in these soils (Hashem, 2001). In Vietnam, local PGPB strains were mixed with an unsterile carrier made of rice husks, broth culture, clay soil, water, and sugar. These bacterial cultures are blended with land soil right before rice seedling inoculation during transplantation, leading to higher rice yield (Nguyen et al., 2003). Using a clay soil inoculant mix containing sulfur powder, sulfur-oxidizing bacteria inoculation, and rhizobia resulted in synergistic interactions that increased groundnut yield as well as oil content in soils deficient in sulfur (Anandham et al., 2007).

Only a few additional inorganic materials are employed as inoculants. Talc-based formulations are a type of inoculant that is exclusively found in India. Its effectiveness against pests, illnesses, and nematodes has been established in numerous horticultural and agricultural crops grown in shadehouses and in the field. Despite the fact that this formulation appears to have short storage, significant contamination, and poor field performance, it is nevertheless employed experimentally (Bharathi et al., 2004).

Two popular plant production substrates, vermiculite, and perlite, were investigated as inoculant components. Perlite outperformed peat in terms of preserving the survival of various PGPB and rhizobia and promoting soybean growth (Albareda et al., 2008). When it comes to maximizing the usage of a seed formulation, inoculum preparation, and formulation drying are critical. *Pseudomonas fluorescens* biocontrol bacteria survival was investigated using a 1:1 pelleting material EB-vermiculite mix, nutrients, and sugar beet seed. This increased the strain's survivability in the formulations and on the seeds during storage. Slower drying formulations improved the strain's survivability throughout storage. Partial biological control of *Pythium* spp. was achieved (~10% more healthy plants) (Moënne-Loccoz et al., 1999).

1.5.4 Liquid inoculants

To solve some of the constraints described previously, liquid inoculants are an improved derivation of no-formulation inoculants. They are essentially microbial suspensions or cultures that have been modified with chemicals that may increase adhesion, stability, surfactant, and dispersion characteristics (Singleton et al., 2002). The primary benefit of liquid inoculants over solid inoculants is their convenience of handling. Liquid formulations, as opposed to solid carrier-based inoculants, permit the producer to incorporate appropriate cell protectants, nutrient amounts, and inducers responsible for cyst/spore/cell production to improve certain characteristics. While standard solid carrier-based inoculants have a storage life of 6 months, a liquid formulation has a shelf life of up to 2 years. They have a few main drawbacks: (1) Some products have a limited shelf life, (2) freezing or cool conditions are essential for extended shelf life, and (3) higher prices limit their usage to developed countries and preclude use in the majority of underdeveloped countries (Stephens and Rask, 2000). Because of their large cell numbers, they became popular for legume inoculation in industrialized nations. Liquid inoculant with a concentration of 2×10^9 cells/mL is the most common, allowing for increased inoculant efficiency and lower application rates (Schulz and Thelen, 2008). Furthermore, it is stated that these inoculants are contamination-free, have more extended storage for some formulations, are more resistant to environmental challenges, and have higher field efficacy than peat-based inoculants (Singleton et al., 2002). They are compatible with machinery used on large farms, including seed augers and air seeders. Finally, these are advantageous for small-scale inoculant makers in the tropics that cannot handle carriers such as peat.

1.5.5 Primitive inoculants

Interestingly, the oldest technique of inoculating plants and seeds using a bacterial culture solution is still in use today. It is generally used in research facilities (Bashan et al., 2014). Various research studies have demonstrated that, under close supervision (most likely not relevant to commercial producers), good positive inoculation outcomes may be obtained utilizing PGPB without any formulation. Clearly, despite the enormous number of articles confirming researchers' recurrent success, "no formulation" inoculants are not a feasible inoculation method for farmers. This is mostly due to the fact that the level of knowledge required for appropriate distribution of the bacteria under ideal circumstances is frequently beyond the farmers' skill and time constraints. It is further impracticable for wide-scale use due to the large volume of liquid involved and the refrigeration and incubation facilities required to maintain culture media with no formulation, making it doubtful for fulfilling commercial and economic demands. This technique will mostly be used for bench and greenhouse preliminary testing at research centers.

1.6 Role of formulation in microbial community

In many circumstances, results acquired in the laboratory or greenhouse are challenging to translate to the field, especially when working with nonsporulating, gram-negative bacteria. They are more vulnerable to harmful effects that occur during processing and field application since they do not generate spores. As a result, they need appropriate, protective formulations to improve their efficacy at the target site and to make practical usage by farmers easier. Numerous researchers have written in-depth evaluations of formulation development (Catroux et al., 2001; Malusá et al., 2012; Herrmann and Lesueur, 2013; Bashan et al., 2014). However, one of the biggest barriers to the effective and widespread use of microbial inoculants is the shortage of suitable formulations and consequent low inoculant quality (Stephens and Rask, 2000). A high number of viable cells sent to the plant is required to achieve a suitable colonization rate, which subsequently improves the intended impact in the field. Inoculant viability may be compromised at several stages before and during application. First, the product must have suitably extended storage, which characterizes the stability of the product during the manufacturing process, packing, storage, and transit conditions (Arora et al., 2011). Through further field application, inoculants are exposed to additional conditions that are adverse to their viability. UV radiation, especially when applied to above-ground plant portions, variable soil parameters including texture, pH, temperature (Arora et al., 2011), along with recurrent drying-rewetting cycles based on precipitation frequency are some examples. The intrinsic seed coat toxicity of inoculants applied directly to seeds can be hazardous (Deaker et al., 2012). Moreover, biotic interaction with natural macrofauna and microflora provides a significant barrier to any strains used. Cell counts of an imported strain frequently decrease after application to nonsterile soil because they are outcompeted by indigenous bacteria or reduced by predators like protozoa (Bashan, 1998; Arora et al., 2011). Apparently, preapplication stress factors occurring throughout the manufacturing process are worsening the problem; fewer viable cells transported to the field have a less successful establishment at the target sites. As a result, a gentle formulation procedure is critical.

1.7 Conclusions and future prospects

Technological advancements in inoculation have been hindered by the slow development of novel and effective formulations for rhizobia and PGPB inoculation. It is difficult to create novel formulations in practical microbiology. Formulation advancements are critical to the creation of improved high-end inoculants. According to literature searches, identifying novel isolates with PGPB capacity is not always difficult, and several are discovered each year. However, the majority of PGPB strains halt there, never making it to the formulation step.

The following topics should be the main priorities for researching new or better rhizobia and PGPB delivery methods:

- Thorough evaluation of carriers. Every now and then, a new formulation utilizing an already known carrier, for example, liquid inoculant having an adhesive or alginate, is promoted as a solution for all problems associated with earlier carriers. However, analyses of the benefits along with drawbacks of every formulation are rarely conducted. As dozens of carrier materials have been presented in recent decades of inoculation procedures, honest and fair appraisals are needed of the most frequent formulations, polymerics, organic wastes, and liquids, with the exception of peat formulations that have already hit their peak.
- Enhancement of a formulation that demonstrates favorable field outcomes by fine-tuning key constituents in the formulations or enhancements in the manufacturing process. Since it is assumed that the industry evaluates, precise details are not available.
- Boosting the microorganism's survival in the inoculants. Lowering the decrease of injected bacteria population throughout formulation and administration should be a significant focus of study. If the mortality rate can be greatly lowered, the cell numbers applied per seed might be increased by two to three orders of magnitude. These include the cell's physiological age, water activity, and relative humidity throughout storage, which varies by species. The lower the water activity, the higher the

chance of survival. This is due to the fact that water activity is a critical aspect in survival, especially in hot environments. The rehydration rate should be improved as well. Slow-drying fabrics are typically better than fast-drying fabrics (Date, 2001).

- Longevity. This is a critical issue since application time in the field is less, but large-quantity manufacturing in factories takes time. The existing shelf life is rather short. Two years is said to be the ideal amount of time. Nonetheless, different experimental formulations, like those utilizing alginates, demonstrate that bacteria may survive for several years without losing potency.
- Benefits of multistrain inoculants. Many types of research have demonstrated the benefits of various combinations, which are normally done in the laboratory without any formulation. While the interactions of the partners within these formulations are mostly unknown, the formulation of numerous microorganisms does not create major technological problems compared to the formation of one bacterium. Explorations into (1) similarity between microbial populations, (2) interaction or symbiosis with plants, (3) effectiveness leading to plant growth-promoting effects, (4) growth rates, (5) potential formation of biofilms, and (6) technical glitches in microorganism culturing in fermenters, in which everyone has various nutritional demands, or other in-vitro growth conditions, required for the development of such inoculants. Because multistrain applications represent the field's current frontier, the effective formulation should accommodate various combinations. However, research on formulations with additives has been conducted and is essentially empirical. Their manner of operation is barely understood. This aspect of the formulation process is new and should be given more attention.
- Inoculants made of polymers. Despite several indicators that this may be the inoculant's future, no formulation of rhizobia or PGPB has cleared the industrial approval threshold, as has happened in other sectors. Bioencapsulation is a rapidly growing developing subject in pharmaceuticals, nanotechnology, aquaculture, medicine, and cosmetics. Many effective encapsulation techniques have been developed for this purpose (Schoebitz et al., 2013). Except for the direct polymerization of a few polymers, hardly any were studied in the inoculant sector. Most of these developing innovations from other industries are worthy of consideration in the agricultural inoculant industry. Farming techniques are unlikely to alter considerably, even to accommodate a technology that produces high-quality inoculants. As a result, the objective should be

to develop formulations that are more than just farmer-friendly, current inoculants. Aside from acquiring "enhanced seeds" with the necessary microorganisms, the best option will be a formulation that does not require further farmer input in application.

- Controlling costs. Inoculants with expensive transport materials or high manufacturing costs cannot sustain agriculture. As a result of the vast volumes of inoculants needed and the major crops that utilize inoculants rather than cash crops, any technique must be developed to reduce its costs consistently. It is improbable that an excellent formulation at a high price will establish a niche in this market. Low-cost solutions for prolonging inoculant shelf life at the farm level in underdeveloped nations require further research. Low-cost technologies must be developed to improve inoculant efficacy. Because not all PGPB strains perform best in every agricultural setting, local strains should optimize performance. As the efficiency of inoculation depends on various factors, such as soil, targeted plant species, and climatic conditions, inoculants must be identified and matched correctly. Because financial incentives mandate keeping the diversity of products modest, this necessity complicates the process of supplying effective inoculants. This industrial necessity contradicts the actuality of crop species, soil biotic communities, and inoculant species, all of which support the distribution and production of various inoculants. Local development and dissemination of chosen native PGPB-rhizobia isolates might address a few of the present challenges hampering inoculant field performance (Selvamukilan et al., 2006). Particular formulas for use in less-than-ideal farming circumstances should be developed. These include soil salinity, acid soils, high temperatures, drought, somewhat degraded soil, and soil with limited access for plant culture upkeep. Local PGPB-rhizobia strains that are resistant or tolerant to chemical or physical stressors, for example, salt-tolerant strains, can be employed to address such issues. Nursery inoculation is demanding and typically produces superior results for transplanted crops. Compared to field treatment, the cost of the product and application will be far lower. This is due to the limited amount of the infected substrate and the ease of controlling growth conditions. As a result, greater research and technology transfer to farmers should focus on plants grown in the nursery. Microbial inoculant-dipped seedling roots, for example, are successful and simple for microbial inoculation of transplanted crops (Choudhury and Kennedy, 2004). This approach should be tried for various transplanted crops.

- Inoculants composed of active PGPB fragments in the absence of live cells.
- Except for the additional public exposure for the company, formal literature publications, including reviews of proprietary formulas, are usually ineffective. The farmer's field experience and market conditions will determine if such formulations are worthwhile to repurchase in the next growing season. Such articles just add to the uncertainty in this sector.

Formulation and field application of inoculants are purely technical platforms that rely on fundamental principles in microbiology and material science. The convergence of these sectors yields beneficial goods that are and will remain an important contribution to sustainable agricultural and environmental practices.

References

Adholeya, A., Das, M., 2012. Biofertilizers: potential for crop improvement under stressed conditions. In: Improving Crop Productivity in Sustainable Agriculture, pp. 183–200.

Ahmad, M., Pataczek, L., Hilger, T.H., Zahir, Z.A., Hussain, A., Rasche, F., Schafleitner, R., Solberg, S.Ø., 2018. Perspectives of microbial inoculation for sustainable development and environmental management. Front. Microbiol. 9, 2992.

Albareda, M., Rodríguez-Navarro, D.N., Camacho, M., Temprano, F.J., 2008. Alternatives to peat as a carrier for rhizobia inoculants: solid and liquid formulations. Soil Biol. Biochem. 40 (11), 2771–2779.

Anandham, R., Sridar, R., Nalayini, P., Poonguzhali, S., Madhaiyan, M.J.M.R., 2007. Potential for plant growth promotion in groundnut (*Arachis hypogaea* L.) cv. ALR-2 by co-inoculation of sulfur-oxidizing bacteria and Rhizobium. Microbiol. Res. 16, 139–153.

Anith, K.N., Vaishakhi, A.S., Viswanathan, A., Varkey, S., Aswini, S., 2017. Population dynamics and efficiency of coconut water based liquid formulation of *Pseudomonas fluorescens* AMB-8. J. Trop. Agric. 54 (2), 184.

Arora, N.K., Khare, E., Maheshwari, D.K., 2011. Plant growth promoting rhizobacteria: constraints in bioformulation, commercialization, and future strategies. In: Plant Growth and Health Promoting Bacteria. Springer, Berlin, pp. 97–116.

Arora, N.K., Khare, E., Naraian, R., Maheshwari, D.K., 2008. Sawdust as a superior carrier for production of multipurpose bioinoculant using plant growth promoting rhizobial and pseudomonad strains and their impact on productivity of *Trifolium repens*. Curr. Sci., 90–94.

Baez-Rogelio, A., Morales-García, Y.E., Quintero-Hernández, V., Muñoz-Rojas, J., 2017. Next generation of microbial inoculants for agriculture and bioremediation. Microb. Biotechnol. 10, 19–21.

Bashan, Y., 1986. Alginate beads as synthetic inoculant carriers for slow release of bacteria that affect plant growth. Appl. Environ. Microbiol. 51 (5), 1089–1098.

Bashan, Y., 1998. Inoculants of plant growth-promoting bacteria for use in agriculture. Biotechnol. Adv. 16 (4), 729–770.

Bashan, Y., de Bashan, L.E., Prabhu, S.R., Hernandez, J.P., 2014. Advances in plant growth-promoting bacterial inoculant technology: formulations and practical perspectives (1998–2013). Plant Soil 378 (1), 1–33.

Bashan, Y, 2011. Development of two culture media for mass cultivation of *Azospirillum* spp. and for production of inoculants to enhance plant growth. Biol. Fertil. Soils 47 (8), 963–969.

Bashan, Y., de Bashan, L.E., Prabhu, S.R., 2016. Superior polymeric formulations and emerging innovative products of bacterial inoculants for sustainable agriculture and the environment. In: Agriculturally Important Microorganisms. Springer, Singapore, pp. 15–46.

Berg, G., 2009. Plant–microbe interactions promoting plant growth and health: perspectives for controlled use of microorganisms in agriculture. Appl. Microbiol. Biotechnol. 84 (1), 11–18.

Berger, B., Patz, S., Ruppel, S., Dietel, K., Faetke, S., Junge, H., Becker, M., 2018. Successful formulation and application of plant growth-promoting *Kosakonia radicincitans* in maize cultivation. Biomed. Res. Int. 2018, e6439481.

Bernabeu, P.R., Garcia, S.S., López, A.C., Vio, S.A., Carrasco, N., Boiardi, J.L., Luna, M.F., 2018. Assessment of bacterial inoculant formulated with *Paraburkholderia tropica* to enhance wheat productivity. World J. Microbiol. Biotechnol. 34 (6), 1–10.

Berninger, T., Mitter, B., Preininger, C., 2017. Zeolite-based, dry formulations for conservation and practical application of *Paraburkholderia phytofirmans* PsJN. J. Appl. Microbiol. 122 (4), 974–986.

Bharathi, R., Vivekananthan, R., Harish, S., Ramanathan, A., Samiyappan, R., 2004. Rhizobacteria-based bio-formulations for the management of fruit rot infection in chillies. Crop Prot. 23 (9), 835–843.

Bharti, N., Suryavanshi, M., 2021. Quality control and regulations of biofertilizers: Current scenario and future prospects. In: Biofertilizers. Woodhead Publishing, pp. 133–141.

Bhattacharyya, J., Jha, D.K., 2012. Plantgrowth-promotingrhizobacteria (PGPR): Emergencein agriculture. World J. Microbiol. Biotechnol. 28, 1327–1350.

Bisutti, I.L., Hirt, K., Stephan, D., 2015. Influence of different growth conditions on the survival and the efficacy of freeze-dried *Pseudomonas fluorescens* strain Pf153. Biocontrol Sci. Tech. 25 (11), 1269–1284.

Bonaterra, A., Cabrefiga, J., Camps, J., Montesinos, E., 2007. Increasing survival and efficacy of a bacterial biocontrol agent of fire blight of rosaceous plants by means of osmoadaptation. FEMS Microbiol. Ecol. 61 (1), 185–195.

Broeckx, G., Vandenheuvel, D., Claes, I.J., Lebeer, S., Kiekens, F., 2016. Drying techniques of probiotic bacteria as an important step towards the development of novel pharmabiotics. Int. J. Pharm. 505 (1–2), 303–318.

Cabrefiga, J., Francés, J., Montesinos, E., Bonaterra, A., 2011. Improvement of fitness and efficacy of a fire blight biocontrol agent via nutritional enhancement combined with osmoadaptation. Appl. Environ. Microbiol. 77 (10), 3174–3181.

Cabrefiga, J., Francés, J., Montesinos, E., Bonaterra, A., 2014. Improvement of a dry formulation of *Pseudomonas fluorescens* EPS 62e for fire blight disease biocontrol by combination of culture osmoadaptation with a freeze-drying lyoprotectant. J. Appl. Microbiol. 117 (4), 1122–1131.

Catroux, G., Hartmann, A., Revellin, C., 2001. Trends in rhizobial inoculant production and use. Plant Soil 230 (1), 21–30.

Costa, E., Usall, J., Teixido, N., Garcia, N., Vinas, I., 2000. Effect of protective agents, rehydration media and initial cell concentration on viability of Pantoeaagglomerans strain CPA-2 subjected to freeze-drying. J. Appl. Microbiol. 89 (5), 793–800.

Choudhury, A.T.M.A., Kennedy, I.R., 2004. Prospects and potentials for systems of biological nitrogen fixation in sustainable rice production. Biol. Fertil. Soils 39, 219–227.

Clayton, G.W., Rice, W.A., Lupwayi, N.Z., Johnston, A.M., Lafond, G.P., Grant, C.A., Walley, F., 2004b. Inoculant formulation and fertilizer nitrogen effects on field pea: nodulation, N2 fixation, and nitrogen partitioning. Can. J. Plant Sci. 84 (1), 79–88.

Clayton, G.W., Rice, W.A., Lupwayi, N.Z., Johnston, A.M., Lafond, G.P., Grant, C.A., Walley, F., 2004a. Inoculant formulation and fertilizer nitrogen effects on field pea: crop yield and seed quality. Can. J. Plant Sci. 84 (1), 89–96.

Costa, E., Teixidó, N., Usall, J., Fons, E., Gimeno, V., Delgado, J., Viñas, I., 2002a. Survival of Pantoeaagglomerans strain CPA-2 in a spray-drying process. J. Food Prot. 65 (1), 185–191.

Costa, E., Usall, J., Teixido, N., Torres, R., Vinas, I., 2002b. Effect of package and storage conditions on viability and efficacy of the freeze-dried biocontrol agent Pantoeaagglomerans strain CPA-2. J. Appl. Microbiol. 92 (5), 873–878.

Date, R.A., 2001. Advances in inoculant technology: a brief review. Aust. J. Exp. Agric. 41 (3), 321–325.

Deaker, R., Hartley, E., Gemell, G., 2012. Conditions affecting shelf-life of inoculated legume seed. Agriculture 2 (1), 38–51.

Deaker, R., Roughley, R.J., Kennedy, I.R., 2004. Legume seed inoculation technology—a review. Soil Biol. Biochem. 36 (8), 1275–1288.

Dey, A., 2021. Liquid biofertilizers and their applications: an overview. In: Environmental and Agricultural Microbiology: Applications for Sustainability, pp. 275–292.

El-Tarabily, K.A., 2004. Suppression of *Rhizoctonia solani* diseases of sugar beet by antagonistic and plant growth-promoting yeasts. J. Appl. Microbiol. 96 (1), 69–75.

Fu, N., Chen, X.D., 2011. Towards a maximal cell survival in convective thermal drying processes. Food Res. Int. 44 (5), 1127–1149.

Hashem, M.A., 2001. Problems and prospects of cyanobacterial biofertilizer for rice cultivation. Funct. Plant Biol 28 (9), 881–888.

Herrmann, L., Lesueur, D., 2013. Challenges of formulation and quality of biofertilizers for successful inoculation. Appl. Microbiol. Biotechnol. 97 (20), 8859–8873.

Hynes, R.K., Jans, D.C., Bremer, E., Lupwayi, N.Z., Rice, W.A., Clayton, G.W., Collins, M.M., 2001. Rhizobium population dynamics in the pea rhizosphere of rhizobial inoculant strain applied in different formulations. Can. J. Microbiol. 47 (7), 595–600.

Jambhulkar, P.P., Sharma, P., Yadav, R., 2016. Delivery systems for introduction of microbial inoculants in the field. In: Microbial Inoculants in Sustainable Agricultural Productivity. Springer, New Delhi, pp. 199–218.

John, R.P., Tyagi, R.D., Brar, S.K., Surampalli, R.Y., Prévost, D., 2011. Bio-encapsulation of microbial cells for targeted agricultural delivery. Crit. Rev. Biotechnol. 31 (3), 211–226.

Krishnaiah, D., Nithyanandam, R., Sarbatly, R., 2014. A critical review on the spray drying of fruit extract: effect of additives on physicochemical properties. Crit. Rev. Food Sci. Nutr. 54 (4), 449–473.

Kumar, V., Rajauria, G., Sahai, V., Bisaria, V.S., 2012. Culture filtrate of root endophytic fungus *Piriformospora indica* promotes the growth and lignan production of *Linum album* hairy root cultures. Process Biochem. 47 (6), 901–907.

Lee, S.K., Lur, H.S., Lo, K.J., Cheng, K.C., Chuang, C.C., Tang, S.J., Yang, Z.W., Liu, C. T., 2016. Evaluation of the effects of different liquid inoculant formulations on the survival and plant-growth-promoting efficiency of *Rhodopseudomonas palustris* strain PS3. Appl. Microbiol. Biotechnol. 100 (18), 7977–7987.

Liffourrena, A.S., Lucchesi, G.I., 2018. Alginate-perlite encapsulated *Pseudomonas putida* A (ATCC 12633) cells: preparation, characterization and potential use as plant inoculants. J. Biotechnol. 278, 28–33.

Loján, P., Demortier, M., Velivelli, S.L., Pfeiffer, S., Suárez, J.P., De Vos, P., Prestwich, B. D., Sessitsch, A., Declerck, S., 2017. Impact of plant growth-promoting rhizobacteria on root colonization potential and life cycle of *Rhizophagus irregularis* following co-entrapment into alginate beads. J. Appl. Microbiol. 122 (2), 429–440.

Malusá, E., Sas-Paszt, L., Ciesielska, J., 2012. Technologies for beneficial microorganisms inocula used as biofertilizers. Sci. World J. 2012, 491206. https://doi.org/10.1100/2012/491206.

Malusà, E., Pinzari, F., Canfora, L., 2016. Efficacy of biofertilizers: challenges to improve crop production. In: Microbial Inoculants in Sustainable Agricultural Productivity. Springer, New Delhi, pp. 17–40.

Marcelino, P.R.F., Milani, K.M.L., Mali, S., Dos Santos, O.J.A.P., de Oliveira, A.L.M., 2016. Formulations of polymeric biodegradable low-cost foam by melt extrusion to deliver plant growth-promoting bacteria in agricultural systems. Appl. Microbiol. Biotechnol. 100 (16), 7323–7338.

Maron, P.A., Sarr, A., Kaisermann, A., Lévêque, J., Mathieu, O., Guigue, J., Karimi, B., Bernard, L., Dequiedt, S., Terrat, S., Chabbi, A., 2018. High microbial diversity promotes soil ecosystem functioning. Appl. Environ. Microbiol. 84 (9), e02738-17.

Mendes, G.D.O., da Silva, N.M.R.M., Anastácio, T.C., Vassilev, N.B., Ribeiro Jr., J.I., da Silva, I.R., Costa, M.D., 2015. Optimization of *Aspergillus niger* rock phosphate solubilization in solid-state fermentation and use of the resulting product as a P fertilizer. Microb. Biotechnol. 8 (6), 930–939.

Mitter, B., Pfaffenbichler, N., Flavell, R., Compant, S., Antonielli, L., Petric, A., Berninger, T., Naveed, M., Sheibani-Tezerji, R., von Maltzahn, G., Sessitsch, A., 2017. A new approach to modify plant microbiomes and traits by introducing beneficial bacteria at flowering into progeny seeds. Front. Microbiol. 8, 11.

Moënne-Loccoz, Y., Naughton, M., Higgins, P., Powell, J., O'connor, B., O'gara, F., 1999. Effect of inoculum preparation and formulation on survival and biocontrol efficacy of *Pseudomonas fluorescens* F113. J. Appl. Microbiol. 86 (1), 108–116.

Morgan, C.A., Herman, N., White, P.A., Vesey, G., 2006. Preservation of micro-organisms by drying; a review. J. Microbiol. Methods 66 (2), 183–193.

Nguyen, H.T., Deaker, R., Kennedy, I.R., Roughley, R.J., 2003. The positive yield response of field-grown rice to inoculation with a multi-strain biofertiliser in the Hanoi area, Vietnam. Symbiosis. 35, 1–15.

Oliveira, A.L., Santos, O.J., Marcelino, P.R., Milani, K.M., Zuluaga, M.Y., Zucareli, C., Gonçalves, L.S., 2017. Maize inoculation with *Azospirillum brasilense* Ab-V5 cells enriched with exopolysaccharides and polyhydroxybutyrate results in high productivity under low N fertilizer input. Front. Microbiol. 8, 1873.

Pastor-Bueis, R., Mulas, R., Gómez, X., González-Andrés, F., 2017. Innovative liquid formulation of digestates for producing a biofertilizer based on *Bacillus siamensis*: field testing on sweet pepper. J. Plant Nutr. Soil Sci. 180 (6), 748–758.

Prabhu, S.R., Thomas, G.V., 2002. Biological conversion of coir pith into a value-added organic resource and its application in agri-horticulture: current status, prospects and perspective. J. Plant. Crops 30 (1), 1–17.

Pusey, P.L., 2002. Biological control agents for fire blight of apple compared under conditions limiting natural dispersal. Plant Dis. 86 (6), 639–644.

Riddech, N., Phibunwatthanawong, T., Sarin, P., 2021. Suitable formulation of microbial inoculants as a bio-fertilizer for promoting growth of hairy-leafed apitong (*Dipterocarpus alatus*). In: Waste and Biomass Valorization, pp. 1–13.

Romano, I., Ventorino, V., Ambrosino, P., Testa, A., Chouyia, F.E., Pepe, O., 2020. Development and application of low-cost and eco-sustainable bio-stimulant containing a new plant growth-promoting strain *Kosakonia pseudosacchari* TL13. Front. Microbiol. 11.

Rouissi, T., Mahmoudi, A., Tyagi, R.D., Brar, S.K., Prèvost, D., Surampalli, R.Y., 2013. Optimisation of spray drying by response surface methodology for the production of *Sinorhizobium meliloti* powder formulation by using starch industry wastewater. Biosyst. Eng. 114 (3), 334–343.

Sahu, P.K., Brahmaprakash, G.P., 2016. Formulations of biofertilizers–approaches and advances. In: Microbial Inoculants in Sustainable Agricultural Productivity. Springer, New Delhi, pp. 179–198.

Santivarangkna, C., Kulozik, U., Foerst, P., 2007. Alternative drying processes for the industrial preservation of lactic acid starter cultures. Biotechnol. Prog. 23 (2), 302–315.

Scheidt, W., dos Santos Pedroza, I.C.P., Fontana, J., da Cruz Meleiro, L.A., de Barros Soares, L.H., Reis, V.M., 2020. Optimization of culture medium and growth conditions of the plant growth-promoting bacterium *Herbaspirillum seropedicae* BR11417 for its use as an agricultural inoculant using response surface methodology (RSM). Plant Soil 451 (1), 75–87.

Schisler, D.A., Slininger, P.J., Olsen, N.L., 2016. Appraisal of selected osmoprotectants and carriers for formulating Gram-negative biocontrol agents active against fusarium dry rot on potatoes in storage. Biol. Control 98, 1–10.

Schoebitz, M., López, M.D., Roldán, A., 2013. Bioencapsulation of microbial inoculants for better soil–plant fertilization. A review. Agron. Sustain. Dev. 33 (4), 751–765.

Schulz, T.J., Thelen, K.D., 2008. Soybean seed inoculant and fungicidal seed treatment effects on soybean. Crop Sci. 48 (5), 1975–1983.

Selvamukilan, B., Selvamukilan, B., Rengalakshmi, S., Tamizoli, P., Nair, S., 2006. Village-level production and use of biocontrol agents and biofertilizers. In: Biological Approaches to Sustainable Soil Systems. CRC Press, Boca Raton, pp. 647–653.

Shilev, S., Azaizeh, H., Vassilev, N., Georgiev, D., Babrikova, I., 2019. Interactions in soil-microbe-plant system: adaptation to stressed agriculture. In: Microbial Interventions in Agriculture and Environment, pp. 131–171.

Singleton, P., Keyser, H., Sande, E., 2002. Development and evaluation of liquid inoculants. In: Inoculants and Nitrogen Fixation of Legumes in Vietnam. ACIAR Proceedings 109e, Canberra, pp. 52–66.

Soto-Muñoz, L., Torres, R., Usall, J., Viñas, I., Solsona, C., Teixidó, N., 2015. DNA-based methodologies for the quantification of live and dead cells in formulated biocontrol products based on *Pantoea agglomerans* CPA-2. Int. J. Food Microbiol. 210, 79–83.

Stamenković, S., Beškoski, V., Karabegović, I., Lazić, M., Nikolić, N., 2018. Microbial fertilizers: a comprehensive review of current findings and future perspectives. Span. J. Agric. Res. 16 (1).

Stephan, D., Da Silva, A.P.M., Bisutti, I.L., 2016. Optimization of a freeze-drying process for the biocontrol agent *Pseudomonas* spp. and its influence on viability, storability and efficacy. Biol. Control 94, 74–81.

Stephens, J.H.G., Rask, H.M., 2000. Inoculant production and formulation. Field Crop Res. 65 (2–3), 249–258.

Swarnalakshmi, K., Yadav, V., Tyagi, D., Dhar, D.W., Kannepalli, A., Kumar, S., 2020. Significance of plant growth promoting rhizobacteria in grain legumes: growth promotion and crop production. Plan. Theory 9 (11), 1596.

Temprano, F., Albareda, M., Camacho, M., Daza, A., Santamaria, C., Rodríguez-Navarro, N.D., 2002. Survival of several Rhizobium/Bradyrhizobium strains on different inoculant formulations and inoculated seeds. Int. Microbiol. 5 (2), 81–86.

Trifonova, R., Postma, J., Van Elsas, J.D., 2009. Interactions of plant-beneficial bacteria with the ascomycete *Coniochaeta ligniaria*. J. Appl. Microbiol. 106 (1), 1859–1866.

Tuomi, T., Heino, M., Rosenqvist, H., Nordström, K., Laakso, S., 2001. Fiber fractions from processing of barley in production and conservation of a biologic control agent. Appl. Biochem. Biotechnol. 94, 135–145.

Valetti, L., Angelini, J.G., Taurian, T., Ibáñez, F.J., Muñoz, V.L., Anzuay, M.S., Ludueña, L. M., Fabra, A., 2016. Development and field evaluation of liquid inoculants with native Bradyrhizobial strains for peanut production. Afr. Crop. Sci. J. 24 (1), 1–13.

Vassilev, N., de Oliveira Mendes, G., 2018. Solid-state fermentation and plant-beneficial microorganisms. In: Current Developments in Biotechnology and Bioengineering. Elsevier, pp. 435–450.
Vassilev, N., Requena, A.R., Nieto, L.M., Nikolaeva, I., Vassileva, M., 2009. Production of manganese peroxidase by Phanerochaetechrisosporium grown on medium containing agro-wastes/rock phosphate and biocontrol properties of the final product. Ind. Crop. Prod. 30 (1), 28–32.
Vassilev, N., Eichler-Löbermann, B., Flor-Peregrin, E., Martos, V., Reyes, A., Vassileva, M., 2017. Production of a potential liquid plant bio-stimulant by immobilized *Piriformospora indica* in repeated-batch fermentation process. AMB Express 7 (1), 1–7.
Vassileva, M., Malusà, E., Sas-Paszt, L., Trzcinski, P., Galvez, A., Flor-Peregrin, E., Shilev, S., Canfora, L., Mocali, S., Vassilev, N., 2021. Fermentation strategies to improve soil bio-inoculant production and quality. Microorganisms 9 (6), 1254.
Xavier, I.J., Holloway, G., Leggett, M., 2004. Development of rhizobial inoculant formulations. Crop Manag. 3 (1), 1–6.
Yánez-Mendizábal, V., Viñas, I., Usall, J., Torres, R., Solsona, C., Abadias, M., Teixidó, N., 2012. Formulation development of the biocontrol agent *Bacillus subtilis* strain CPA-8 by spray-drying. J. Appl. Microbiol. 112 (5), 954–965.
Zhou, J., Deng, B., Zhang, Y., Cobb, A.B., Zhang, Z., 2017. Molybdate in rhizobial seed-coat formulations improves the production and nodulation of alfalfa. PLoS One 12 (1), e0170179.

CHAPTER 2

Application of microbial inoculants as an alternative to chemical products for decomposition of organic wastes

Mohammad Haris[a], Touseef Hussain[a,b], Atirah Tauseef[a], Amir Khan[a], Abrar Ahmad Khan[a], and Yasheshwar[c]
[a]Plant Pathology and Nematology Section, Department of Botany, Aligarh Muslim University, Aligarh, India
[b]Division of Plant Pathology, ICAR-Indian Agricultural Research Institute, New Delhi, India
[c]Department of Botany, Acharya Narendra Dev College, Govindpuri, Kalka ji, University of Delhi, New Delhi, India

2.1 Introduction

Environmental contamination has the potential to pose a major threat to the survival of all living organisms. The excessive use of chemical fertilizers and pesticides can contribute to environmental degradation (Kaosol, 2009). Various chemicals used in agriculture not only disrupt traditional farming practices but are also responsible for the biological, physical, and chemical deterioration of arable lands (Pretty and Bharucha, 2014; Dhankar et al., 2021). The use of chemical fertilizers in agriculture leads to degradation of the health of the soil and vegetation (Haris et al., 2019). The systematic use of these agrochemicals can harm soil biodiversity, resulting in the loss of nutritional ingredients and unwanted chemical intermediaries in the food chain (Lal, 2015). As a result, several countries are taking steps to minimize the use of pesticides and fertilizers in agricultural production (FAO, 2017). Using beneficial microorganisms as a substitute for toxic chemical fertilizers and pesticides can be a viable option. Soil microbes play an essential role in the improvement of soil management and crop productivity, in plant growth and development, and in nutrient cycling (Yan et al., 2015; Mukherjee et al., 2022). Soil microbial dynamics also determine the productivity potential of soil crops. Every year, about 700 million tonnes of organic wastes are generated in India, as a result of which issues for their safe disposal

Microbial Inoculants
https://doi.org/10.1016/B978-0-323-99043-1.00006-2

arise; usually, these wastes are either burned or landfilled (Bhiday, 1994; Nagavallemma et al., 2006; Zeinhom et al., 2010). However, numerous naturally existing microbes can transform organic wastes into valuable resources like plant nutrients and decrease the C:N ratio to promote soil quality. Different microbes break down different types of organic matter (Schulz et al., 2013; Husain et al., 2022). These microorganisms are also crucial for maintaining the flow of nutrients from one system to another and for minimizing ecological imbalances (Umsakul et al., 2010; Novinsak et al., 2008). Decomposition of organic wastes by applying microbial inoculants is a preferred biological and ecological process by which organic wastes are reduced to organic fertilizers. Various parameters, such as moisture content, pH, composting temperature, C:N ratio, and the existence of potential pathogens, including coliform bacteria, are used to ensure the stability and quality of the compost during the composting process (Sanmanee et al., 2011; Fourti et al., 2011; Al-Turki, 2010; Erickson et al., 2009; Steger et al., 2007; Wu and Ma, 2002). Farmers are typically unaware of the relevance of microbial resources to surface and underground soils as well as their beneficial effects on farmlands. However, they do collect knowledge regarding the technology of decomposition through microbial mediation and compost enrichments with profitable microorganisms. This chapter examines the importance of microbe-mediated organic waste bioconversion practices and their reuse to strengthen soils. We also describe microbial diversity, the potential benefits of microbe-mediated compost as agriculture inputs, the role of decomposers in soil fertility, the biodecomposition of agricultural wastes, the benefits and limitations of biological fertilizers, and the challenges for decomposition of organic wastes.

2.2 Potential benefits of microbe-mediated compost as farm inputs

In addition to being farmer-friendly and replicable, composting technologies also create useful farm inputs that greatly enhance agricultural productivity and profitability. Microbe-mediated activities promote efficient composting processes. These help reintroduce organic matter into soils, and beneficial microbes assist soils by improving their nutrient content for plant growth and development. Microorganisms as bioconversion agents have the potential to convert lignocellulosic and cellulosic wastes into organic matter, bioremediate environmental contaminants, and interact with the rhizosphere to enhance soil structure and plant growth

(Sanchez, 2009; Huang et al., 2010). Microorganisms are the main actors involved in maintaining the flow of nutrients from residues to agricultural soils (Erickson et al., 2009). They are essential for the management of natural resources in farmers' fields. The microbiological process of composting breaks down all types of agricultural residues in an appropriate and timely manner, producing low-cost and high-value bioorganic agricultural inputs (Singh and Prabha, 2017; Sudharmaidevi et al., 2017; Singh and Nain, 2015; Ahmad et al., 2007). Excessive use of chemicals causes severe imbalances in the natural soil ecosystem and poses a threat to fertility, soil structure, and function, leads to crop poisoning and loss of productivity, and damages the harmony of soil–crop interactions (Aktar et al., 2009). In such a scenario, farming systems better increase the use of both agricultural residual resources and low external inputs. Such systems will captivate farmers with locally available sources and use their resources to obtain better results and minimize reliance on high external input costs. For this reason, microbial technological interventions must essentially drive farming communities to produce better functional foods, improve the organic matter of soil through the application of low-cost, self-produced compost, and contribute microbiologically enriched farm inputs to strengthen field soils (Fig. 1).

Thus, the process of fast composting can assist farmers in the application of compost and contribution of enriched bioorganic farm inputs of desired quality, in a timely manner, to meet the needs of not only organic farming but also farming of advanced value-added commercial crops such as flowers, fruits, and vegetables (Seyedbagheri, 2010; Hoornweg et al., 2000). If farmers need composting materials that are high in biopesticides to control pathogenic fungi transmitted by soil, seeds, or seedlings in the field, then they can use bioagents to biofortify the raw compost (Ng et al., 2016; Siddiqui et al., 2008;). Similarly, a consortium of microbes that fix nitrogen solubilize both zinc and phosphorus. They also mobilize potassium, which can be used to strengthen the waste materials of the compost to obtain the

Fig. 1 Organic waste bioconversion by beneficial microorganisms.

desired quality under appropriate conditions of temperature and humidity. This can produce potential bioorganic inputs, improved by collecting and recycling N, P, K, and Zn (Pallavi et al., 2017; Kamran et al., 2017; Baig et al., 2012; Pugliese et al., 2011; Mandal et al., 2021). The complete process depends on the needs of farmers, their experience, the availability of indigenous resources, local conditions, and existing human resources.

2.3 Agrowaste biodecomposition

One of the main reasons identified for the decrease in the sustainability of agriculture is the poor condition of the soil due to the low application of organic matter in farms. Mineralization results in the availability of soluble plant nutrients by the biodecomposition of organic matter in the soil. Table 1 lists several microorganisms, their potential components contributing to rapid crop residue bioconversion, biodegradation, decomposition, and other valuable products. Plant materials, particularly crop residues, are high in lignocellulosic biomass, but their crystalline structures are encrusted with suberin, silica, lignin, and different polymeric components, which hinder the mild microbial degradation composting process (Huber and Praznik, 2004). So, pretreatment of lignocellulosic biomass with acids, alkalis, urea, steam, and hydrolytic enzymes is recommended for substantial decomposition of complex components to smooth the composting process (Mosier et al., 2005; Table 1). Potential microorganisms show increased enzymatic capacity for fast degradation of recalcitrant lignin (Table 1) (Varma et al., 2017; Perez et al., 2014). Such organic compounds contain complex nested fractions, and their biomass recovery is tenacious and resistant to solubilization and hydrolysis (Kumar and Sharma, 2017). Therefore, rather than a single pretreatment process, chemical and multiple biological and physical steps are needed in an integrated manner to decrease unwanted inhibitors (Shrestha et al., 2017).

Agricultural food wastes are interesting composting materials, which are converted into manures and are used for producing high value-added crops (Rubio et al., 2013). The process of composting, which ensures high nutrient conditions, adequate carbon classification, humification of organic matter, and adequate volume, has been widely considered necessary to reduce nitrogen losses and exceed production costs (Bernal et al., 2009). Findings have confirmed that composting aids in the detoxification and degradation of phytotoxic compounds in waste materials. As a result, it provides a

Table 1 Degradation and decomposition of agricultural wastes into compost by various microbes.

Microbes	Activities of biodegradation	Nature of organic wastes	References
Pleurotus flabellatus	Lignocellulose degradation	Straw of rice and leaves of sisal	Mshandete and Cuff (2008)
Pleurotus eryngii	Lignocellulose degradation and activity of laccase enzymes	Agricultural wastes	Yildirim et al. (2015)
Aspergillus niger	Production of cellulase and xylanase	Predecomposition of sugarcane bagasse and organic matter	Romero et al. (2007), Singh and Sharma (2002)
Phanerochaete chrysosporium	Manganese peroxydase Lignin peroxydase and glyoxal oxidase	Lignin-containing biomass such as wood shavings and agrowastes	Zhang et al. (2013), Kersten and Cullen (2007), Martinez (2002)
Xylaria hypoxylon	Laccase, glucosidase, xylanase, and esterase	Woody matters	Liers et al. (2006)
Trametes versicolor	Laccase	Woody substrates and agrowastes	Cabuk et al. (2006)
Aspergillus awamori	Cellulases	Agrowastes	Pleissner et al. (2013), Gaind and Nain (2007)
Paecilomyces marquandii	Keratinase	Poultry wastes (feather wastes)	Vesela and Friedrich (2009)
Phanerochaete chrysosporium	The humification degree of humic acid increases	Agrowastes	Huang et al. (2009)
Bacillus spp. and *Paenibacillus* spp.	Lignin degradation	Pulp paper waste degradation	Chandra et al. (2007)
Aneurinibacillus aneurinilyticus	Lignin degradation	Pulp paper waste degradation	Raj et al. (2007)
Pseudomonas putida	Laccase and manganese peroxydase	Agrowastes	Ahmad et al. (2010)

Continued

Table 1 Degradation and decomposition of agricultural wastes into compost by various microbes—cont'd

Microbes	Activities of biodegradation	Nature of organic wastes	References
Pseudomonas aeruginosa	Laccase, lipid peroxidase, and manganese peroxidase	Agrowastes	Bholay et al. (2012)
Serratia marcescens	Laccase, manganese peroxidase, and lipid peroxidase	Agrowastes	Chandra et al. (2012)
Citrobacter freundii	Lignin degradation and manganese peroxidase	Agrowastes (saw dust)	Ali et al. (2017)
Streptomyces spp.	Acetylesterase, cellulase, xylanase, and xylosidase	Agrowastes	Benimelia et al. (2007)
Mono- and cocultures of *P. ostreatus* and *B. subtilis*	Cellulases	Cereal wastes mixed with plum and apple wastes	Petre et al. (2014)
Bacillus cereus and *Bacillus megaterium*	Hemicellulose and cellulose breakdown	Organic substrates	Ribeiro et al. (2017)

cost-effective technique to recycle wastes in terms of high-value products (Pampuro et al., 2016).

2.4 Microbial inoculants: Their diversity and potential use in long-term crop production

Microbial inoculants, often known as biofertilizers, are organically generated products that include specific bacteria obtained from plant roots and the rhizoplane. Their application has demonstrated promising outcomes in increasing yield and stimulating plant development by 10%–40% (Kawalekar, 2013). When applied to plant surfaces, seeds, or soil, these bacteria inhabit the rhizosphere and enter the plants, thus modulating plant growth (Raghuwanshi, 2012; Khan et al., 2022). Microbial inoculants protect plants from pests and diseases in addition to improving the fertility of poor soils and enhancing crop output.

The use of microbial inoculants stimulates strong root growth, improves stability, degrades hazardous compounds, increases seedling survival, and shortens flowering time (Youssef and Eissa, 2014). Another significant benefit is that after 3–4 years of continuous application of biofertilizers, there is no need for them because the parental inoculum is sufficient for multiplication and growth (Bumandalai and Tserennadmid, 2019).

Plants require 17 critical elements to maintain their growth and development. Nitrogen (N), potassium (K), and phosphorous (P) are three of the most sought-after elements (Mishra and Dash, 2014). Numerous microorganisms, such as *Cyanobacteria*, nitrogen-fixing soil bacteria, and phosphate-solubilizing bacteria, are widely used as biofertilizers in combination with various types of fungi and molds (Umesha et al., 2018). Similarly, bacteria that produce phytohormones are used in biofertilizer formulations. They provide plants with growth-promoting chemicals such as vitamins, amino acids, and indole acetic acid (IAA), increase soil productivity and fertility, and improve crop output (Parikh and James, 2012; Haris et al., 2020).

2.5 Different forms of biofertilizers

Biofertilizers are classified into different categories based on their roles and modes of action. Plant growth-promoting *Rhizobacteria* (PGPR), nitrogen fixers (N-fixers), phosphorus solubilizers (P-solubilizers), and potassium solubilizers (K-solubilizers) are some of the most commonly utilized biofertilizers (Mahdi et al., 2010). Bacterial growth in soil is influenced by the physical and chemical properties of the soil and the amount of phosphorus, organic matter, and cultural activities. Microbes also play a role in the ecosystem's nutrient cycles (Hussain et al., 2020a, 2020b).

2.6 N-fixing biofertilizers

One of the most limiting dietary factors for plant growth is nitrogen (Gupta et al., 2012). Although the amount of nitrogen in the atmosphere is more than 80%, plants are unable to utilize it in this situation. To make nitrogen available to plants, a colony of bacteria must fix atmospheric nitrogen into nitrogenous salts. Biological nitrogen fixers (BNFs) are bacteria that fix nitrogen in the environment.

They convert inert nitrogen into an organic form that plants can use (Reed et al., 2011). Nitrogen fixation can provide 300–400 kg N/ha/year and increase crop yield by 10%–50%. Plants get approximately a quarter of their nitrogen from nitrogen fixation.

Exudates from plant roots are released into the soil, attracting bacteria and facilitating colonization and nitrogen fixation inside the plant rhizosphere. Free-living bacteria (*Azotobacter* and *Azospirillum*), blue-green algae, and symbionts (*Rhizobium*, *Frankia*, and *Azolla*) are the three types. *Rhizobium*, *Bradyrhizobium*, *Mesorhizobium*, *Sinorhizobium*, and *Azorhizobium* are N_2-fixing bacteria found in legumes. *Achromobacter*, *Arthrobacter Alcaligenes*, *Acetobacter*, *Azomonas*, *Bacillus*, *Beijerinckia*, *Clostridium*, *Desulfovibrio*, *Enterobacter*, *Erwinia*, *Derxia*, *Klebsiella*, *Corynebacterium*, *Herbaspirillum*, *Campylobacter*, *Lignobacter*, *Mycobacterium*, *Rhodospirillum*, *Rhodo-pseudomon* (Meena et al., 2017) and different species of *Azospirillum* and *Azotobacter* isolated from the rhizosphere have been widely studied in the field for increasing legume and grain yields (Bhat et al., 2015). The most common N_2-fixing bacteria are subsequently listed.

2.6.1 Rhizobium

Rhizobium, one of the best examples of symbiotic nitrogen fixation, is a member of the Rhizobiaceae bacterial family. Both leguminous and nonleguminous crops can use it to fix nitrogen. The rate of nitrogen fixation by *Rhizobium* in several leguminous crops approaches 300 kg N/ha/year (Pindi, 2012). The bacteria enter the roots of leguminous plants and create nodules, within which molecular nitrogen is reduced to ammonia, which is then used by the plant to synthesize vitamins, proteins, and other nitrogen-containing substances. As a result, these root nodules serve as ammonium-ion storage sites (Flores-Félix et al., 2013). *Rhizobium* increases growth in nonleguminous crops by changing their root shape and growth physiology. Furthermore, *Rhizobium* spraying boosts crop productivity by increasing plant height, seed germination, nitrogen content, and leaf chlorophyll (Sara et al., 2013). Rice seed inoculation with various *Rhizobium* strains at increasing levels of nitrogen increases the content of straw by 4%–19% and the yield of rice grain by 8%–22% (Sammauria et al., 2020). *Rhizobia* include bacteria such as *Rhizobium*, *Bradyrhizobium*, *Azorhizobium*, *Mesorhizobium*, and *Sinorhizobium*. They can work both directly by fixing nitrogen and inducing plant hormone synthesis and indirectly by reducing pathogen inhibitory effects (Mabrouk et al., 2018). *Rhizobium* maintains soil fertility while also increasing agricultural yields (Arora et al., 2017).

2.6.2 Azotobacter

Azotobacter is a nonparasitic, nitrogen-fixing diazotrophic bacterium that, due to its high metabolic activity, is a significant component of the nitrogen cycle. It has the ability to create vitamins such as thiamine and riboflavin (Sahoo et al., 2014). It belongs to the Azotobacteriaceae family and is used to fertilize nonleguminous crops such as cotton, rice, vegetables, sugarcane, sweet potato, and sweet sorghum. *Azotobacter* inoculation boosts seed yield in rapeseed and mustard, according to Datta et al. (2015). It fixes about 30 kg of nitrogen per year and is largely used in the sugarcane industry since it increases cane production by 25–50 t per hectare and sugar content by 10%–15%. *Azotobacter* can be found in both alkaline and acidic soils. The dominant species in the soil is *A. chroococcum*, but other species such as *A. vinelandii*, *A. macrocytogenes*, *A. insignis*, and *A. beijerinckii* have also been detected. *Azotobacter* produces antibiotics and antifungal chemicals that minimize seedling mortality by inhibiting the growth of several harmful fungi in the rhizosphere (Wani et al., 2016).

2.6.3 Azospirillum

This particular bacterium is especially significant since it has the ability to fix a significant amount of nitrogen in the soil. In nonleguminous crops such as oilseeds, cotton, cereals, millets, and sorghum, it is found in the rhizosphere and fixes 20–40 kg N/ha. It frequently forms symbiotic relationships with a variety of plants. Various studies have demonstrated *Azospirillum*'s ability to improve a variety of crops (Leite et al., 2019; Galindo et al., 2020).

2.6.4 Anabaena azollae

A. azollae is a symbiotically related bacterium that fixes nitrogen in the atmosphere, primarily in rice plants (Yadav et al., 2014). Many countries, including Vietnam, China, Thailand, and the Philippines, use *Azolla* as a biofertilizer. Another benefit of employing this biofertilizer is its ability to tolerate metals; as a result, it can be used in heavy metal-polluted areas (Akhtar et al., 2020).

2.7 Blue-green algae (*Cyanobacteria*)

Cyanobacteria that fix nitrogen are one of the most common N_2 fixers on the planet. *Aulosira, Nostoc, Oscillatoria, Anabaena,* and *Lyngbya* are prokaryotes that belong to the *Cyanobacteria* group (Sharma et al., 2011). They are important for providing nitrogen to the soil and for delivering vitamin B complex and growth-enhancing hormones, including auxins and gibberellic acid, which modulate plant growth. When applied at a rate of 10 kg/ha, they can fix almost 20–30 kg/N/ha in water-logged rice fields and increase crop output by 10%–15%. The availability of nitrogen to plants was reported to be increased following the treatment of *Cyanobacteria* in agriculture, primarily in rice fields (Singh et al., 2016; Kumar et al., 2022).

2.8 Phosphate-solubilizing and phosphate-mobilizing biofertilizers

On a dry weight basis, plants contain about 0.2% of phosphorus, which is an essential macronutrient for plant growth and development. Phosphorus is the least mobile macronutrient available to plants developing in soil conditions when compared to other macronutrients. Phosphate in insoluble form must be converted to soluble form by microorganisms. Phosphate solubilization is aided by a small number of fungal groups and a large number of bacteria (Kalayu, 2019). Phosphate-solubilizing bacteria (PSB) use a variety of processes to convert insoluble phosphates like HPO_4 and H_2PO_4 into soluble forms, including organic acid generation, ion exchange reactions, and chelation. Phosphate-solubilizing bacteria account for 1%–50% of all phosphate-solubilizing activities, whereas fungi account for only 0.1%–0.5% (Sharma et al., 2013). PSB create metabolites with hydroxyl (gluconic) and carboxyl (ketogluconic) groups that chelate the cation connected to the phosphate and convert it to a soluble form that the plants may use.

Furthermore, organic acids lower the pH of the soil and dissolve the phosphate connected to it, thus making it available to plants (Itelima et al., 2018).

2.9 Mycorrhizal fungi

Certain fungal groupings form a symbiotic relationship with the host plant. It is, without a doubt, the most important symbiosis on the planet (DeLuca et al., 2019). This symbiotic relationship provides the host plant with essential nutrients, such as phosphorus and growth-promoting hormones, and regulates plant growth. They also increase root surface area, which aids in nutrient absorption from the soil and confers resistance to diseases in plants (Dwivedi et al., 2015).

2.10 VAM fungi (Endomycorrhiza)

The symbiotic interaction between particular Phycomycetes fungi and angiosperm roots is known as vesicular arbuscular mycorrhiza (VAM). Nearly 80% of plant families have symbiotic fungi living in their roots (Sadhana, 2014). They use specialized structures called vesicles and arbuscules to help allocate nutrients from the soil to the roots. The fungal hyphae increase root and shoot length by enhancing the intake of phosphorus and other nutrients.

VAM has the potential to improve plants' ability to withstand various biotic and abiotic stresses, to replace the need for fertilizers, and to reduce the present use of chemical fertilizers (Abbasi et al., 2015). Antibiotically improved twin-fold mobilization of less mobile nutrients such as Ca, P, Mg, S, Zn, and Cu from the rhizosphere was achieved by VAM fungi (Vala and Desai, 2020).

2.11 Potassium-solubilizing and potassium-mobilizing biofertilizers

After nitrogen and phosphorus, potassium (K) is the second-most prevalent and necessary nutrient in plants. Despite the fact that K is a plentiful element in soil, only 1%–2% of it is accessible to plants, with the rest existing as a mineral form of potassium that is not readily available to plants. As a result, regular K replenishment of the soil solution is required (Meena et al., 2014). It is essential for the growth and development of plants. If not enough is applied, then plants will grow slowly, have weak roots, and produce few seeds, thus lowering yield (Prajapati and Modi, 2012).

2.12 Sulfur-oxidizing biofertilizers

Sulfur is a micronutrient that plants require. It has been shown to play an important role in the refinement of various biological and physical features of soil. Sulfur is well-known for its soil-buffering abilities, which enable it to maintain high pH levels. It also boosts the efficacy of nitrogen and phosphorus fertilizers as well as the ability of crops to absorb micronutrients, according to previous research. *Thiobacillus* spp. is an example of a sulfur-oxidizing bacterium; *Thiobacillus thioparus* and *T. thiooxidans* catalyze the oxidation of sulfur to sulfates, which are easily absorbed by plants and aid in plant nutrition (Riaz et al., 2020).

2.13 Zinc biofertilizers

Zinc is an essential element that plants require in relatively low quantities (5–100 mg/kg) in their tissues to promote growth and development. Zinc deficiency is widespread in soil, and it is caused by improper fertilizer application and poor soil health.

If current motivations are not addressed, then zinc deficiency is predicted to increase from 42% to 63% by 2025. Zinc is a key component in the production of growth hormones. Zinc deficiency in plants can cause stunted shoot growth, altered membrane integrity, decreased leaf size, chlorophyll loss, and increased vulnerability to heat, light, and fungal diseases as well as affect grain production, root development, pollen formation, and water intake and transport (Tavallali et al., 2010).

2.14 *Rhizobacteria* that promote plant growth (PGPR)

PGPR refer to a collection of free-living bacteria found in soil, which colonize plant roots and have favorable impacts on plant growth (Timmusk et al., 2014). They act as biofertilizers by promoting plant growth and development, improving biotic and abiotic stress tolerance, and aiding in soil mineralization through decomposition of organic materials. Plants benefit from PGPR inoculation in a variety of ways. They improve drought tolerance (Ilyas et al., 2020), salinity tolerance (Bharti et al., 2013), and biotic stress tolerance (Bharti et al., 2013; Verma et al., 2016). They stimulate the synthesis of phytohormones like auxins, IAA, gibberellic acid, ethylene, and others, which boost seed germination, improve soil fertility, and promote growth (Kumari et al., 2016). They can affect plant secondary metabolites and help in heavy metal and other pollution bioremediation (Kumar et al., 2021; Siddiqua et al., 2021). *Alcaligenes*, *Agrobacterium*, *Azotobacter*, *Arthrobacter*, and *Actinoplanes* are among the genera represented by PGPR. *Bacillus*, *Frankia*, *Rhizobium*, *Pseudomonas*, *Micrococcus*, *Streptomyces*, *Enterobacter*, *Xanthomonas*, *Cellulomonas*, *Thiobacillus*, *Serratia*, *Flavobacterium*, and other bacteria also included in the PGPR representation (Yadav et al., 2017).

2.15 The role of decomposers in soil fertility

Decomposers are microorganisms that feed on dead or decaying organisms in the soil, whose abundance and decomposition undergo major changes with increasing soil depth (Preusser et al., 2019). Abundance decreases regularly from the uppermost to the lowermost layer of the soil and is differentially distributed in the subsoil than in the topsoil (Taylor et al., 2002). It has been found that the lowermost layer of soil has low microbial diversity both genetically and metabolically compared to topsoil (Will et al., 2010; Eilers et al., 2012). Therefore, several types of biodecomposers, viz., fungi, bacteria, and protozoa, are able to break down cellulose with the help of cellulase enzymes, which hydrolyze lignocelluloses (Iqbal et al., 2020). Arbuscular mycorrhizal fungi (AMF), *Humicola*, *Trichoderma*, *Penicillium*, and *Aspergillus* are the common types of biological decomposers. AMF establish themselves in the soil and plant roots by producing a large extra-radical network, which protects the plant from harmful soil-borne pathogens and enhances the plant's water and mineral uptake from the soil. The key product for decomposition is soil organic matter (SOM), which can be seen as an integrator of many processes such as nutrient cycling and other ecosystem services. SOM help build and maintain soil structure and health, which are essential for water infiltration and storage, and aerobic soil atmosphere, which is conductive to nutrient uptake and plant growth, and regulate trace gas emissions (Palm and Swift, 2002). Physical, chemical, and biological decompositions are the main events in which large substances are converted into smaller, utilizable forms by plants. They also help enhance the nutrient composition of soil, and, among these, decomposition by living microorganisms is the main and effective process in which fungi, bacteria, and their spores increase the decomposition rate under both anaerobic and aerobic conditions (Iqbal et al., 2020). Biodecomposition increases nutrient quantity by phosphorous solubilization, nitrogen fixing, and cellulose decomposition (Manna et al., 1997).

Soil quality, fertility, and its capacity to enhance agricultural plants is decided by three interacting and mutually dependent components: chemical, physical, and biological fertility. Chemical fertility constitutes both nutrient quality and quantity that decide the chemical composition of soil, whether alkaline, acidic, or saline, which may be toxic to the plants (Ansari et al., 2022; Saxena et al., 2022). Physical fertility is attributed to the physical condition of the soil like soil texture, water-holding capacity, and absorption, whereas biological fertility reveals that some organisms live in the soil, interact with other components, and perform critical functions in the nutrient and carbon cycles, whereas some are harmful soil pathogens (Johns, 2017; Ahamd et al., 2019). Fungi and bacteria are known to boost soil texture and play an important role in soil fertility, nutrient cycling, and crop productivity, to encourage the production of soil pores and aggregates (Miller and Jastrow, 2000; Degens, 1997; Dhankar et al., 2020), and to sustain soil microorganisms that are important for optimal crop yield and soil fertility (Garbeva et al., 2004; Luo et al., 2016). Fungal cells secret mucilaginous substances composed of cell wall polysaccharides, surface polysaccharides, and intracellular or somatic polysaccharides present within the cytoplasmic membrane. Bacteria secret exopolysaccharides, which form organo-mineral compounds, and hold soil aggregates, which are important for improving both aeration and porosity in soil (Degens, 1997). Gupta and Germida (2015) observed that soil fertility not only depends on the mineral contents of soil but also results from the presence of soil microorganisms. Besides this, secretions from fungi, bacteria, decomposed cells, and animal and plant residues, especially in soils managed by organic inputs, are also responsible for improving the organic components of the soil, which, in turn, improve the fertility, quality, structure, and function of the soil. Treatments of soil with fertilizers also affect its fertility, chemical composition, structure, and microbial community. According to Zhang et al. (2012), organic amendments enhance the relative abundance of fungi and bacteria and lower the presence of *Actinomycetes*. However, balanced fertilization, organics, and low amounts of chemicals are beneficial for optimal microbial communities and their decomposition to improve both rice field fertility and rice quality (Liu et al., 2009). Decomposers break down proteins, starches, and other complex organic molecules inside the soil, convert compounds containing nitrogen (N), phosphorous (P), calcium (Ca), and sulfur (S) into forms that can be utilized by plants, and enhance soil fertility

(Irawan and Sutopo, 2017). There is also evidence that the growth of fungi in soils has an extremely important role in soil carbon (C) and nitrogen (N) turnover (D'Annibale et al., 2006). Jeffries et al. (2003) observed that the hyphae of AMF serve as a direct conduit for host C into the soil and directly contribute to its C pools, thus bypassing the decomposition process. Due to this, the activity and amount of other soil microorganisms are affected; however, this seems to be a selective method, since it particularly stimulates those microbes that exhibit antagonistic activity against soil-borne pathogens (Linderman, 2000). The reason for this phenomenon is unknown, but the observation clearly shows that AMF could be a useful biological tool for maintaining soil fertility and structure (Jeffries et al., 2003). Examining the mechanisms of bacterial and fungal inocula together with organic fertilizers could be an extremely useful tool for increasing soil fertility (Song et al., 2015) and aggregation. Such a strategy may help in reducing the utilization of chemical fertilizers, in managing eco-friendly integrated soil nutrients, and in sustaining soil fertility. A combined inoculation of both fungi and bacteria with organics could be an efficient method for management of crop production and soil fertility (Rillig et al., 2002; Minerdi et al., 2001; Mortimer et al., 2008; Ceaser-Tonhat et al., 2014) in strongly degraded soil. Several fungal and bacterial species play a crucial role in increasing soil fertility. These microorganisms enhance soil organic matter that increase the availability of N, P, K, and iron in soil (Egamberdiyeva and Höflich, 2004; Leifheit et al., 2015; Ceaser-Tonhat et al., 2014). Such microorganisms are also involved in the formation of certain essential organic acids, which enhance the uptake and mobilization of nutrients in the soil, and, due to this, the dependency on chemical fertilizers in agroecosystems will greatly reduce (Figueiredo et al., 2010). However, researchers are engaged in finding out the mechanism by which the inocula of microbes promote soil fertility, although it is believed that microbes use various direct and indirect mechanisms for improving soil fertility and structure (Glick, 2012; Nishat et al., 2022).

References

Abbasi, H., Akhtar, A., Sharf, R., 2015. Vesicular arbuscular mycorrhizal (VAM) fungi: a tool for sustainable agriculture. Am. J. Plant Nutr. Fertil. Technol. 5, 40–49.

Ahamd, G., Nishat, Y., Haris, M., Danish, M., Hussain, T., 2019. Efficiency of soil, plant and microbes for the healthy plant immunity and sustainable agricultural system. In: Varma, A., Tripathi, S., Prasad, R. (Eds.), Plant-Microbe Interface. Springer, Cham. https://doi.org/10.1007/978-3-030-19831-2_15.

Ahmad, R., Jilani, G., Arshad, M., Zahir, Z.A., Khalid, A., 2007. Bio-conversion of organic wastes for their recycling in agriculture: an overview of perspectives and prospects. Ann. Microbiol. 57, 471–479. https://doi.org/10.1007/BF031 75343.

Ahmad, M., Taylor, C.R., Pink, D., Burton, K.S., Eastwood, D.C., Bending, G.D., 2010. Development of novel assays for lignin degradation: comparative analysis of bacterial and fungal lignin degraders. Mol. BioSyst. 6, 815–821. https://doi.org/10.1007/BF03175343.

Akhtar, M., Sarwar, N., Ashraf, A., Ejaz, A., Ali, S., Rizwan, M., Science, S., 2020. Beneficial role of Azolla sp. in paddy soils and their use as bioremediators in polluted aqueous environments: implications and future perspectives. Arch. Agron., 1–14.

Aktar, M.W., Sengupta, D., Chowdhury, A., 2009. Impact of pesticides use in agriculture: their benefits and hazards. Interdiscip. Toxicol. 2, 1–12. https://doi.org/10.2478/v10102-009-0001-7.

Ali, S.S., Abomohra, A.E.F., Sun, J., 2017. Effective bio-pretreatment of sawdust waste with a novel microbial consortium for enhanced biomethanation. Bioresour. Technol. 238, 425–432. https://doi.org/10.1016/j.biortech.2017.03.187.

Al-Turki, A.I., 2010. Quality assessment of commercially produced composts in Saudi Arabia market. Int. J. Agric. Res. 5, 70–79.

Ansari, M.S., Atirah, T., Haris, M., Khan, A., Hussain, T., Khan, A.A., 2022. Effects of heavy metals present in sewage sludge, their impact on soil fertility, soil microbial activity, and environment. In: Shah, M.P., Rodriguez-Couto, S., Banerjee, R. (Eds.), Development in Waste Water Treatment Research and Processes Treatment and Reuse of Sewage Sludge: An Innovative Approach for Wastewater Treatment. Elsevier, pp. 197–214. https://doi.org/10.1016/B978-0-323-85584-6.00013-3.

Arora, N.K., Verma, M., Mishra, J., 2017. Rhizobial bioformulations: past, present and future. In: Rhizotrophs: Plant Growth Promotion to Bioremediation. Springer, Berlin/Heidelberg, Germany, pp. 69–99.

Baig, K.S., Arshad, M., Shaharoona, B., Khalid, A., Ahmed, I., 2012. Comparative effectiveness of Bacillus spp. possessing either dual or single growth-promoting traits for improving phosphorus uptake, growth and yield of wheat (*Triticum aestivum* L.). Ann. Microbiol. 62, 1109–1119. https://doi.org/10.1007/s13213-011-0352-0.

Benimelia, C.S., Castroa, G.R., Chailec, A.P., Amoroso, M.J., 2007. Lindane uptake and degradation by aquatic Streptomyces sp. strain M7. Int. Biodeterior. Biodegrad. 59, 148–155. https://doi.org/10.1016/j.ibiod.2006.07.014.

Bernal, M.P., Alburquerque, J.A., Moral, R., 2009. Composting of animal manures and chemical criteria for compost maturity assessment. A review. Bioresour. Technol. 10, 5444–5453. https://doi.org/10.1016/j.biort ech.2008.11.027.

Bharti, N., Yadav, D., Barnawal, D., Maji, D., Kalra, A., 2013. *Exiguobacterium oxidotolerans*, a halotolerant plant growth promoting rhizobacteria, improves yield and content of secondary metabolites in *Bacopa monnieri* (L.) Pennell under primary and secondary salt stress. J. World J. Microbiol. Biotechnol. 29, 379–387.

Bhat, T.A., Ahmad, L., Ganai, M.A., Khan, O., 2015. Nitrogen fixing biofertilizers; mechanism and growth promotion: a review. J. Pure Appl. Microbiol. 9, 1675–1690.

Bhiday, M.R., 1994. Earthworms in agriculture. Indian Farm 43, 31–34.

Bholay, A., Borkhataria, B.V., Jadhav, P.U., Palekar, K.S., Dhalkari, M.V., Nalawade, P.M., 2012. Bacterial lignin peroxidase: a tool for biobleaching and biodegradation of industrial effluents. Univers J Environ Res Technol 2, 58–64.

Bumandalai, O., Tserennadmid, R., 2019. Effect of *Chlorella vulgaris* as a biofertilizer on germination of tomato and cucumber seeds. Int. J. Aquat. Biol. 7, 95–99.

Cabuk, A., Unal, A.T., Kolankaya, N., 2006. Biodegradation of cyanide by a white rot fungus, *Trametes versicolor*. Biotechnol. Lett. 28, 1313–1317. https://doi.org/10.1007/s10529-006-9090-y.

Ceaser-Tonhat, T.C., Stevens, W.B., Sainju, U.M., Ceaser, A.J., West, M., Gaskin, J.F., 2014. Soil-aggregating bacterial community as affected by irrigation, tillage, and cropping system in the northern great plains. Soil Sci. 179 (1), 11–20.

Chandra, R., Raj, A., Purohit, H.J., Kapley, A., 2007. Characterization and optimization of three potential aerobic bacterial strains for kraft lignin degradation from pulp paper waste. Chemosphere 67, 839–846. https://doi.org/10.1016/j.chemosphere.2006.10.011.

Chandra, R., Singh, R., Yadav, S., 2012. Effect of bacterial inoculum ratio in mixed culture for decolourization and detoxification of pulp paper mill effluent. J. Chem. Technol. Biotechnol. 87, 436–444. https://doi.org/10.1002/jctb.2758.

D'Annibale, A., Rosetto, F., Leonardi, V., Federici, F., Petruccioli, M., 2006. Role of autochthonous filamentous fungi in bioremediation of a soil historically contaminated with aromatic hydrocarbons. Appl. Environ. Microbiol. 72 (1), 28–36.

Datta, A., Singh, R.K., Tabassum, S., 2015. Isolation, characterization and growth of Rhizobium strains under optimum conditions for effective biofertilizer production. Int. J. Pharm. Sci. Rev. Res. 32, 199–208.

Degens, B.P., 1997. Macro-aggregation of soils by biological bonding and binding mechanisms and the factors affecting these: a review. Soil Res. 35 (3), 431–460.

DeLuca, T., Pingree, M., Gao, S., 2019. Assessing soil biological health in forest soils. In: Developments in Soil Science. Vol. 36. Elsevier, Amsterdam, The Netherlands, pp. 397–426.

Dhankar, R., Goyal, S., Kumar, K., Hussain, T., 2021. Bacterial community response to pesticides polluted soil. In: Mandal, S., et al. (Eds.), Recent Advancement in Microbial Biotechnology—Agricultural and Industrial Approach. Elsevier, pp. 339–355. https://doi.org/10.1016/B978-0-12-822098-6.00010-0.

Dhankar, R., Tyagi, P., Kamble, S.S., Gupta, D., Hussain, T., 2020. Advances in fungi: rejuvenation of polluted sites. In: Sharma, V.K., Shah, M.P., Parmar, S., Kumar, A. (Eds.), Fungi Bio-Prospects in Sustainable Agriculture, Environment and Nano-Technology. vol. 2. Academic Press, UK, pp. 251–275. https://doi.org/10.1016/B978-0-12-821925-6.00012-5.

Dwivedi, S., Sangeeta, G.R., Gopal, R., 2015. Role of mycorrhizae as biofertilizer and bioprotectant. Int J Pharm. Bio. Sci 6, 1014–1026.

Egamberdiyeva, D., Höflich, G., 2004. Effect of plant growth-promoting bacteria on growth and nutrient uptake of cotton and pea in a semi-arid region of Uzbekistan. J. Arid Environ. 56 (2), 293–301.

Eilers, K.G., Debenport, S., Anderson, S., Fierer, N., 2012. Digging deeper to find unique microbial communities: the strong effect of depth on the structure of bacterial and archaeal communities in soil. Soil Biol. Biochem. 50, 58–65.

Erickson, M.C., Liao, J., Ma, L., Jiang, X., Doyle, M.P., 2009. Inactivation of *Salmonella* spp. In crow manure composts formulated to different initial C:N ratios. Bioresour. Technol. 100, 5898–5903.

FAO, 2017. The future of food and agriculture-trends and challenges. FAO, Rome. http://www.fao.org/3/a-i6583e.pdf. (Accessed 18 December 2018).

Figueiredo, M.D.V.B., Seldin, L., de Araujo, F.F., Mariano, R.D.L.R., 2010. Plant growth promoting rhizobacteria: fundamentals and applications. In: Plant Growth and Health Promoting bacteria. Springer, Berlin, Heidelberg, pp. 21–43.

Flores-Félix, J.D., Menéndez, E., Rivera, L.P., Marcos-García, M., Martínez-Hidalgo, P., Mateos, P.F., Martínez-Molina, E., Velázquez, M.D.L.E., García-Fraile, P., Rivas, R., 2013. Use of *Rhizobium leguminosarum* as a potential biofertilizer for *Lactuca sativa* and *Daucus carota* crops. J. Plant Nutr. Soil Sci. 176, 876–882.

Fourti, O., Jedidi, N., Hassen, A., 2011. Comparison of methods for evaluating stability and maturity of co-composting of municipal solid wastes and sewage sludge in semi-arid pedo-climatic condition. Nat. Sci. 3, 124–135.

Gaind, S., Nain, L., 2007. Chemical and biological properties of wheat soil in response to paddy straw incorporation and its biodegradation by fungal inoculants. Biodegradation 18, 495–503. https://doi.org/10.1007/s10532-006-9082-6.

Galindo, F.S., Teixeira Filho, M.C.M., Buzetti, S., Rodrigues, W.L., Fernandes, G.C., Boleta, E.H.M., Neto, M.B., de Azambuja Pereira, M.R., Rosa, P.A.L., Pereira, Í. T., 2020. Influence of *Azospirillum brasilense* associated with silicon and nitrogen fertilization on macronutrient contents in corn. Open Agric. 5, 126–137.

Garbeva, P.V., Van Veen, J.A., Van Elsas, J.D., 2004. Microbial diversity in soil: selection of microbial populations by plant and soil type and implications for disease suppressiveness. Annu. Rev. Phytopathol. 42, 243–270.

Glick, B.R., 2012. Plant growth-promoting bacteria: mechanisms and applications. Scientifica 2012.

Gupta, V.V., Germida, J.J., 2015. Soil aggregation: influence on microbial biomass and implications for biological processes. Soil Biol. Biochem. 80, A3–A9.

Gupta, G., Panwar, J., Akhtar, M.S., Jha, P.N., 2012. Endophytic nitrogen-fixing bacteria as biofertilizer. In: Sustainable Agriculture Reviews. Springer, Berlin/Heidelberg, Germany, pp. 183–221.

Haris, M., Ahmad, G., Shakeel, A., Khan, A.A., 2019. Utilization of fly ash to improve the growth and the management of root-knot nematode on carrot. Haya Saudi J. Life Sci. 4, 221–226.

Haris, M., Shakeel, A., Ansari, M.A., Hussain, T., Khan, A.A., Dhankar, R., 2020. Sustainable crop production and improvement through bio-prospecting of fungi. In: Sharma, V. K., Shah, M.P., Parmar, S., Kumar, A. (Eds.), Fungi Bio-Prospects in Sustainable Agriculture, Environment and Nano- Technology. Vol. 1. Academic Press, UK, pp. 407–428. https://doi.org/10.1016/B978-0-12-821394-0.00016-0.

Hoornweg, D., Thomas, L., Otten, L., 2000. Composting and its applicability in developing countries. In: Urban Waste Management. Published for the Urban Development Division the World Bank, Washington, DC. Working Paper Series. 8.

Huang, H.L., Zeng, G.M., Jiang, R.Q., Yuan, X.Z., Yu, M., 2009. Fluorescence spectroscopy characteristics of humic acid by inoculating white-rot fungus during different phases of agricultural waste composting. J. Cent. S. Univ. Technol. 16, 440–443. https://doi.org/10.1007/s1177 1-009-0074-7.

Huang, D.L., Zeng, G.M., Feng, C.L., et al., 2010. Changes of microbial population structure related to lignin degradation during lignocellulosic waste composting. Bioresour. Technol. 101, 4062–4067. https://doi.org/10.1016/j.biort ech.2009.12.145.

Huber, A., Praznik, W., 2004. Identification and quantification of renewable crop materials. In: Stevens, C.V., Verhe, R. (Eds.), Renewable Bioresources: Scope and Modifications for Non-food Applications. Wiley, England.

Husain, R., Vikram, N., Pandey, S., et al., 2022. Microorganisms: an eco-friendly tools for the waste management and environmental safety. In: Arora, S., Kumar, A., Ogita, S., Yau, Y.Y. (Eds.), Biotechnological Innovations for Environmental Bioremediation. Springer, Singapore. https://doi.org/10.1007/978-981-16-9001-3_36.

Hussain, T., Akhtar, N., Aminedi, R., Danish, M., Nishat, Y., Patel, S., 2020a. Role of the potent microbial based bioagent and their emerging strategies for the eco-friendly management of Agricultural Phytopathogens. In: Singh, J., Yadav, A.N. (Eds.), Natural Bioactive Products in Sustainable Agriculture. Springer, Singapore, pp. 45–66.

Hussain, T., Singh, S., Danish, M., Pervez, R., Hussain, K., Husain, R., 2020b. Natural metabolites an eco-friendly approach to manage plant diseases and for better agricultural farming. In: Singh, J., Yadav, A.N. (Eds.), Natural Bioactive Products in Sustainable Agriculture. Springer, Singapore, pp. 1–13. https://doi.org/10.1007/978-981-15-3024-1_1.

Ilyas, N., Mumtaz, K., Akhtar, N., Yasmin, H., Sayyed, R., Khan, W., Enshasy, H.A.E., Dailin, D.J., Elsayed, E.A., Ali, Z., 2020. Exopolysaccharides producing bacteria for the amelioration of drought stress in wheat. J. Sustain. 12, 8876.

Iqbal, N., Agrawal, A., Dubey, S., Kumar, J., 2020. Role of decomposers in agricultural waste management. In: Biomass. Intech Open.

Irawan, B., Sutopo, H.A.D.I., 2017. Effects of saprophytic microfungi application on soil fertility based on their decomposition properties. J. Appl. Biol. Sci. 11 (2), 15–19.

Itelima, J., Bang, W., Onyimba, I., Oj, E., 2018. A review: biofertilizer; a key player in enhancing soil fertility and crop productivity. J. Microbiol. Biotechnol. Rep. 2, 22–28.

Jeffries, P., Gianinazzi, S., Perotto, S., Turnau, K., Barea, J.M., 2003. The contribution of arbuscular mycorrhizal fungi in sustainable maintenance of plant health and soil fertility. Biol. Fertil. Soils 37 (1), 1–16.

Johns, C., 2017. The role of carbon in promoting healthy soils. In: Strategic Analysis Paper. Future Directions International, p. 10.

Kalayu, G., 2019. Phosphate solubilizing microorganisms: promising approach as biofertilizers. Int. J. Agron., 1–7.

Kamran, S., Shahid, I., Baig, D.N., Rizwan, M., Malik, K.A., Mehnaz, S., 2017. Contribution of zinc solubilizing bacteria in growth promotion and zinc content of wheat. Front. Microbiol. 8 (2593). https://doi.org/10.3389/fmicb.2017.02593.

Kaosol, T., 2009. Sustainable solutions for municipal solid waste management in Thailand. World Acad. Sci. Eng. Technol. 60, 665–670.

Kawalekar, J.S., 2013. Role of biofertilizers and biopesticides for sustainable agriculture. J. Bio. Innov. 2, 73–78.

Kersten, P., Cullen, D., 2007. Extracellular oxidative systems of the lignin-degrading basidiomycete *Phanerochaete chrysosporium*. Fungal Genet. Biol. 44, 77–87. https://doi.org/10.1016/j.fgb.2006.07.007.

Khan, A., Anwar Ansari, M.S., Hussain, T., Khan, A.A., 2022. Chapter 7: Role of beneficial microbes for plant growth improvement. In: Soni, R., Suyal, D.C., Goel, R. (Eds.), Plant Protection: From Chemicals to Biologicals. De Gruyter, Berlin, Boston, pp. 141–176. https://doi.org/10.1515/9783110771558-007.

Kumar, A., Chaurasia, U., Elshobary, M.E., Kumari, S., Hussain, T., Bharti, A.P., Maurya, D.K., Samntha, L., El-Sheekh, M.M., 2022. Utilization of algae in crop improvement and crop protection for a better agricultural system. In: El-Sheekh, M.M., Abdullah, N., Ahmad, I. (Eds.), Handbook of Research on Algae as a Sustainable Solution for Food, Energy, and the Environment. IGI Global Publisher, USA, pp. 442–470. https://doi.org/10.4018/978-1-6684-2438-4.ch018.

Kumar, A., Hussain, T., Susmita, C., Maurya, D.K., Danish, M., Farooqui, S.A., 2021. Microbial remediation and detoxification of heavy metals by plants and microbes. In: Shah, M., et al. (Eds.), The Future of Effluent Treatment Plants – Biological Treatment Systems. Elsevier, UK, pp. 589–614. https://doi.org/10.1016/B978-0-12-822956-9.00030-1.

Kumar, A.K., Sharma, S., 2017. Recent updates on different methods of pretreatment of lignocellulosic feedstocks: a review. Bioresour. Bioprocess. 4 (7). https://doi.org/10.1186/s40643-017-0137-9.

Kumari, B., Mallick, M., Hora, A., 2016. Plant growth-promoting rhizobacteria (PGPR): their potential for development of sustainable agriculture. Bioexploit. Sustain. Agric., 1–19.

Lal, R., 2015. Restoring soil quality to mitigate soil degradation. Sustainability 7, 5875–5895. https://doi.org/10.3390/su7055875.

Leifheit, E.F., Verbruggen, E., Rillig, M.C., 2015. Arbuscular mycorrhizal fungi reduce decomposition of woody plant litter while increasing soil aggregation. Soil Biol. Biochem. 81, 323–328.

Leite, R.D.C., dos Santos, J.G., Silva, E.L., Alves, C.R., Hungria, M., Leite, R.D.C., dos Santos, A.C., 2019. Productivity increase, reduction of nitrogen fertiliser use and drought-stress mitigation by inoculation of Marandu grass (*Urochloa brizantha*) with *Azospirillum brasilense*. J. Crop. Pasture Sci. 70, 61–67.

Liers, C., Ullrich, R., Steffen, K.T., Hatakka, A., Hofrichter, M., 2006. Mineralization of 14C-labelled synthetic lignin and extracellular enzyme activities of the wood-colonizing ascomycetes *Xylaria hypoxylon* and *Xylaria polymorpha*. Appl. Microbiol. Biotechnol. 69, 573–579. https://doi.org/10.1007/s0025 3-005-0010-1.

Linderman, R.G., 2000. Effects of mycorrhizas on plant tolerance to diseases. In: *Arbuscular mycorrhizas*: Physiology and Function. Springer, Dordrecht, pp. 345–365.

Liu, M., Hu, F., Chen, X., Huang, Q., Jiao, J., Zhang, B., Li, H., 2009. Organic amendments with reduced chemical fertilizer promote soil microbial development and nutrient availability in a subtropical paddy field: the influence of quantity, type and application time of organic amendments. Appl. Soil Ecol. 42 (2), 166–175.

Luo, X., Fu, X., Yang, Y., Cai, P., Peng, S., Chen, W., Huang, Q., 2016. Microbial communities play important roles in modulating paddy soil fertility. Sci. Rep. 6 (1), 1–12.

Mabrouk, Y., Hemissi, I., Salem, I.B., Mejri, S., Saidi, M., Belhadj, O., 2018. Potential of rhizobia in improving nitrogen fixation and yields of legumes. Symbiosis, 107–122.

Mahdi, S.S., Hassan, G., Samoon, S., Rather, H., Dar, S.A., Zehra, B., 2010. Bio-fertilizers in organic agriculture. J. Phytol. 2, 42–54.

Mandal, S.D., Sonali, Singh, S., Hussain, K., Hussain, T., 2021. Plant microbe association for the mutual benefits for plant growth and soil health. In: Yadav, A.N., et al. (Eds.), Current Trends in Microbial Biotechnology for Sustainable Agriculture, Environmental and Microbial Biotechnology. Springer Nature, Singapore, pp. 95–121. https://doi.org/10.1007/978-981-15-6949-4_5.

Manna, M.C., Hazra, J.N., Sinha, N.B., Ganguly, T.K., 1997. Enrichment of compost by bioinoculants and mineral amendments. J. Indian Soc. Soil Sci. 45 (4), 831–833.

Martinez, A.T., 2002. Molecular biology and structure–function of lignin-degrading heme peroxidases. Enzym. Microb. Technol. 30 (425–432). https://doi.org/10.1016/S0141-0229(01)00521-X.

Meena, V.S., Maurya, B., Verma, J.P., 2014. Does a rhizospheric microorganism enhance K + availability in agricultural soils? Microbiol. Res. 169, 337–347.

Meena, V.S., Mishra, P.K., Bisht, J.K., Pattanayak, A., 2017. Agriculturally Important Microbes for Sustainable Agriculture: Volume 2: Applications in Crop Production and Protection. Springer, Berlin/Heidelberg, Germany.

Miller, R.M., Jastrow, J.D., 2000. Mycorrhizal fungi influence soil structure. In: Arbuscular Mycorrhizas: Physiology and Function. Springer, Dordrecht, pp. 3–18.

Minerdi, D., Fani, R., Gallo, R., Boarino, A., Bonfante, P., 2001. Nitrogen fixation genes in an endosymbiotic *Burkholderia* strain. Appl. Environ. Microbiol. 67 (2), 725–732.

Mishra, P., Dash, D., 2014. Rejuvenation of biofertilizer for sustainable agriculture and economic development. Consilience, 41–61.

Mortimer, P.E., Pérez-Fernández, M.A., Valentine, A.J., 2008. The role of arbuscular mycorrhizal colonization in the carbon and nutrient economy of the tripartite symbiosis with nodulated *Phaseolus vulgaris*. Soil Biol. Biochem. 40 (5), 1019–1027.

Mosier, N., Wyman, C.E., Dale, B.E., Elander, R., Lee, Y.Y., Holtzapple, M., Ladisch, M., 2005. Features of promising technologies for pretreatment of lignocellulosic biomass. Bioresour. Technol. 96, 673–686. https://doi.org/10.1016/j.biortech.2004.06.025.

Mshandete, A.M., Cuff, J., 2008. Cultivation of three types of indigenous wild edible mushrooms: *Coprinus cinereus*, *Pleurotus flabellatus* and *Volvariella volvacea* on composted sisal decortications residue in Tanzania. Afr. J. Biotechnol. 7, 4551–4562.

Mukherjee, S., Pandey, V., Parvez, A., Qi, X., Hussain, T., 2022. Bacillus as a versatile tool for crop improvement and agro-industry. In: Islam, M.T., Rahman, M., Pandey,

P. (Eds.), Bacilli in Agrobiotechnology. Bacilli in Climate Resilient Agriculture and Bioprospecting. Springer, Cham, pp. 429–452. https://doi.org/10.1007/978-3-030-85465-2_19.

Nagavallemma, K.P., Wani, S.P., Stephane, L., Padmaja, V.V., Vineela, C., Babu, Rao, M., Sahrawat, K.L., 2006. Vermicomposting: recycling wastes into valuable organic fertilizer. ICRISAT 2, 16.

Ng, L.C., Sariah, M., Radziah, O., Zainal Abidin, M.A., Sariam, O., 2016. Development of microbial-fortified rice straw compost to improve plant growth, productivity, soil health, and rice blast disease management of aerobic rice. Compost. Sci. Util. 24 (86097).

Nishat, Y., Mohammad, D., Siddiqui, N., Hussain, T., 2022. Removal of emerging pollutants from the environment through microbes. In: Samuel, J., Kumar, A., Singh, J. (Eds.), Relationship Between Microbes and the Environment for Sustainable Ecosystem Services. Microbial Mitigation of Waste for Sustainable Ecosystem Services, vol. 2. Elsevier, pp. 181–203. https://doi.org/10.1016/B978-0-323-89937-6.00002-4.

Novinsak, A., Surette, C., Allain, C., Filion, M., 2008. Application of molecular technologies to monitor the microbial content of biosolids and composted biosolids. Water Sci. Technol. 57, 471–477.

Pallavi, Chandra, D., Sharma, A.K., 2017. Commercial microbial products: exploiting beneficial plant-microbe interaction. In: Singh, D.P., Singh, H.B., Prabha, R. (Eds.), Plant–Microbe Interactions in Agro-Ecological Perspectives. Springer, Singapore, https://doi.org/10.1007/978-981-10-6593-4_2.

Palm, C.A., Swift, M.J., 2002. Soil fertility as an ecosystem concept. In: Accomplishments and Changing Paradigm Towards the 21st Century. Proceedings of the 17th world Congress of Soil Science, Bangkok, Thailand, pp. 14–21.

Pampuro, N., Dinuccio, E., Balsari, P., Cavallo, E., 2016. Evaluation of two composting strategies for making pig slurry solid fraction suitable for pelletizing. Atmos. Pollut. Res. 7, 288–293. https://doi.org/10.1016/j.apr.2015.10.001.

Parikh, S.J., James, B.R., 2012. Soil: the foundation of agriculture. Nat. Educ. Knowl. 3, 2.

Perez, J., Munoz-Dorado, J., de la Rubia, T., Martinez, J., 2014. Biodegradation and biological treatments of cellulose, hemicellulose and lignin: an overview. Int. Microbiol. 5, 53–63. https://doi.org/10.1007/s10123-002-0062-3.

Petre, M., Petre, V., Rusea, I., 2014. Microbial composting of fruit tree wastes through controlled submerged fermentation. Ital. J. Agron. 9, 152–156. https://doi.org/10.4081/ija.2014.610.

Pindi, P.K., 2012. Liquid microbial consortium – a potential tool for sustainable soil health. J. Fertil. Pestic. 3, 124.

Pleissner, D., Kwan, T.H., Lin, C.S.K., 2013. Fungal hydrolysis in submerged fermentation for food waste treatment and fermentation feedstock preparation. Bioresour. Technol. 158, 48–54. https://doi.org/10.1016/j.biortech.2014.01.139.

Prajapati, K., Modi, H., 2012. The importance of potassium in plant growth – a review. Indian J. Plant Sci. 1, 177–186.

Pretty, J., Bharucha, Z.P., 2014. Sustainable intensification in agricultural systems. Ann. Bot. 114, 1571–1596. https://doi.org/10.1093/aob/mcu205.

Preusser, S., Poll, C., Marhan, S., Angst, G., Mueller, C.W., Bachmann, J., Kandeler, E., 2019. Fungi and bacteria respond differently to changing environmental conditions within a soil profile. Soil Biol. Biochem. 137, 107543.

Pugliese, M., Liu, B., Gullino, M.L., Garibaldi, A., 2011. Microbial enrichment of compost with biological control agents to enhance suppressiveness to four soil-borne diseases in greenhouse. J. Plant Dis. Prot. 118, 45–50. https://doi.org/10.1007/BF03356380.

Raghuwanshi, R., 2012. Opportunities and challenges to sustainable agriculture in India. Nebio 3, 78–86.

Raj, A., Chandra, R., Reddy, M.M.K., Hemant, J.P., Kapley, A., 2007. Biodegradation of kraft lignin by a newly isolated bacterial strain, *Aneurinibacillus aneurinilyticus* from the sludge of a pulp paper mill. World J. Microbiol. Biotechnol. 23, 793–799. https://doi.org/10.1007/s1127 4-006-9299-x.

Reed, S.C., Cleveland, C.C., Townsend, A.R., 2011. Functional ecology of free-living nitrogen fixation: a contemporary perspective. Annu. Rev. Ecol. Evol. Syst. 42, 489–512.

Riaz, U., Mehdi, S.M., Iqbal, S., Khalid, H.I., Qadir, A.A., Anum, W., Ahmad, M., Murtaza, G., 2020. Bio-fertilizers: eco-friendly approach for plant and soil environment. In: Bioremediation and Biotechnology. Springer, Berlin/Heidelberg, Germany, pp. 189–213.

Ribeiro, N.D.Q., Souza, T.P., Costa, L.M.A.S., Castro, C.P.D., Dias, E.S., 2017. Microbial additives in the composting process. Ciência e Agrotecnologia 41, 159–168. https://doi.org/10.1590/1413-7054201741 20382 16.

Rillig, M.C., Wright, S.F., Eviner, V.T., 2002. The role of arbuscular mycorrhizal fungi and glomalin in soil aggregation: comparing effects of five plant species. Plant Soil 238 (2), 325–333.

Romero, E., Esperanza, M., Garcia-Guinea, J., Martinez, A.T., Martínez, M.J., 2007. An anamorph of the white-rot fungus *Bjerkandera adusta* capable of colonizing and degrading compact disc components. FEMS Microbiol. Lett. 275, 122–129. https://doi.org/10.1111/j.1574-6968.2007.00876.x.

Rubio, R., Perez-Murcia, M.D., Agullo, E., Bustamante, M.A., Sanchez, C., Paredes, C., Moral, R., 2013. Recycling of agro-food wastes into vineyards by composting: agronomic validation in field conditions. Commun. Soil Sci. Plant Anal. 44, 502–516. https://doi.org/10.1080/00103624.2013.744152.

Sadhana, B., 2014. Arbuscular Mycorrhizal Fungi (AMF) as a biofertilizer – a review. Int. J. Curr. Microbiol. App. Sci. 3, 384–400.

Sahoo, R.K., Ansari, M.W., Dangar, T.K., Mohanty, S., Tuteja, N., 2014. Phenotypic and molecular characterisation of efficient nitrogen fixing Azotobacter strains from rice fields for crop improvement. Protoplasma 251, 511–523.

Sammauria, R., Kumawat, S., Kumawat, P., Singh, J., Jatwa, T.K., 2020. Microbial inoculants: potential tool for sustainability of agricultural production systems. Arch. Microbiol. 202, 677–693.

Sanchez, C., 2009. Lignocellulosic residues: biodegradation and bioconversion by fungi. Biotechnol. Adv. 27, 185–194. https://doi.org/10.1016/j.biotechadv.2008.11.001.

Sanmanee, N., Panishkan, K., Obsuwan, K., Dharmvanij, S., 2011. Study of compost maturity during humification process using UV spectroscopy. World Acad. Sci. Eng. Technol. 80, 403–405.

Sara, S., Morad, M., Reza, C.M., 2013. Effects of seed inoculation by Rhizobium strains on chlorophyll content and protein percentage in common bean cultivars (*Phaseolus vulgaris* L.). Int. J. Biosci. 3, 1–8.

Saxena, K., Hussain, T., Dhanker, R., Jain, P., Goyal, S., 2022. Phytoremediation: a sustainable solution to combat pollution. In: Arora, S., Kumar, A., Ogita, S., Yau, Y.Y. (Eds.), Biotechnological Innovations for Environmental Bioremediation. Springer, Singapore. https://doi.org/10.1007/978-981-16-9001-3_11.

Schulz, S., Brankatschk, R., Dumig, A., Kogel-Knabner, I., Schloter, M., 2013. The role of microorganisms at different stages of ecosystem development for soil formation. Biogeosciences 10 (6), 3983–3996.

Seyedbagheri, M.M., 2010. Compost: Production, Quality, and Use in Commercial Agriculture. CIS 1175. University of Idaho, USA. http://cals.uidaho.edu/edcomm/pdf/CIS/CIS1175.pdf. (Accessed 18 December 2018).

Sharma, N.K., Tiwari, S.P., Tripathi, K., Rai, A.K., 2011. Sustainability and cyanobacteria (blue-green algae): facts and challenges. J. Appl. Phycol. 23, 1059–1081.

Sharma, S.B., Sayyed, R.Z., Trivedi, M.H., Gobi, T.A., 2013. Phosphate solubilizing microbes: sustainable approach for managing phosphorus deficiency in agricultural soils. Springerplus 2, 587.

Shrestha, S., Fonoll, X., Khanal, S.K., Raskin, L., 2017. Biological strategies for enhanced hydrolysis of lignocellulosic biomass during anaerobic digestion: current status and future perspectives. Bioresour. Technol. 245 (Pt-A), 1245–1257. https://doi.org/10.1016/j.biort ech.2017.08.089.

Siddiqua, K.S., Farooqui, S.A., Hussain, T., Mohamed, H.I., 2021. Microbial enzymes and their role in phytoremediation. In: Mohamed, H.I., El-Beltagi, H.E.D.S., Abd-Elsalam, K.A. (Eds.), Plant Growth-Promoting Microbes for Sustainable Biotic and Abiotic Stress Management. Springer, Cham. https://doi.org/10.1007/978-3-030-66587-6_22.

Siddiqui, Y., Meon, S., Ismail, M.R., Ali, A., 2008. Trichoderma-fortified compost extracts for the control of choanephora wet rot in okra production. Crop Prot. 27, 385–390. https://doi.org/10.1016/j.cropro.2007.07.002.

Singh, S., Nain, L., 2015. Microorganisms in the conversion of agricultural wastes to compost. Proc. Indian Natl. Sci. Acad. 80, 473–481. https://doi.org/10.16943/ptinsa/2014/v80i2/7.

Singh, D.P., Prabha, R., 2017. Bioconversion of agricultural wastes into high value biocompost: a route to livelihood generation for farmers. Adv. Recycl. Waste Manag. 2, 3. https://doi.org/10.4172/2475-7675.1000137.

Singh, A., Sharma, S., 2002. Composting of a crop residue through treatment with microorganisms and subsequent vermicomposting. Bioresour. Technol. 85, 107–111.

Singh, J.S., Kumar, A., Rai, A.N., Singh, D.P., 2016. Cyanobacteria: a precious bio-resource in agriculture, ecosystem, and environmental sustainability. Front. Microbiol. 7, 529.

Song, X., Liu, M., Wu, D., Griffiths, B.S., Jiao, J., Li, H., Hu, F., 2015. Interaction matters: synergy between vermicompost and PGPR agents improves soil quality, crop quality and crop yield in the field. Appl. Soil Ecol. 89, 25–34.

Steger, K., Sjogren, A.M., Jarvis, A., Jansson, J.K., Sundh, I., 2007. Development of compost maturity and Actinobacteria populations during full-scale composting of organic household waste. J. Appl. Microbiol. 103, 487–498.

Sudharmaidevi, C.R., Thampatt, K.C.M., Saifudeen, N., 2017. Rapid production of organic fertilizer from degradable waste by thermochemical processing. Int. J. Recycl. Org. Waste Agric. 6, 1–11. https://doi.org/10.1007/s40093-016-0147-1.

Tavallali, V., Rahemi, M., Eshghi, S., Kholdebarin, B., Ramezanian, A., 2010. Zinc alleviates salt stress and increases antioxidant enzyme activity in the leaves of pistachio (*Pistacia vera* L. 'Badami') seedlings. Turk. J. Agric. For. 34, 349–359.

Taylor, J.P., Wilson, B., Mills, M.S., Burns, R.G., 2002. Comparison of microbial numbers and enzymatic activities in surface soils and subsoils using various techniques. Soil Biol. Biochem. 34 (3), 387–401.

Timmusk, S., Abd El-Daim, I.A., Copolovici, L., Tanilas, T., Kännaste, A., Behers, L., Nevo, E., Seisenbaeva, G., Stenström, E., Niinemets, Ü., 2014. Drought-tolerance of wheat improved by rhizosphere bacteria from harsh environments: enhanced biomass production and reduced emissions of stress volatiles. PLoS ONE 9, e96086.

Umesha, S., Singh, P.K., Singh, R.P., 2018. Microbial biotechnology and sustainable agriculture. In: Biotechnology for Sustainable Agriculture. Elsevier, Amsterdam, The Netherlands, pp. 185–205.

Umsakul, K., Dissara, Y., Srimuang, N., 2010. Chemical physical and microbiological changes during composting of the water hyacinth. Pak. J. Biol. Sci. 13, 985–992.

Vala, Y., Desai, N.B., 2020. Biofertilizers: an approach towards saving fertilizers. Available online: https://www.academia.edu/download/65194961/AEN_Issue3_Nov2020.pdf#page=17. (Accessed 8 February 2021).

Varma, V.S., Das, S., Sastri, C.V., Kalamdhad, A.S., 2017. Microbial degradation of lignocellulosic fractions during drum composting of mixed organic waste. Sustain. Environ. Res. 27, 265–272. https://doi.org/10.1016/j.serj.2017.05.004.

Verma, P., Yadav, A.N., Khannam, K.S., Kumar, S., Saxena, A.K., Suman, A., 2016. Molecular diversity and multifarious plant growth promoting attributes of Bacilli associated with wheat (*Triticum aestivum* L.) rhizosphere from six diverse agro-ecological zones of India. J. Basic Microbiol. 56, 44–58.

Vesela, M., Friedrich, J., 2009. Amino acid and soluble protein cocktail from waste keratin hydrolysed by a fungal keratinase of *Paecilomyces marquandii*. Biotechnol. Bioprocess Eng. 14, 84–90. https://doi.org/10.1007/s12257-008-0083-7.

Wani, S.A., Chand, S., Wani, M.A., Ramzan, M., Hakeem, K.R., 2016. *Azotobacter chroococcum*–A potential biofertilizer in agriculture: an overview. J. Soil Sci. Agric. Environ. Prospect., 333–348.

Will, C., Thürmer, A., Wollherr, A., Nacke, H., Herold, N., Schrumpf, M., et al., 2010. Horizon-specific bacterial community composition of German grassland soils, as revealed by pyrosequencing-based analysis of 16S rRNA genes. Appl. Environ. Microbiol. 76 (20), 6751–6759.

Wu, L., Ma, L.Q., 2002. Relationship between compost stability and extractable organic carbon. J. Environ. Qual. 31, 1323–1328.

Yadav, R., Abraham, G., Singh, Y., Singh, P., 2014. Advancements in the utilization of Azolla-Anabaena system in relation to sustainable agricultural practices. Proc. Indian Natl. Sci. Acad. 80, 301–316.

Yadav, A., Verma, P., Singh, B., Chauahan, V., 2017. Plant growth promoting bacteria: biodiversity and multifunctional attributes for sustainable agriculture. Adv. Biotechnol. Microbiol. 5, 1–16.

Yan, N., Marschner, P., Cao, W., Zuo, C., Qin, W., 2015. Influence of salinity and water content on soil microorganisms. Int. Soil Water Conserv. Res. 3 (4), 316–323.

Yildirim, N., Yildirim, N.C., Yildiz, A., 2015. Laccase enzyme activity during growth and fruiting of *Pleurotus eryngii* under solid state fermentation medium containing agricultural wastes. Int. J. Pure Appl. Sci. 1, 64–71.

Youssef, M., Eissa, M., 2014. Biofertilizers and their role in management of plant parasitic nematodes. A review. J. Biotechnol. Pharm. Res. 5, 1–6.

Zeinhom, E.A., Elhadary, R., Elashry, A., 2010. Integrating GIS and MCDM to deal with landfill site selection. Int. J. Eng. Technol. 10, 32–42.

Zhang, W., Xu, M., Wang, X., Huang, Q., Nie, J., Li, Z., Lee, K.B., 2012. Effects of organic amendments on soil carbon sequestration in paddy fields of subtropical China. J. Soils Sediments 12 (4), 457–470.

Zhang, J., Zeng, G., Chen, Y., et al., 2013. Impact of *Phanerochaete chrysosporium* inoculation on indigenous bacterial communities during agricultural waste composting. Appl. Microbiol. Biotechnol. 97, 3159–3169. https://doi.org/10.1007/s0025 3-012-4124-y.

CHAPTER 3

Microbial inoculants and their potential application in salinity management

Giselle Gomes Monteiro Fracetto[a], Felipe José Cury Fracetto[a], Edivan Rodrigues de Souza[a], Victor Lucas Vieira Prudêncio de Araújo[b], Midouin Lidelias[a], José Petrônio Mendes Júnior[a], and Cintia Caroline Gouveia da Silva[a]
[a]Federal Rural University of Pernambuco, Recife, Pernambuco, Brazil
[b]Luiz de Queiroz College of Agriculture, University of São Paulo, Piracicaba, São Paulo, Brazil

3.1 Soil salinity: General aspects

Soil salinity is a serious threat to agricultural production and affects ca. 1 billion hectares worldwide. This area is expected to increase by roughly 50% by 2050 due to increased food demands, continuous soil mismanagement, and global climate change (Negacz et al., 2021; Ivushkin et al., 2019). This problem is mostly caused by natural processes, such as soils from source materials high in salts, land use, particularly lands under irrigation, and increasingly high sea levels, leading to ocean water creeping into water tables (Monteiro et al., 2021).

Saline and sodic soils, also known as salt-affected soils, contain soluble salts and exchangeable sodium in each pedogenetic horizon. The increase in soluble salt concentration in the soil (salinity) and the percentage of exchangeable sodium (sodicity) affect healthy plant development (Ribeiro et al., 2016). In natural or primary salinization of soils, the main sources of soluble salts are primary minerals subjected to chemical weathering, releasing ions into water. Silicates are minerals that predominate in lithosphere rocks, including feldspars (58%), pyroxenes and amphiboles (13%), quartz (11%), micas, chlorites, and clay minerals (10%). Soils that originate from sedimentary rocks, containing calcite or dolomite, can release, in addition to the constituent cations, carbonates (CO_3^{2-}), bicarbonates (HCO_3^{-}), gypsum, and sulfates (SO_4^{2-}) (Ribeiro et al., 2016).

Microbial Inoculants
https://doi.org/10.1016/B978-0-323-99043-1.00013-X

In environments with humid and subhumid climates, soluble salts with Na^+ ions tend to percolate to the water table or be dragged by surface runoff water to the drainage system before reaching the ocean. In arid or semiarid tropical regions, due to low rainfall, poor irrigation system, high evapotranspiration rate, and relief topography, salts tend to accumulate inside the soil profile due to low leaching (Shaygan and Baumgartl, 2022). Groundwater is also responsible for soil salinity, particularly because this water rises by capillarity, causing salinization of the soil surface layers when evaporation occurs. In irregular relief and poor drainage systems, water evaporates and promotes the accumulation of salts both on the surface and subsurface of the soil profile (Shankar and Evelin, 2019).

Although soils affected by salts occur extensively under natural conditions, soils that were previously productive but have now become saline can worsen this global issue, mainly due to inadequate management of irrigation systems (i.e., secondary or anthropogenic salinization). Furthermore, the indiscriminate use of fertilizers and inputs with a high salinity composition increase the osmotic pressure in the soil solution, often harming the germination of seeds and the development of plants (Litalien and Zeeb, 2020).

3.2 Plant and soil microbiota responses to salinity

It is necessary to know how plant growth is limited by the osmotic effect of salts in the soil and by the toxic effects of salts to understand the physiological mechanisms responsible for salinity tolerance. The response of plants to salinity stress, with a consequent reduction in growth, occurs in two steps: (i) a rapid response, due to a decrease in the environmental osmotic potential (the osmotic phase), and (ii) a slower response, due to the accumulation of salts, especially Na^+, in the leaves (the ionic phase) (Lins et al., 2018) (Fig. 1).

The osmotic phase begins immediately after the concentration of salts around the roots exceeds the salinity threshold (i.e., one that starts causing damage to plants). For most plants, this value is $4\,dS\,m^{-1}$ (40 mM of NaCl) in the soil saturation paste extract (Richards, 1954). Fig. 1A illustrates the effect of this phase on the growth of the shoot, where it particularly refers to an increase in dry mass or leaf area over time. Leaf growth is reduced, as is the speed of emergence of new leaves and side branches. Interestingly, shoot growth is more salt-sensitive than is root growth, similar to what occurs in water-deficient soils. According to Munns and Tester (2008), this condition is due to the reduction of leaf superficial area, causing a decrease in water

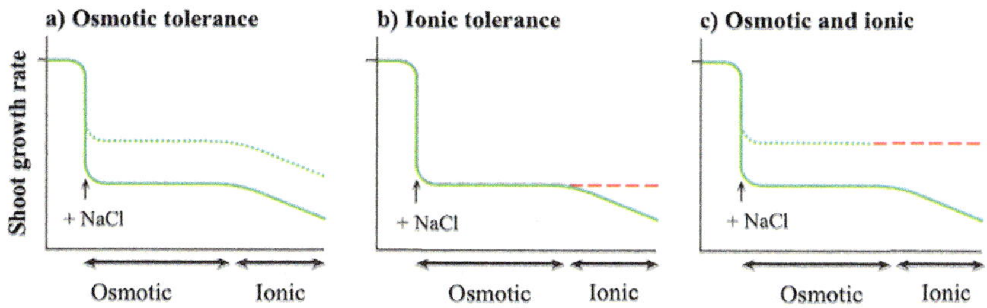

Fig. 1 The growth response to salinity stress occurs in two steps: a rapid response, due to a decrease in the environmental osmotic potential (the osmotic phase), and a slower response due to the accumulation of Na^+ in leaves (the ionic phase). The *solid green line* represents the change in the growth rate after the addition of NaCl. (A) The *broken green line* represents the hypothetical response of a plant with an increased tolerance to the osmotic component of salinity stress. (B) The *broken red line* represents the response of a plant with an increased tolerance to the ionic component of salinity stress. (C) The *green and red line* represents the response of a plant with increased tolerance to both the osmotic and ionic components of salinity stress.

consumption by the plant and, consequently, conserving soil moisture around the roots.

The osmotic effect (Fig. 1A) is responsible for most of the damage caused to the plant, mainly due to the reduction of water availability. Only at high salinity levels or in salt-sensitive species does the ionic effect overlap with the osmotic one, due to the absence of Na^+ transport control. The isolated tolerance of the osmotic effect is represented by the dashed line in Fig. 1A. The increased tolerance of both stresses (osmotic and ionic) favors plant growth at reasonable levels, sufficient to complete the life cycle (Fig. 1C).

The specific ion toxicity phase (Fig. 1B) initially affects older leaves as they do not have the same rate of leaf area expansion as do younger leaves, and this causes leaf death due to the higher concentration of salts. If the death rate of old leaves is high, then the photosynthetic capacity will not be sufficient to meet the carbohydrate requirement of young leaves, resulting in reduced plant growth. The decrease in leaf growth after increasing salinity is due to the osmotic effect around the roots. Even though the loss of water from cells and the decrease in turgor are controlled by the osmotic adjustment, the plant growth rate is affected. Over time, decreased rates of cell division and elongation lead to reduced leaf size and early flowering with a reduced number of flowers (Munns and Tester, 2008). The most evident response of a plant to salinity is a decrease in stomatal opening, which is affected by the osmotic effect, and this response is probably regulated by root signaling.

Soil microorganisms immediately respond to variations in salinity due to their high surface-to-volume ratio and the presence of permeable cell membranes (Morrissey et al., 2014). In general, salinity has a negative effect on soil microbial communities because of osmotic stress, which can cause cell drying and lysis (Leogrande and Vitti, 2019; Rath and Rousk, 2015). The estimated microbial biomass, basal soil respiration, and the activity of metabolic enzymes decrease with increasing salinity (Min et al., 2016; Yan and Marschner, 2012; Ghollarata and Raiesi, 2007; Yuan et al., 2007). Although high levels of salinity in agricultural systems decrease prokaryote diversity, the densities of Proteobacteria, Bacteroidetes, and Firmicutes increase, along with microbial interactions, due to the evolutionary mechanisms of adaptation of these bacteria (Zheng et al., 2017).

3.3 Rhizobial inoculants

Soil- and plant-associated microorganisms can be used as an efficient strategy to improve plant health and growth under salinity stress. Several studies have indicated the use of legumes to recover saline soils, due to their ability to perform symbiosis with nitrogen-fixing bacteria, known as *Rhizobia* (Abiala et al., 2018). Indeed, legumes have long been recognized as "natural soil fertilizers," as they promote nutrient cycling, increase carbon storage, and minimize soil erosion while also helping to establish other plant species (Barros et al., 2021; Lira Junior et al., 2020). In addition to their potential to remove toxic ions, legumes also increase soil nitrogen content, as observed for *Hedysarum carnosum*, a pastoral legume capable of increasing Na^+ ions in roots and maintaining a high efficiency of symbiotic N_2 fixation under high salinity (100 mM NaCl) (Kouas et al., 2010).

Rhizobium-legume symbiosis is a complex developmental process that involves colonization of the host roots by bacteria and formation of a symbiotic nitrogen-fixing structure. Salinity stress can trigger several physiological, biochemical, and molecular changes in *Rhizobia*, negatively affecting both symbiotic partners, such as structural defects in the infection cord, total inhibition of nodulation, limitations in root colonization, reductions in the quantity, weight, and size of nodules, imbalance in the oxygen levels within nodules, and reduction in nitrogenase enzyme activity (Chakraborty and Harris, 2022). In addition, changes in flavonoids and proteins occur in plants, also impairing symbiosis (Sharma et al., 2020).

However, plants inoculated with *Rhizobium* produce light-molecular-weight organic solutes, such as betaine, glycine, proline, polyamines, and ectoines, which protect plant cells, by stabilizing the structure and conformation of proteins, as well as cell membranes from water stress and desiccation. Furthermore, these solutes increase photosynthetic and biomass productivity and the activities of several enzymatic and nonenzymatic antioxidants (Wdowiak-Wróbel et al., 2013; Yurgel et al., 2013; Reina-Bueno et al., 2012). Inoculation with *Rhizobia* improves plant tolerance to different abiotic stresses, due to a more efficient absorption of nutrients, providing a broad response in various physiological activities such as attenuation of ionic imbalance, facilitation of water absorption, and decreasing generation of reactive oxygen species. In chickpea grown under salinity stress, inoculation with *Rhizobium* before sowing was demonstrated to be highly advantageous as it improved mineral absorption, reduced leakage of electrolytes that directly influences photosynthesis, and improved yield attributes (Mushtaq et al., 2021).

Rhizobial symbiosis can improve leaf water status in the host plant under salinity stress conditions. Inoculation of *Rhizobium leguminosarum* in common bean (*Phaseolus vulgaris*) under salinity stress has been shown to improve leaf water status, root osmotic water flow, and nitrogen content. Furthermore, there was a decrease in sodium concentration in the root tissues and in the osmotic potential of the xylem and an increase in the amount of aquaporins, favoring a greater osmotic flow of water into the roots (Franzini et al., 2019). Inoculation with *Rhizobia* can also improve the activity of antioxidant enzymes, such as superoxide dismutase, catalase, peroxidase, and ascorbate peroxidase, and increase osmotic solutes, such as glutathione and soluble sugars, under salinity stress conditions, as was observed in alfalfa (*Medicago sativa* L.) inoculated with *Rhizobium meliloti* (Wang et al., 2016).

The combination of cultivars and salinity stress-tolerant *Rhizobia* can result in synergistic advantages in legume life cycle, and this can be a strategy to improve the productivity in these environments. For example, salinity stress-halotolerant alfalfa inoculated with the Rm1521 salt-tolerant rhizobial strain maintained a high water content and showed a pronounced abundance of pinitol, sucrose, and starch in the nodules. In addition, this strain benefited alfalfa by maintaining nodule drain strength for compatible solutes and nodule metabolism (Bertrand et al., 2015).

Subsequently, it was shown that cultivars may differ in their response to recurrent selection: under 160 mM NaCl stress, aromatic and branched-chain amino acids (BCAAs) increased in B-ST3 nodules (Bridgeview cultivar populations obtained after three cycles of recurrent selection for salt tolerance) compared to B-TS0 (initial cultivar), whereas these accumulations were not observed in H-TS3 (Halo cultivar populations obtained after three cycles of recurrent selection for salt tolerance). BCAAs are known to control the development of *Bacteroides*, and their accumulation under severe stress may have contributed to the high nodulation of B-TS3 (Bertrand et al., 2020).

In addition to using salt-tolerant cultivars and *Rhizobia*, it is extremely important to evaluate the compatibility of different cultivars with their different symbionts under saline conditions to choose the inoculant. Soybean (*Glycine* max (L.) Merrill), one of the most important crops worldwide, whose annual production is roughly 350 billion kg (Foreign Agricultural Service USDA, 2020), establishes a symbiotic relationship with *Bradyrhizobium* and *Sinorhizobium* (*Ensifer*). Under saline conditions, the CNS soybean variety presented a higher number of nodules by *S. fredii*. However, symbiotic efficiency and growth rates were generally higher in *Bradyrhizobium* than those in *Sinorhizobium* until treatment with 50 mM NaCl (Nitawaki et al., 2021). Interestingly, when soybean was inoculated with a mix of five inoculants (*B. elkanii* USDA 31, *B. diazoefficiens* USDA 110T, and *S. fredii* USDA 191, 192, and 193), the nodule occupancy rate with the USDA 191 strain significantly increased along with the NaCl concentration, whereas the USDA 110T strain showed the opposite response. These results partially describe the ecological advantage of *S. fredii Rhizobia* for possible dominance under saline soil conditions (Nitawaki et al., 2021).

3.4 Plant growth-promoting bacterial inoculants

Plant growth-promoting bacteria (PGPB) have been extensively studied and employed for relieving plant salinity stress (Kumar et al., 2020; Numan et al., 2018; Kumar, 2022). PGPB are mainly found in rhizospheric soil, in addition to inhabiting plant tissues (endosphere), shoots, and stems (phyllosphere), constituting the plant microbiome, and may be linked to both direct (i.e., related to plant nutrition) and indirect functional roles (i.e., related to plant health) (Compant et al., 2019).

PGPB have numerous growth-promoting mechanisms, such as indole-3-acetic acid (IAA) production, phosphate and potassium solubilization, and biological nitrogen fixation (Pathak et al., 2017; Cohen et al., 2015). Furthermore, some mechanisms can induce salinity tolerance, which involves several biochemical and physiological changes in the plant (Ma et al., 2020; Yang et al., 2009). Among these mechanisms, biofilm formation

(Timmusk et al., 2014), exopolysaccharide production (Rolli et al., 2015), and osmotic adjustment (Sarma and Saikia, 2014) stand out.

Salinity stress can inhibit the activities of PGPB, making it difficult to apply these bacteria in agriculture. Currently, the prospection and study of salt-tolerant PGPB that can alleviate this stress and, at the same time, promote plant growth, deserve greater scientific attention (Verma et al., 2021; Gupta et al., 2022). With the prospection of PGPB in saline soils in China, it was possible to isolate *Bacillus* and *Brevibacterium*. *B. frigoritolerans* presented growth-promoting mechanisms, such as biological nitrogen fixation, IAA production, phosphate solubilization, and ACC deaminase production at high salt concentrations (Wang et al., 2021b). Furthermore, inoculation of this bacterium promoted wheat growth under salinity stress when combined with *Bacillus velezensis* and *Bacillus thuringiensis* (Wang et al., 2021b).

Exopolysaccharide and biofilm production are important survival mechanisms for colonization of bacteria in rhizospheric soil and root surfaces, especially under salinity stress (Dutta et al., 2022; Kasim et al., 2016). The plant-growth promoting bacterium *Bacillus pumilus* FAB10, which is a producer of biofilms, exopolysaccharides, IAA, phosphate solubilization, and ACC deaminase activity in vitro colonized the rhizosphere and formed a biofilm on the surface of wheat roots in the presence of up to 250 mM of NaCl (Ansari et al., 2019). Furthermore, inoculation significantly improved wheat growth and other biochemical attributes related to photosynthesis, transpiration, and proline content in plant tissues (Ansari et al., 2019). Similar results were also found with the inoculation of biofilm-producing PGPB (*Bacillus licheniformis* AP6 and *Pseudomonas plecoglossicida*) in sunflower, allowing an increase in photosynthetic pigments, gas exchange activities, and nutrient absorption (Yasmeen et al., 2020).

Responses to salinity stress by plants are mediated by the production of phytohormones (Choudhary et al., 2021), including endogenous ones, in which the plant increases the biosynthesis of abscisic acid (ABA) and ethylene (Riyazuddin et al., 2020; Mulholland et al., 2003). The protective role of ABA is imperative in plant development because it induces stomatal closure that reduces water loss and ion exchange activity (Osakabe et al., 2014), in addition to controlling the development of arbuscular mycorrhiza (Fracetto et al., 2017).

Endogenous ABA content increases in plants in salinity stress soils (Albacete et al., 2008); however, inoculation of the plant growth-promoting bacterium *Bacillus amyloliquefaciens* RWL-1 in rice significantly reduced the endogenous ABA content in the plant, increasing the endogenous salicylic acid content and improving the plant's physiological responses to salinity stress (Shahzad et al., 2017). These studies demonstrated the importance of evaluating the ability to produce phytohormones by PGPB, to increase the resistance of plants to that stress.

In addition to ABA, plants under salinity stress accumulate ethylene in their tissues, causing damage to cell elongation and activation of the plant defense system (Tao et al., 2015; Fracetto et al., 2013). A mechanism that has been extensively studied is the production of the enzyme ACC deaminase by PGPB to prevent salinity stress in several crops (Orozco-Mosqueda et al., 2020). The limiting step of ethylene biosynthesis is the conversion of AdoMet (S-adenosylmethionine) to ACC (1-aminocyclopropane-1-carboxylic acid), which is catalyzed by the ACC synthase enzyme (Lin et al., 2009). The enzymatic activity of ACC deaminase results in the production of α-ketobutyrate and ammonia, which prevent excessive increases in ethylene synthesis, and this is one of the most efficient mechanisms to induce plant tolerance to stress (Bharti and Barnawal, 2019; Gupta and Pandey, 2019).

PGPB ACC deaminase-producing *Aneurinibacillus aneurinilyticus* and *Paenibacillus* sp. isolated from *Allium sativum* rhizospheres can produce more than 1500 nmol of α-ketobutyrate mg protein^{-1} h^{-1}. The inoculation of these consorted bacteria in *P. vulgaris* seeds significantly decreased (60%) the levels of ethylene. Furthermore, they alleviated the negative effects of salinity stress and increased root length (110%), root fresh weight (45%), shoot length (60%), shoot fresh weight (255%), root biomass (220%), shoot biomass (425%), and the total chlorophyll content (57%) of French bean seedlings (Gupta and Pandey, 2019).

3.5 Mycorrhizal inoculants

There is evidence that arbuscular mycorrhizal fungi (AMF) cause variations in soil–water-plant relationships and improve tolerance to biotic and abiotic stresses (López-Ráez, 2016). Arbuscular mycorrhiza (AM) symbiosis represents the most widespread beneficial associations between plants and microorganisms, and this is more than 460 million years old and is considered a fundamental step in the evolution of land plants (Smith and Read, 2010; Redecker et al., 2000). AMF are must-have symbiotic fungi that phylogenetically belong to the Glomeromycotina subphylum and the Mucoromycota phylum (Bonfante and Venice, 2020; Spatafora et al., 2016) and can establish a relationship with more than 80% of land plants (Kaur and Suseela,

2020; Smith and Read, 2010), thus improving the acquisition of phosphorus and other nutrients by plants, soil aggregation performance by production of glycoprotein glomalin, etc. (Singh et al., 2022; Mardukhi et al., 2011). AMF influence the atmospheric fixation of CO_2 through the host plant that drives photoassimilates straight to the roots (Begum et al., 2019); in addition, fungal hyphae intensify the decomposition of soil organic matter, thus aiding in soil C storage (Paterson et al., 2016).

Under salinity stress conditions, it is known that the formation of AM symbiosis improves nutrient acquisition and water absorption, maintains ionic homeostasis, keeps the osmotic balance in plants, induces the antioxidant system to prevent damage by reactive oxygen species, increases photosynthetic efficiency, and modulates the phytohormone profile to minimize the effects of salt on the growth and development of plants (Santander et al., 2021; Evelin et al., 2019).

A high concentration of Na^+ and Cl^- in the soil solution competes with the absorption of essential ions such as Ca^{2+}, P, K^+, and Mg^{2+} and changes the right proportions of salts in the soil solution, affecting the acquisition of nutrients by the plants and hence restricting plant growth and biomass production (Sagar et al., 2021). However, AMF enhance the ability of symbiosis to use the soil and assimilate nutrients, improving the resilience of plants under salinity stress (Diao et al., 2022; Evelin et al., 2019). As a matter of fact, extraradical arbuscular mycorrhizal fungal hyphae can supply roughly 80, 25, 10, 25, and 60% of P, N, K, Zn, and Cu, respectively (Evelin et al., 2019; Marschner and Dell, 1994).

Increases in P uptake by mycorrhizal plants in saline soils can reduce the negative impact of Na^+ and Cl^- by maintaining the integrity of the vacuolar membrane, which promotes compartmentalization within the vacuoles and the selective uptake of ions, which do not interfere with growth metabolic pathways (Dar et al., 2022). In addition, the activity of nitrate reductase, involved in the assimilation of NO_3^-, increases in plants inoculated with AMF, and this is correlated with a better tolerance of mycorrhizal plants under salinity stress conditions (Dar et al., 2022; Azcón et al., 2013).

Another mechanism that AMF use to promote plant growth under salinity stress is homeostasis or ionic balance. Particularly, plants tend to absorb excessive amounts of Na^+, resulting in inhibition of K^+ uptake (Ma et al., 2021). A high Na^+: K^+ ratio creates an ionic imbalance in the cytoplasm, and Na^+ ions compete with K^+ for binding sites essential for various cellular functions (Dar et al., 2022; Giri et al., 2007), and, hence, this interrupts various enzymatic processes and synthesis of proteins (Sagar et al., 2021).

AMF maintain a high K^+: Na^+ ratio in plants exposed to salinity stress conditions by facilitating K^+ uptake and limiting Na^+ uptake and translocation to shoot tissues (Dar et al., 2022; Selvakumar and Thamizhiniyan, 2011).

Arbuscular mycorrhizal fungal inoculation can upregulate the antiporter SOS1 (Salt Overly Sensitive) located in the plasma membrane and involved in the exclusion of Na^+ from the cytosol to the apoplast, in control of long-distance Na^+ transport to load xylem (Shi et al., 2002), and in the vacuolar antiporter NHXs (sodium: hydrogen exchange), responsible for separating Na^+ into vacuoles, minimizing Na^+ toxicity, and, simultaneously, assisting in osmotic adjustment (Chen et al., 2017), as observed in *Robinia pseudoacacia* (Santander et al., 2021; Chen et al., 2017).

AMF are known to improve the water absorption by plants, due to the expansion of the extraradical hyphal network in the soil (Sagar et al., 2021). Mycorrhizal plants of *R. pseudoacacia* and *Malus hupehensis* showed a higher relative water content and maintained greater leaf turgidity and lower leaf osmotic potential compared to nonmycorrhizal plants when subjected to salinity stress (Chen et al., 2017; Yang et al., 2014). AMF improve the functioning of water channel proteins and aquaporins, triggering transport of water within cells and maintaining cellular osmoregulation (Cheng et al., 2021; Santander et al., 2021). Furthermore, mycorrhizal plants have a higher osmotic potential than do nonmycorrhizal plants due to the accumulation of osmolytes such as proline, betaine, polyamines, sugars, organic acids, amino acids, and trehalose in salinity stress (Alqarawi et al., 2014; Navarro et al., 2014).

Salinity stress reduces photosynthetic efficiency, decreasing the number of active PSII reaction centers and inhibiting electron transport from PSII to PSI (Jajoo and Mathur, 2021; Mehta et al., 2010). Due to better water status in mycorrhizal plants, resulting in the retention of greater leaf area and greater stomatal conductance, thus increasing CO_2 assimilation, these impacts are reduced (Dar et al., 2022; Chen et al., 2017; Wu et al., 2015). The symbiotic association between AMF and plants upregulates the expression of the chloroplast genes *Rppsb*A and *Rppsb*D, resulting in a higher PSII efficiency and thus increasing photosynthetic capacity (Sagar et al., 2021; Gamalero et al., 2010).

Salt tolerance measured by AMF is also attributed to their ability to detoxify reactive oxygen species (ROS) and maintain balance with antioxidants. Colonization by AMF restricts excessive ROS generation by increasing the activities of antioxidant enzymes such as superoxide dismutase (SOD), catalase, peroxidase (POX), ascorbate peroxidase, and glutathione reductase and antioxidant molecules (carotenoids, ascorbic acid, glutathione, tocopherols) (Jajoo and Mathur, 2021; Santander et al., 2020). The benefits of AM symbiosis in containing ROS levels and increasing the activity of antioxidant enzymes have already been reported in several studies (Wang et al., 2021a; Li et al., 2020; Santander et al., 2020; Ye et al., 2019; Chang et al., 2018), demonstrating that inoculation with AMF results in a well-established defense mechanism against salinity, in which the mitigation of oxidative stress can be crucial.

Plant hormones are extremely important in the process of tolerance to salinity stress (Fahad et al., 2015). ABA allows plant growth under salinity stress (Basu et al., 2017). ABA performs the accumulation of K^+, Ca^{2+}, and compatible solutes such as proline and sugars in root cells, preventing the uptake of Na^+ and Cl^- (Atia et al., 2018). The production of ABA is essential for the proper establishment and development of mycorrhizal colonization in the roots, and it has a direct impact on the formation and functioning of arbuscules (Fracetto et al., 2017; Pozo et al., 2015). Under salinity stress, ABA concentration in plants increases due to leaf water potential, indicating that endogenous ABA production is a result of the water deficit (Hussain et al., 2022; Zhang et al., 2006).

3.6 Dark septate endophytic fungal inoculants

In addition to mycorrhizae, there is a group of root endophytic fungi, known as dark septate endophytes (DSEs), directly involved in host resistance to environmental stress. DSEs are characterized by intense dark pigmentation and the formation of septate and melanized hyphae and microsclerotia (Yuan et al., 2016; Knapp et al., 2015). Studies about the role of DSE fungi in the ecosystem are limited (Gupta et al., 2021), and they have low host specificity. The beneficial effects of DSEs vary according to the fungal species and may be limited under low soil organic matter content, but they are more evident at moderate salinity levels (Malicka et al., 2022).

Some studies have shown the potential of DSEs to promote plant growth in saline environments through mechanisms like those used by arbuscular mycorrhizae. Santos et al. (2017) observed that inoculation with DSEs promoted growth and decreased oxidative stress of two rice varieties. Inoculation of bean plants with DSEs *Sordariomycetes* sp1-B′2 and *Melanconiella elegans*-21W2 improved the N and P nutrition of beans under salinity stress, favoring plant growth and net photosynthetic rate (Farias et al., 2020). The inoculation of DSE *Curvularia* sp. significantly promoted the growth of poplar (*Populus tomentosa*), resulting in an increased production of antioxidant enzymes, mainly superoxide dismutase and ascorbate peroxidase (Pan et al., 2018). In addition, the photosystem II-based electron transport rate, actual quantum yield in the light-adapted steady state (ϕPSII), and photochemical quenching values (qP) were significantly higher in colonized plants (Pan et al., 2018). These studies demonstrate the potential use of endophytes as inoculants to improve the salt tolerance of different plants.

3.7 Final remarks and prospects

Crop decline caused by abiotic stresses has been one of the main obstacles to world food security. Soil salinity can damage plant growth and affect large-scale agricultural productivity in arid and semiarid tropical regions. Therefore, studies about salinity stress are expanding to find solutions to this environmental damage. Currently, it is well-accepted in the scientific community that most of these solutions are due to plant-associated microorganisms.

Microorganisms increase the tolerance of plants to salinity stress and promote their growth through production of ACC deaminase and phytohormones (IAA, cytokinins, ABA). These organisms can produce exopolysaccharides and biofilms, maintain ionic homeostasis and absorption of nutrients, and boost the efficiency of water use and osmosis protection, hence ensuring both photosynthetic efficiency and antioxidant metabolism (Fig. 2). Countless studies have been conducted on the prospection and characterization of these mechanisms in several experiments under controlled conditions; however, further research is needed to assess the efficiency of these microorganisms under natural field conditions.

Studies about the tripartite interactions between AMF, PGPB, and plants, aiming to mitigate the adverse effects caused by salinity stress in tropical arid and semiarid environments, are insufficient and need to be further explored. Most studies that use bioinoculants are performed using *Rhizobia*, PGPB, AMF, or DSEs, disregarding the synergistic effects of joint inoculation. Interactions between different types of microorganisms can improve the growth and tolerance of plants under soil salinity conditions (Sagar et al., 2021; Moreira et al., 2020; Krishnamoorthy et al., 2016; Pereira et al., 2016;). We believe that the synergism between AMF and PGPB can be an excellent strategy for producing an inoculant in salinity management (Fig. 3). Inoculation of ACC deaminase-producing PGPB inhibits ethylene production and this could promote the development of AM, thus

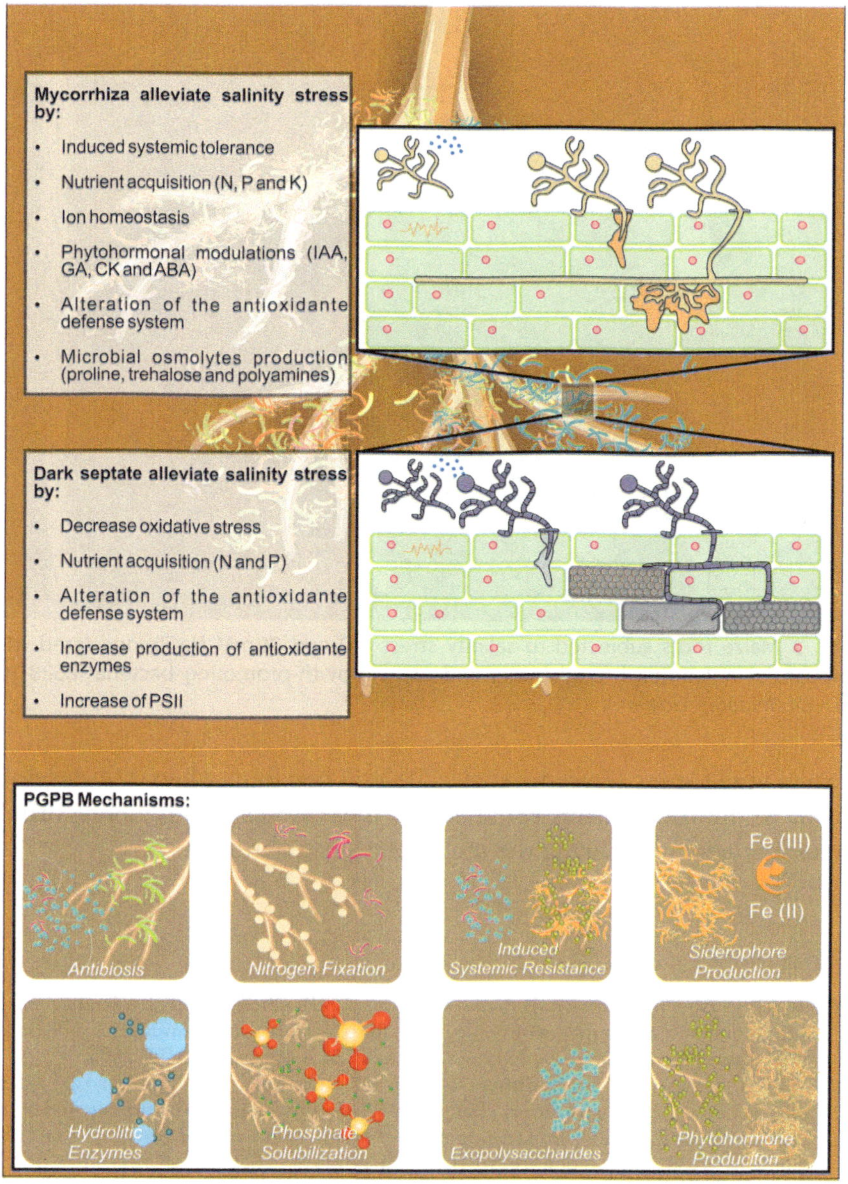

Fig. 2 Interactions between mycorrhizas, dark septate endophytic fungi, and plant growth-promoting bacteria (PGPB) confer salinity stress tolerance. *IAA*, indole-3-acetic acid; *GA*, gibberellin; *CK*, cytokinin; *ABA*, abscisic acid.

Fig. 3 Maize roots subjected to salinity stress (40 and 80 mM NaCl) inoculated with arbuscular mycorrhizal fungi (AMF) and plant growth-promoting bacteria (BS28-10). *(Credit: Midouin Lidelias.)*

reducing salt stress (Fracetto et al., 2013). Therefore, when AM development is stimulated, plants could absorb more water and nutrients, mainly phosphorus, thus reducing the use of phosphate fertilizers.

References

Abiala, M.A., Abdelrahman, M., Burritt, D.J., Tran, L.S.P., 2018. Salt stress tolerance mechanisms and potential applications of legumes for sustainable reclamation of salt-degraded soils. Land Degrad. Dev. 29, 3812–3822. https://doi.org/10.1002/ldr.3095.

Albacete, A., Ghanem, M.E., Martínez-Andújar, C., Acosta, M., Sánchez-Bravo, J., Martínez, V., et al., 2008. Hormonal changes in relation to biomass partitioning and shoot growth impairment in salinized tomato (*Solanum lycopersicum* L.) plants. J. Exp. Bot. 59, 4119–4131. https://doi.org/10.1093/jxb/ern251.

Alqarawi, A.A., Abd Allah, E.F., Hashem, A., 2014. Alleviation of salt-induced adverse impact via mycorrhizal fungi in *Ephedra aphylla* Forssk. J. Plant Interact. 9, 802–810. https://doi.org/10.1080/17429145.2014.949886.

Ansari, F.A., Ahmad, I., Pichtel, J., 2019. Growth stimulation and alleviation of salinity stress to wheat by the biofilm forming *Bacillus pumilus* strain FAB10. Appl. Soil Ecol. 143, 45–54. https://doi.org/10.1016/j.apsoil.2019.05.023.

Atia, A., Barhoumi, Z., Debez, A., Hkiri, S., Abdelly, C., Smaoui, A., et al., 2018. Plant hormones: Potent targets for engineering salinity tolerance in plants. In: Salinity

Responses and Tolerance in Plants. Vol. 1. Springer, Cham, pp. 159–184, https://doi.org/10.1007/978-3-319-75671-4_6.

Azcón, R., Medina, A., Aroca, R., Ruiz-Lozano, J.M., 2013. Abiotic stress remediation by the arbuscular mycorrhizal symbiosis and rhizosphere bacteria/yeast interactions. Mol. Microb. Ecol. Rhizosphere 1, 991–1002. https://doi.org/10.1002/9781118297674.ch93.

Barros, F.M.R., Fracetto, F.J.C., Junior, M.A.L., Bertini, S.C.B., Fracetto, G.G.M., 2021. Spatial and seasonal responses of diazotrophs and ammonium-oxidizing bacteria to legume-based silvopastoral systems. Appl. Soil Ecol. 158, 103797. https://doi.org/10.1016/j.apsoil.2020.103797.

Basu, S., Rabara, R., Negi, S., 2017. Towards a better greener future-an alternative strategy using biofertilizers. I: Plant growth promoting bacteria. Plant Gene 12, 43–49. https://doi.org/10.1016/j.plgene.2017.07.004.

Begum, N., Qin, C., Ahanger, M.A., Raza, S., Khan, M.I., Ashraf, M., et al., 2019. Role of arbuscular mycorrhizal fungi in plant growth regulation: implications in abiotic stress tolerance. Front. Plant Sci. 10, 1068. https://doi.org/10.3389/fpls.2019.01068.

Bertrand, A., Dhont, C., Bipfubusa, M., Chalifour, F.P., Drouin, P., Beauchamp, C.J., 2015. Improving salt stress responses of the symbiosis in alfalfa using salt-tolerant cultivar and rhizobial strain. Appl. Soil Ecol. 87, 108–117. https://doi.org/10.1016/j.apsoil.2014.11.008.

Bertrand, A., Gatzke, C., Bipfubusa, M., Lévesque, V., Chalifour, F.P., Claessens, A., et al., 2020. Physiological and biochemical responses to salt stress of alfalfa populations selected for salinity tolerance and grown in symbiosis with salt-tolerant rhizobium. Agronomy 10, 569. https://doi.org/10.3390/agronomy10040569.

Bharti, N., Barnawal, D., 2019. Amelioration of salinity stress by PGPR: ACC deaminase and ROS scavenging enzymes activity. In: PGPR Amelioration in Sustainable Agriculture. Woodhead Publishing, pp. 85–106, https://doi.org/10.1016/B978-0-12-815879-1.00005-7.

Bonfante, P., Venice, F., 2020. Mucoromycota: going to the roots of plant-interacting fungi. Fungal Biol. Rev. 34, 100–113. https://doi.org/10.1016/j.fbr.2019.12.003.

Chakraborty, S., Harris, J.M., 2022. At the crossroads of salinity and rhizobium-legume symbiosis. Mol. Plant-Microbe Interact. 35, 540–553. https://doi.org/10.1094/MPMI-09-21-0231-FI.

Chang, W., Sui, X., Fan, X.X., Jia, T.T., Song, F.Q., 2018. Arbuscular mycorrhizal symbiosis modulates antioxidant response and ion distribution in salt-stressed *Elaeagnus angustifolia* seedlings. Front. Microbiol. 9, 652. https://doi.org/10.3389/fmicb.2018.00652.

Chen, J., Zhang, H., Zhang, X., Tang, M., 2017. Arbuscular mycorrhizal symbiosis alleviates salt stress in black locust through improved photosynthesis, water status, and K+/Na+ homeostasis. Front. Plant Sci. 1739. https://doi.org/10.3389/fpls.2017.01739.

Cheng, X.F., Wu, H.H., Zou, Y.N., Wu, Q.S., Kuča, K., 2021. Mycorrhizal response strategies of trifoliate orange under well-watered, salt stress, and waterlogging stress by regulating leaf aquaporin expression. Plant Physiol. Biochem. 162, 27–35. https://doi.org/10.1016/j.plaphy.2021.02.026.

Choudhary, P., Pramitha, L., Rana, S., Verma, S., Aggarwal, P.R., Muthamilarasan, M., 2021. Hormonal crosstalk in regulating salinity stress tolerance in graminaceous crops. Physiol. Plant. 173, 1587–1596. https://doi.org/10.1111/ppl.13558.

Cohen, A.C., Bottini, R., Pontin, M., Berli, F.J., Moreno, D., Boccanlandro, H., et al., 2015. *Azospirillum brasilense* ameliorates the response of *Arabidopsis thaliana* to drought mainly via enhancement of ABA levels. Physiol. Plant. 153, 79–90. https://doi.org/10.1111/ppl.12221.

Compant, S., Samad, A., Faist, H., Sessitsch, A., 2019. A review on the plant microbiome: ecology, functions, and emerging trends in microbial application. J. Adv. Res. 19, 29–37. https://doi.org/10.1016/j.jare.2019.03.004.

Dar, M., Razvi, S.M., Singh, N., Mushtaq, A., Dar, S., 2022. Arbuscular mycorrhizal fungi—alleviator for salinity stress: a review. Pedosphere. https://doi.org/10.1016/j.pedsph.2022.06.027.

Diao, F., Jia, B., Wang, X., Luo, J., Hou, Y., Li, F.Y., Guo, W., 2022. Proteomic analysis revealed modulations of carbon and nitrogen by arbuscular mycorrhizal fungi associated with the halophyte *Suaeda salsa* in a moderately saline environment. Land Degrad. Dev. 33, 1933–1943. https://doi.org/10.1002/ldr.4274.

Dutta, A., Banerjee, S., Dinda, S., Chowdhury, I., Haldar, S., Bandyopadhyay, S., 2022. A critical analysis on the roles of exopolysaccharides and ACC deaminase in salinity stress tolerance in crop plants. Biocatal. Agric. Biotechnol., 102372. https://doi.org/10.1016/j.bcab.2022.102372.

Evelin, H., Devi, T.S., Gupta, S., Kapoor, R., 2019. Mitigation of salinity stress in plants by arbuscular mycorrhizal symbiosis: current understanding and new challenges. Front. Plant Sci. 10, 470. https://doi.org/10.3389/fpls.2019.00470.

Fahad, S., Nie, L., Chen, Y., Wu, C., Xiong, D., Saud, S., et al., 2015. Crop plant hormones and environmental stress. Sustain. Agric. Rev., 371–400. https://doi.org/10.1007/978-3-319-09132-7_10.

Farias, G.C., Nunes, K.G., Soares, M.A., de Siqueira, K.A., Lima, W.C., Neves, A.L.R., de Lacerda, C.F., 2020. Dark septate endophytic fungi mitigate the effects of salt stress on cowpea plants. Braz. J. Microbiol. 51, 243–253. https://doi.org/10.1007/s42770-019-00173-4.

Foreign Agricultural Service USDA, 2020. World Agricultural Production. Circular Series WAP 4-20. April, 2020.

Fracetto, G.G., Peres, L.E., Mehdy, M.C., Lambais, M.R., 2013. Tomato ethylene mutants exhibit differences in arbuscular mycorrhiza development and levels of plant defense-related transcripts. Symbiosis 60, 155–167. https://doi.org/10.1007/s13199-013-0251-1.

Fracetto, G.G.M., Peres, L.E.P., Lambais, M.R., 2017. Gene expression analyses in tomato near isogenic lines provide evidence for ethylene and abscisic acid biosynthesis fine-tuning during arbuscular mycorrhiza development. Arch. Microbiol. 199, 787–798. https://doi.org/10.1007/s00203-017-1354-5.

Franzini, V.I., Azcón, R., Ruiz-Lozano, J.M., Aroca, R., 2019. Rhizobial symbiosis modifies root hydraulic properties in bean plants under non-stressed and salinity-stressed conditions. Planta 249, 1207–1215. https://doi.org/10.1007/s00425-018-03076-0.

Gamalero, E., Berta, G., Massa, N., Glick, B.R., Lingua, G., 2010. Interactions between *Pseudomonas putida* UW4 and *Gigaspora rosea* BEG9 and their consequences for the growth of cucumber under salt-stress conditions. J. Appl. Microbiol. 108, 236–245. https://doi.org/10.1111/j.1365-2672.2009.04414.x.

Ghollarata, M., Raiesi, F., 2007. The adverse effects of soil salinization on the growth of *Trifolium alexandrinum* L. and associated microbial and biochemical properties in a soil from Iran. Soil Biol. Biochem. 39, 1699–1702. https://doi.org/10.1016/j.soilbio.2007.01.024.

Giri, B., Kapoor, R., Mukerji, K.G., 2007. Improved tolerance of *Acacia nilotica* to salt stress by arbuscular mycorrhiza, *Glomus fasciculatum* may be partly related to elevated K/Na ratios in root and shoot tissues. Microb. Ecol. 54, 753–760. https://doi.org/10.1007/s00248-007-9239-9.

Gupta, S., Pandey, S., 2019. ACC deaminase producing bacteria with multifarious plant growth promoting traits alleviates salinity stress in French bean (*Phaseolus vulgaris*) plants. Front. Microbiol. 10, 1506. https://doi.org/10.3389/fmicb.2019.01506.

Gupta, S., Schillaci, M., Walker, R., Smith, P., Watt, M., Roessner, U., 2021. Alleviation of salinity stress in plants by endophytic plant-fungal symbiosis: current knowledge, perspectives and future directions. Plant Soil 461, 219–244. https://doi.org/10.1007/s11104-020-04618-w.

Gupta, A., Mishra, R., Rai, S., Bano, A., Pathak, N., Fujita, M., et al., 2022. Mechanistic insights of plant growth promoting bacteria mediated drought and salt stress tolerance in plants for sustainable agriculture. Int. J. Mol. Sci. 23, 3741. https://doi.org/10.3390/ijms23073741.

Hussain, R., Sharma, S., Kumari, I., Ahmed, M., 2022. Plant microbiome: modulation of plant defense and ecological dynamics under stressed environment. In: Mitigation of Plant Abiotic Stress by Microorganisms. Academic Press, pp. 19–40, https://doi.org/10.1016/B978-0-323-90568-8.00002-X.

Ivushkin, K., Bartholomeus, H., Bregt, A.K., Pulatov, A., Kempen, B., De Sousa, L., 2019. Global mapping of soil salinity change. Remote Sens. Environ. 231, 111260. https://doi.org/10.1016/j.rse.2019.111260.

Jajoo, A., Mathur, S., 2021. Role of arbuscular mycorrhizal fungi as an underground saviuor for protecting plants from abiotic stresses. Physiol. Mol. Biol. Plants 27, 2589–2603. https://doi.org/10.1007/s12298-021-01091-2.

Kasim, W.A., Gaafar, R.M., Abou-Ali, R.M., Omar, M.N., Hewait, H.M., 2016. Effect of biofilm forming plant growth promoting rhizobacteria on salinity tolerance in barley. Ann. Agric. Sci. 61, 217–227. https://doi.org/10.1016/j.aoas.2016.07.003.

Kaur, S., Suseela, V., 2020. Unraveling arbuscular mycorrhiza-induced changes in plant primary and secondary metabolome. Meta 10, 335. https://doi.org/10.3390/metabo10080335.

Knapp, D.G., Kovács, G.M., Zajta, E., Groenewald, J.Z., Crous, P.W., 2015. Dark septate endophytic pleosporalean genera from semiarid areas. Persoonia Mol. Phyl. Evol. Fungi 35, 87–100. https://doi.org/10.3767/003158515X687669.

Kouas, S., Slatni, T., Salah, I.B., Abdelly, C., 2010. Eco-physiological responses and symbiotic nitrogen fixation capacity of salt-exposed *Hedysarum carnosum* plants. Afr. J. Biotechnol. 9, 7462–7469. https://doi.org/10.5897/AJB10.211.

Krishnamoorthy, R., Kim, K., Subramanian, P., Senthilkumar, M., Anandham, R., Sa, T., 2016. Arbuscular mycorrhizal fungi and associated bacteria isolated from salt-affected soil enhances the tolerance of maize to salinity in coastal reclamation soil. Agric. Ecosyst. Environ. 231, 233–239. https://doi.org/10.1016/j.agee.2016.05.037.

Kumar, A., 2022. Microbial Biocontrol: Sustainable Agriculture and Phytopathogen Management. 1 Springer Nature Chem, Cham, Switzerland, pp. 1–369, https://doi.org/10.1007/978-3-030-87512-1.

Kumar, A., Singh, S., Gaurav, A.K., Srivastava, S., Verma, J.P., 2020. Plant growth-promoting bacteria: biological tools for the mitigation of salinity stress in plants. Front. Microbiol. 11, 1216. https://doi.org/10.3389/fmicb.2020.01216.

Leogrande, R., Vitti, C., 2019. Use of organic amendments to reclaim saline and sodic soils: a review. Arid Land Res. Manag. 33, 1–21. https://doi.org/10.1080/15324982.2018.1498038.

Li, Z., Wu, N., Meng, S., Wu, F., Liu, T., 2020. Arbuscular mycorrhizal fungi (AMF) enhance the tolerance of *Euonymus maackii* Rupr. at a moderate level of salinity. PLoS One 15, e0231497. https://doi.org/10.1371/journal.pone.0231497.

Lin, Z., Zhong, S., Grierson, D., 2009. Recent advances in ethylene research. J. Exp. Bot. 60, 3311–3336. https://doi.org/10.1093/jxb/erp204.

Lins, C.M.T., de Souza, E.R., de Melo, H.F., Paulino, M.K.S.S., Dourado, P.R.M., de Carvalho Leal, L.Y., Santos, H.R.B., 2018. Pressure-volume (PV) curves in *Atriplex nummularia* Lindl. for evaluation of osmotic adjustment and water status under saline conditions. Plant Physiol. Biochem. 124, 155–159. https://doi.org/10.1016/j.plaphy.2018.01.014.

Lira Junior, M.A., Fracetto, F.J.C., da Silva Ferreira, J., Silva, M.B., Fracetto, G.G.M., 2020. Legume-based silvopastoral systems drive C and N soil stocks in a subhumid tropical environment. Catena 189, 104508. https://doi.org/10.1016/j.catena.2020.104508.

Litalien, A., Zeeb, B., 2020. Curing the earth: a review of anthropogenic soil salinization and plant-based strategies for sustainable mitigation. Sci. Total Environ. 698, 134235. https://doi.org/10.1016/j.scitotenv.2019.134235.

López-Ráez, J.A., 2016. How drought and salinity affect arbuscular mycorrhizal symbiosis and strigolactone biosynthesis? Planta 243, 1375–1385. https://doi.org/10.1007/s00425-015-2435-9.

Ma, Y., Dias, M.C., Freitas, H., 2020. Drought and salinity stress responses and microbe-induced tolerance in plants. Front. Plant Sci. 11, 591911. https://doi.org/10.3389/fpls.2020.591911.

Ma, P., Shi, Z., Diao, F., Hao, L., Zhang, J., Xu, J., et al., 2021. Effects of arbuscular mycorrhizal fungi on growth and Na+ accumulation of *Suaeda glauca* (Bunge) grown in salinized wetland soils. Appl. Soil Ecol. 166, 104065. https://doi.org/10.1016/j.apsoil.2021.104065.

Malicka, M., Magurno, F., Piotrowska-Seget, Z., 2022. Plant association with dark septate endophytes: when the going gets tough (and stressful), the tough fungi get going. Chemosphere 134830. https://doi.org/10.1016/j.chemosphere.2022.134830.

Mardukhi, B., Rejali, F., Daei, G., Ardakani, M.R., Malakouti, M.J., Miransari, M., 2011. Arbuscular mycorrhizas enhance nutrient uptake in different wheat genotypes at high salinity levels under field and greenhouse conditions. Comptes Rendus Biologies 334, 564–571. https://doi.org/10.1016/j.crvi.2011.05.001.

Marschner, H., Dell, B., 1994. Nutrient uptake in mycorrhizal symbiosis. Plant Soil 159 (1), 89–102. https://doi.org/10.1007/BF00000098.

Mehta, P., Jajoo, A., Mathur, S., Bharti, S., 2010. Chlorophyll a fluorescence study revealing effects of high salt stress on photosystem II in wheat leaves. Plant Physiol. Biochem. 48, 16–20. https://doi.org/10.1016/j.plaphy.2009.10.006.

Min, W., Guo, H., Zhang, W., Zhou, G., Ma, L., Ye, J., et al., 2016. Response of soil microbial community and diversity to increasing water salinity and nitrogen fertilization rate in an arid soil. Acta Agric. Scand. Section B Soil Plant Sci. 66, 117–126. https://doi.org/10.1080/09064710.2015.1078838.

Monteiro, D.R., Souza, E.R.D., Dourado, P.R.M., Melo, H., Santos, H., dos Santos, M.A., 2021. Soil water potentials and sweet sorghum under salinity. Commun. Soil Sci. Plant Anal. 52, 1149–1160. https://doi.org/10.1080/00103624.2021.1879114.

Moreira, H., Pereira, S.I., Vega, A., Castro, P.M., Marques, A.P., 2020. Synergistic effects of arbuscular mycorrhizal fungi and plant growth-promoting bacteria benefit maize growth under increasing soil salinity. J. Environ. Manag. 257, 109982. https://doi.org/10.1016/j.jenvman.2019.109982.

Morrissey, E.M., Gillespie, J.L., Morina, J.C., Franklin, R.B., 2014. Salinity affects microbial activity and soil organic matter content in tidal wetlands. Glob. Chang. Biol. 20, 1351–1362. https://doi.org/10.1111/gcb.12431.

Mulholland, B.J., Taylor, I.B., Jackson, A.C., Thompson, A.J., 2003. Can ABA mediate responses of salinity stressed tomato. Environ. Exp. Bot. 50, 17–28. https://doi.org/10.1016/S0098-8472(02)00110-7.

Munns, R., Tester, M., 2008. Mechanisms of salinity tolerance. Annu. Rev. Plant Biol. 59, 651. https://doi.org/10.1146/annurev.arplant.59.032607.092911.

Mushtaq, Z., Faizan, S., Gulzar, B., Hakeem, K.R., 2021. Inoculation of Rhizobium alleviates salinity stress through modulation of growth characteristics, physiological and biochemical attributes, stomatal activities and antioxidant defence in *Cicer arietinum* L. J. Plant Growth Regul. 40, 2148–2163. https://doi.org/10.1007/s00344-020-10267-1.

Navarro, J.M., Pérez-Tornero, O., Morte, A., 2014. Alleviation of salt stress in citrus seedlings inoculated with arbuscular mycorrhizal fungi depends on the rootstock salt tolerance. J. Plant Physiol. 171, 76–85. https://doi.org/10.1016/j.jplph.2013.06.006.

Negacz, K., Bruning, B., Vellinga, P., 2021. Achieving multiple sustainable development goals through saline agriculture. In: Future of Sustainable Agriculture in Saline Environments, pp. 13–28, https://doi.org/10.1201/9781003112327.

Nitawaki, Y., Kitabayashi, H., Mason, M.L.T., Yamamoto, A., Saeki, Y., 2021. Effect of salt stress on soybean growth and nodulation under inoculation with soybean rhizobia. Soil Sci. Plant Nutr. 67, 103–113. https://doi.org/10.1080/00380768.2020.1860644.

Numan, M., Bashir, S., Khan, Y., Mumtaz, R., Shinwari, Z.K., Khan, A.L., et al., 2018. Plant growth promoting bacteria as an alternative strategy for salt tolerance in plants: a review. Microbiol. Res. 209, 21–32. https://doi.org/10.1016/j.micres.2018.02.003.

Orozco-Mosqueda, M.D.C., Glick, B.R., Santoyo, G., 2020. ACC deaminase in plant growth-promoting bacteria (PGPB): an efficient mechanism to counter salt stress in crops. Microbiol. Res. 235, 126439. https://doi.org/10.1016/j.micres.2020.126439.

Osakabe, Y., Yamaguchi-Shinozaki, K., Shinozaki, K., Tran, L.S.P., 2014. ABA control of plant macroelement membrane transport systems in response to water deficit and high salinity. New Phytol. 202, 35–49. https://doi.org/10.1111/nph.12613.

Pan, X., Qin, Y., Yuan, Z., 2018. Potential of a halophyte-associated endophytic fungus for sustaining Chinese white poplar growth under salinity. Symbiosis 76, 109–116. https://doi.org/10.1007/s13199-018-0541-8.

Paterson, E., Sim, A., Davidson, J., Daniell, T.J., 2016. Arbuscular mycorrhizal hyphae promote priming of native soil organic matter mineralisation. Plant Soil 408, 243–254. https://doi.org/10.1007/s11104-016-2928-8.

Pathak, D.V., Kumar, M., Rani, K., 2017. Biofertilizer application in horticultural crops. In: Panpatte, D., Jhala, Y., Vyas, R., Shelat, H. (Eds.), Microorganisms for Green Revolution. Microorganisms for Sustainability. Vol. 6. Springer, Singapore, https://doi.org/10.1007/978-981-10-6241-4_11.

Pereira, S.I., Moreira, H., Argyras, K., Castro, P.M., Marques, A.P., 2016. Promotion of sunflower growth under saline water irrigation by the inoculation of beneficial microorganisms. Appl. Soil Ecol. 105, 36–47. https://doi.org/10.1016/j.apsoil.2016.03.015.

Pozo, M.J., López-Ráez, J.A., Azcón-Aguilar, C., García-Garrido, J.M., 2015. Phytohormones as integrators of environmental signals in the regulation of mycorrhizal symbioses. New Phytol. 205, 1431–1436. https://doi.org/10.1111/nph.13252.

Rath, K.M., Rousk, J., 2015. Salt effects on the soil microbial decomposer community and their role in organic carbon cycling: a review. Soil Biol. Biochem. 81, 108–123. https://doi.org/10.1016/j.soilbio.2014.11.001.

Redecker, D., Kodner, R., Graham, L.E., 2000. Glomalean fungi from the Ordovician. Science 289, 1920–1921. https://doi.org/10.1126/science.289.5486.1920.

Reina-Bueno, M., Argandoña, M., Nieto, J.J., Hidalgo-García, A., Iglesias-Guerra, F., Delgado, M.J., Vargas, C., 2012. Role of trehalose in heat and desiccation tolerance in the soil bacterium *Rhizobium etli*. BMC Microbiol. 12, 1–17. https://doi.org/10.1186/1471-2180-12-207.

Ribeiro, M.R., Ribeiro Filho, M.R., Jakomine, P.K.T., 2016. Origem e classificação dos solos afetados por sais. In: Gheyi, H.R., Dias, N.D.S., de Lacerda, C.F., Gomes Filho, E. (Eds.), Manejo da salinidade na agricultura: Estudos básicos e aplicados. INCTSal, Fortaleza. 504p.

Richards, L.A., 1954. Diagnosis and Improvement of Saline and Alkali Soils. vol. 78 LWW, p. 154.

Riyazuddin, R., Verma, R., Singh, K., Nisha, N., Keisham, M., Bhati, K.K., et al., 2020. Ethylene: a master regulator of salinity stress tolerance in plants. Biomol. Ther. 10, 959. https://doi.org/10.3390/biom10060959.

Rolli, E., Marasco, R., Vigani, G., Ettoumi, B., Mapelli, F., Deangelis, M.L., et al., 2015. Improved plant resistance to drought is promoted by the root-associated microbiome as a water stress-dependent trait. Environ. Microbiol. 17, 316–331. https://doi.org/10.1111/1462-2920.12439.

Sagar, A., Rathore, P., Ramteke, P.W., Ramakrishna, W., Reddy, M.S., Pecoraro, L., 2021. Plant growth promoting rhizobacteria, arbuscular mycorrhizal fungi and their synergistic interactions to counteract the negative effects of saline soil on agriculture: key macromolecules and mechanisms. Microorganisms 9, 1491. https://doi.org/10.3390/microorganisms9071491.

Santander, C., Ruiz, A., García, S., Aroca, R., Cumming, J., Cornejo, P., 2020. Efficiency of two arbuscular mycorrhizal fungal inocula to improve saline stress tolerance in lettuce plants by changes of antioxidant defense mechanisms. J. Sci. Food Agric. 100, 1577–1587. https://doi.org/10.1002/jsfa.10166.

Santander, C., Aroca, R., Cartes, P., Vidal, G., Cornejo, P., 2021. Aquaporins and cation transporters are differentially regulated by two arbuscular mycorrhizal fungi strains in lettuce cultivars growing under salinity conditions. Plant Physiol. Biochem. 158, 396–409. https://doi.org/10.1016/j.plaphy.2020.11.025.

Santos, S.G.D., Silva, P.R.A.D., Garcia, A.C., Zilli, J.É., Berbara, R.L.L., 2017. Dark septate endophyte decreases stress on rice plants. Braz. J. Microbiol. 48, 333–341. https://doi.org/10.1016/j.bjm.2016.09.018.

Sarma, R.K., Saikia, R., 2014. Alleviation of drought stress in mung bean by strain *Pseudomonas aeruginosa* GGRJ21. Plant Soil 377, 111–126. https://doi.org/10.1007/s11104-013-1981-9.

Selvakumar, G., Thamizhiniyan, P., 2011. The effect of the arbuscular mycorrhizal (AM) fungus *Glomus intraradices* on the growth and yield of chilli (*Capsicum annuum* L.) under salinity stress. World Appl. Sci. J. 14, 1209–1214.

Shahzad, R., Khan, A.L., Bilal, S., Waqas, M., Kang, S.M., Lee, I.J., 2017. Inoculation of abscisic acid-producing endophytic bacteria enhances salinity stress tolerance in *Oryza sativa*. Environ. Exp. Bot. 136, 68–77. https://doi.org/10.1016/j.envexpbot.2017.01.010.

Shankar, V., Evelin, H., 2019. Strategies for reclamation of saline soils. In: Microorganisms in Saline Environments: Strategies and Functions. Springer, Cham, pp. 439–449, https://doi.org/10.1007/978-3-030-18975-4_19.

Sharma, V., Bhattacharyya, S., Kumar, R., Kumar, A., Ibañez, F., Wang, J., et al., 2020. Molecular basis of root nodule symbiosis between *Bradyrhizobium* and 'crack-entry' legume groundnut (*Arachis hypogaea* L.). Plan. Theory 9, 276. https://doi.org/10.3390/plants9020276.

Shaygan, M., Baumgartl, T., 2022. Reclamation of salt-affected land: a review. Soil Syst. 6, 61. https://doi.org/10.3390/soilsystems6030061.

Shi, H., Quintero, F.J., Pardo, J.M., Zhu, J.K., 2002. The putative plasma membrane Na^+/H^+ antiporter SOS1 controls long-distance Na^+ transport in plants. Plant Cell 14, 465–477. https://doi.org/10.1105/tpc.010371.

Singh, A.K., Zhu, X., Chen, C., Wu, J., Yang, B., Zakari, S., et al., 2022. The role of glomalin in mitigation of multiple soil degradation problems. Crit. Rev. Environ. Sci. Technol. 52, 1604–1638. https://doi.org/10.1080/10643389.2020.1862561.

Smith, S.E., Read, D.J., 2010. Mycorrhizal Symbiosis. Academic Press.

Spatafora, J.W., Chang, Y., Benny, G.L., Lazarus, K., Smith, M.E., Berbee, M.L., et al., 2016. A phylum-level phylogenetic classification of zygomycete fungi based on genome-scale data. Mycologia 108, 1028–1046. https://doi.org/10.3852/16-042.

Tao, J.J., Chen, H.W., Ma, B., Zhang, W.K., Chen, S.Y., Zhang, J.S., 2015. The role of ethylene in plants under salinity stress. Front. Plant Sci. 6, 1059. https://doi.org/10.3389/fpls.2015.01059.

Timmusk, S., Abd El-Daim, I.A., Copolovici, L., Tanilas, T., Kännaste, A., Behers, L., et al., 2014. Drought-tolerance of wheat improved by rhizosphere bacteria from harsh

environments: enhanced biomass production and reduced emissions of stress volatiles. PLoS One 9, e96086. https://doi.org/10.1371/journal.pone.0096086.
Verma, H., Kumar, D., Kumar, V., Kumari, M., Singh, S.K., Sharma, V.K., Droby, S., Santoyo, G., White, J.F., Kumar, A., 2021. The potential application of endophytes in Management of Stress from drought and salinity in crop plants. Microorganisms 9, 1729. https://doi.org/10.3390/microorganisms9081729.
Wang, Y., Zhang, Z., Zhang, P., Cao, Y., Hu, T., Yang, P., 2016. Rhizobium symbiosis contribution to short-term salt stress tolerance in alfalfa (*Medicago sativa* L.). Plant Soil 402, 247–261. https://doi.org/10.1007/s11104-016-2792-6.
Wang, H., An, T., Huang, D., Liu, R., Xu, B., Zhang, S., et al., 2021a. Arbuscular mycorrhizal symbioses alleviating salt stress in maize is associated with a decline in root-to-leaf gradient of Na^+/K^+ ratio. BMC Plant Biol. 21, 1–15. https://doi.org/10.1186/s12870-021-03237-6.
Wang, R., Wang, C., Feng, Q., Liou, R.M., Lin, Y.F., 2021b. Biological inoculant of salt-tolerant bacteria for plant growth stimulation under different saline soil conditions. J. Microbiol. Biotechnol. 31, 398–407. https://doi.org/10.4014/jmb.2009.09032.
Wdowiak-Wróbel, S., Leszcz, A., Małek, W., 2013. Salt tolerance in *Astragalus cicer* microsymbionts: the role of glycine betaine in osmoprotection. Curr. Microbiol. 66, 428–436. https://doi.org/10.1007/s00284-012-0293-2.
Wu, N., Li, Z., Liu, H., Tang, M., 2015. Influence of arbuscular mycorrhiza on photosynthesis and water status of *Populus cathayana* Rehder males and females under salt stress. Acta Physiol. Plant. 37, 1–14. https://doi.org/10.1007/s11738-015-1932-6.
Yan, N., Marschner, P., 2012. Response of microbial activity and biomass to increasing salinity depends on the final salinity, not the original salinity. Soil Biol. Biochem. 53, 50–55. https://doi.org/10.1016/j.soilbio.2012.04.028.
Yang, J., Kloepper, J.W., Ryu, C.M., 2009. Rhizosphere bacteria help plants tolerate abiotic stress. Trends Plant Sci. 14, 1–4. https://doi.org/10.1016/j.tplants.2008.10.004.
Yang, Y., Tang, M., Sulpice, R., Chen, H., Tian, S., Ban, Y., 2014. Arbuscular mycorrhizal fungi alter fractal dimension characteristics of *Robinia pseudoacacia* L. seedlings through regulating plant growth, leaf water status, photosynthesis, and nutrient concentration under drought stress. J. Plant Growth Regul. 33, 612–625. https://doi.org/10.1007/s00344-013-9410-0.
Yasmeen, T., Ahmad, A., Arif, M.S., Mubin, M., Rehman, K., Shahzad, S.M., et al., 2020. Biofilm forming rhizobacteria enhance growth and salt tolerance in sunflower plants by stimulating antioxidant enzymes activity. Plant Physiol. Biochem. 156, 242–256. https://doi.org/10.1016/j.plaphy.2020.09.016.
Ye, L., Zhao, X., Bao, E., Cao, K., Zou, Z., 2019. Effects of arbuscular mycorrhizal fungi on watermelon growth, elemental uptake, antioxidant, and photosystem II activities and stress-response gene expressions under salinity-alkalinity stresses. Front. Plant Sci. 10, 863. https://doi.org/10.3389/fpls.2019.00863.
Yuan, B.C., Li, Z.Z., Liu, H., Gao, M., Zhang, Y.Y., 2007. Microbial biomass and activity in salt affected soils under arid conditions. Appl. Soil Ecol. 35, 319–328. https://doi.org/10.1016/j.apsoil.2006.07.004.
Yuan, Z.L., Su, Z.Z., Zhang, C.L., 2016. Chapter 11: Understanding the biodiversity and functions of root fungal endophytes: the ascomycete *Harpophora oryzae* as a model case. In: Druzhinina, I., Kubicek, C. (Eds.), Environmental and Microbial Relationships. The Mycota, Vol. IV. Springer, Cham, https://doi.org/10.1007/978-3-319-29532-9_11.
Yurgel, S.N., Rice, J., Mulder, M., Kahn, M.L., Belova, V.S., Roumiantseva, M.L., 2013. Truncated betB2-144 plays a critical role in *Sinorhizobium meliloti* Rm2011 osmoprotection and glycine-betaine catabolism. Eur. J. Soil Biol. 54, 48–55. https://doi.org/10.1016/j.ejsobi.2012.10.004.

Zhang, J., Jia, W., Yang, J., Ismail, A.M., 2006. Role of ABA in integrating plant responses to drought and salt stresses. Field Crop Res. 97, 111–119. https://doi.org/10.1016/j.fcr.2005.08.018.

Zheng, W., Xue, D., Li, X., Deng, Y., Rui, J., Feng, K., Wang, Z.L., 2017. The responses and adaptations of microbial communities to salinity in farmland soils: a molecular ecological network analysis. Appl. Soil Ecol. 120, 239–246. https://doi.org/10.1016/j.apsoil.2017.08.019.

CHAPTER 4

Metal-tolerant microbial inoculants for improved phytoextraction

Pooja Sharma[a,*], Ambreen Bano[b,*], Surendra Pratap Singh[c], and Yen Wah Tong[a,d]

[a]NUS Environmental Research Institute, National University of Singapore, Singapore
[b]IIRC-3, Plant-Microbe Interaction and Molecular Immunology Laboratory, Department of Biosciences, Faculty of Sciences, Integral University, Lucknow, UP, India
[c]Plant Molecular Biology Laboratory, Department of Botany, Dayanand Anglo-Vedic (PG) College, Chhatrapati Shahu Ji Maharaj University, Kanpur, India
[d]Department of Chemical and Biomolecular Engineering, National University of Singapore, Singapore

4.1 Introduction

Human health and environmental protection are increasingly concerned about heavy metal pollution (Teng et al., 2006; Xiao et al., 2015). Furthermore, vanadium ores frequently pollute soils with cadmium and chromium (Wang et al., 2018). Heavy metals associated with vanadium have even spread to farmlands and groundwater (Hao et al., 2018). As harmful metals are transported in the food chain and concentrations are increased throughout as a result of direct exposure, they pose potential threats to both the ecosystem and human health (Xiao et al., 2017; Zou et al., 2019). The use of phytoremediation solutions for soils contaminated with metal(loids) and modified plants is receiving increased attention as a way to improve their effectiveness (Vangronsveld et al., 2009). There are a number of microbes that can improve the performance of plants under stressful conditions, including beneficial bacteria and fungi (Cherian et al., 2012). The use of plant growth-promoting bacteria (PGPB) associated with many plant species is one approach to enhancing metal phytoextraction through increases in shoot biomass and/or shoot metal concentration. Some PGPB can live in plant components as endophytic bacteria (Compant et al., 2010). The colonization process may begin in the root zone, although bacteria from the phyllosphere, atmosphere, and spermosphere may also be involved.

* Authors contributed equally and should be considered as first authors.

Microbial Inoculants
https://doi.org/10.1016/B978-0-323-99043-1.00005-0

Rhizospheric and phyllospheric bacteria are less likely to interact with their hosts than endophytic bacteria (Sturz and Nowak, 2000). Environmentalists are increasingly using biological techniques to safely remove lead (Pb) from the environment. Phytoremediation has been well-researched and is widely regarded as a safe method of removing highly hazardous metals (Gul et al., 2019; Manzoor et al., 2018). Pb phytoextraction is challenging because of its persistence and low mobility. The use of specific hyperaccumulator plants is one of the strategies used to improve phytoremediation efficiency. Therefore, these synthetic chelators are linked to concerns like cost, leaching, and toxicity (Gul et al., 2019; Qiao et al., 2017).

The bacterial rhizobiome, a collection of microbes that colonize a plant's root system, can help in phytoextraction improvement (Xiao et al., 2019; Wang et al., 2020b). Intricate interactions exist between heavy metals, soil, and root-associated microbes. Microbes can aid in increasing heavy metals buildup by promoting plant biomass and metal mobility in polluted soils or by enhancing plant photosynthesis and metal tolerance (Shin et al., 2012). The bioavailability of heavy metals in soils is increased by metal-resistant bacteria isolated from hyperaccumulator rhizospheres, for example, which leads to improved root uptake of heavy metals (Visioli et al., 2015). Heavy metals stress is alleviated, and phytoextraction is improved by endophytic bacteria isolated from hyperaccumulators (Wang et al., 2020a). As a result, there is a pressing need to comprehend the role of microbes in hyperaccumulator plant-microbe interactions. The discovery of the microbial characteristics that increase plant phytoremediation may hinge on the reconfiguration of the extensive network of plant-microbiome interactions under various situations.

The composition and structure of microbiomes associated with hyperaccumulators have been revealed in new studies, suggesting that plants contribute to microbiome evolution in a selective manner (Luo et al., 2019). Plant compartment, i.e., rhizosphere versus endosphere, plant genotype/ecotype, and environmental factors such as the presence of soil contaminants, soil pH, and organic matter content are also determinants of the microbes linked to hyperaccumulating plants (Wu et al., 2020). Heavy metal levels in soil alter the microbiome structure of hyperaccumulators plants, reducing microbial biomass and increasing heavy metal-resistant microbes (Wood et al., 2016). However, there is still a scarcity of field-scale studies on bacteria linked to heavy metal-accumulating plants. Furthermore, we lack an understanding of the relationship between the microbiome structure and plant uptake as well as aboveground heavy metal transfer under various soil types.

4.2 Phytoextraction: Its mechanism

The technique that uses plants for contaminant removal from air, water, and soil is known as phytoremediation (Jeevanantham et al., 2019). In a variety of ways, plants can remediate or clean up contaminated sites. Contaminant absorption by plants predominantly occurs via the root system, which contains the primary mechanisms for toxicity prevention. The root system has a massive surface area that absorbs and collects the nutrients and water required for development and other nonessential pollutants (Edelstein and Ben-Hur, 2018; Kumar and Chopra, 2018). Plant roots alter the soil-root interface by releasing inorganic and organic exudates into the rhizosphere (Bali et al., 2020). Microbes, according to Weyens et al. (2015), can help plants withstand biotic and abiotic challenges by absorbing water and nutrients and by creating plant hormones and siderophores. In addition, phytoremediation relies on microbial biodegradation to break down, detoxify, and sequester contaminants while promoting plant growth (Weyens et al., 2015). As depicted in Fig. 1 (Created with BioRender.com), the phytoremediation approach consists of several steps: phytoextraction, phytostabilization, phytostimulation, phytodegradation, phytovolatilization, rhizodegradation, hydraulic control, and rhizofiltration (Lee et al., 2020). The absorption and transport of metal pollutants by plant roots from the soil into the aboveground sections of the plant referred to as phytoextraction, also known as phytoaccumulation. Plants that are hyperaccumulators absorb large amounts of metals compared to plants that are not hyperaccumulators. These plants are chosen and planted based on the type of metal present and other site circumstances. Plants are harvested after their growing phase and may be burnt or composted to recycle the metals. If required, then this technique can be repeated to reduce soil contamination levels to acceptable levels. The ash from burning plants is disposed of in hazardous waste landfills. If the contaminated soil is dug up for treatment, then the volume of the ash produced will be less than 10% of the total volume produced. The majority of plants that assimilate and absorb abnormally high levels of metals prefer metals such as nickel (Ni), copper (Cu), and zinc (Zn) (Ghori et al., 2016; Talebi et al., 2019).

The phyllosphere and rhizosphere of plants are both involved in phytoextraction since this mechanism extracts pollutants from soil and further transports them to the aboveground sections of the plant (Pandey and Bajpai, 2019). This process is well-known as phytosequestration, phytoabsorption, and phytoaccumulation (Favas et al., 2014). This approach employs hyperaccumulator plants (those that are capable of concentrating

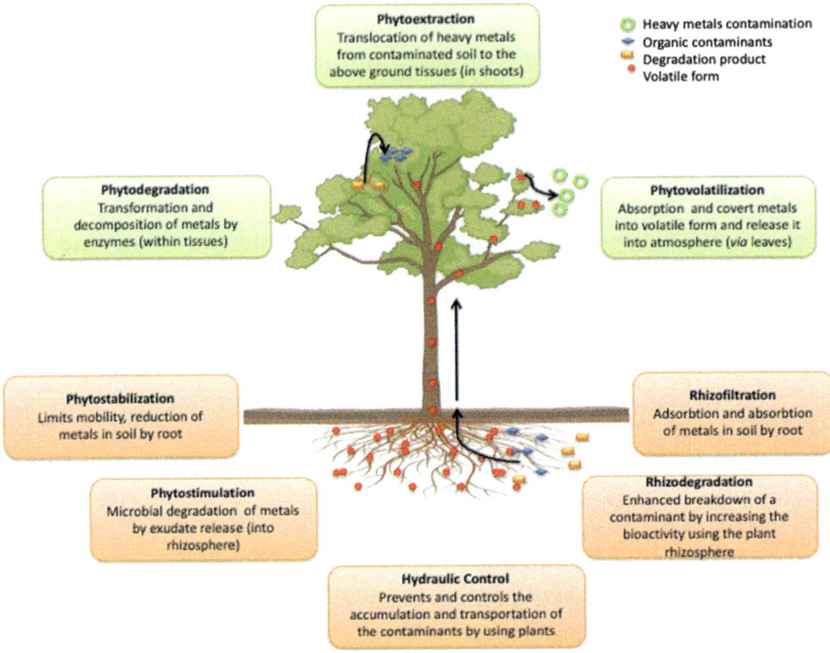

Fig. 1 The phytoremediation approach: phytoextraction, phytostabilization, phytodegradation, phytovolatilization, phytostimulation, rhizodegradation, hydraulic control, and rhizofiltration.

metals a hundred times more than nonaccumulating plants) to absorb contaminants from polluted soils and collect them in the phyllosphere. These plants are chosen based on their biomass, translocation factor, metals tolerance, surface biomass, and ease of harvesting. Incineration is the ultimate step in pollution removal after plants have been gathered from contaminated areas (Pandey and Bajpai, 2019). The capacity of a plant to intercept, accumulate, and absorb metals in shoots is dependent on various factors, including the level of soil pollution, metal availability for absorption by the roots, and the plant's ability to intercept, absorb, and accumulate metals. Finally, soil, metal, and plant interplay determines phytoextraction potential. The intricacy of this interaction, which is influenced by climatic circumstances, suggests that a site-specific phytoremediation method is preferable to a general approach. This emphasizes how critical it is to comprehend the mechanisms and processes that drive metal absorption and accumulation in plants (Fig. 2 [Created with BioRender.com] and Table 1).

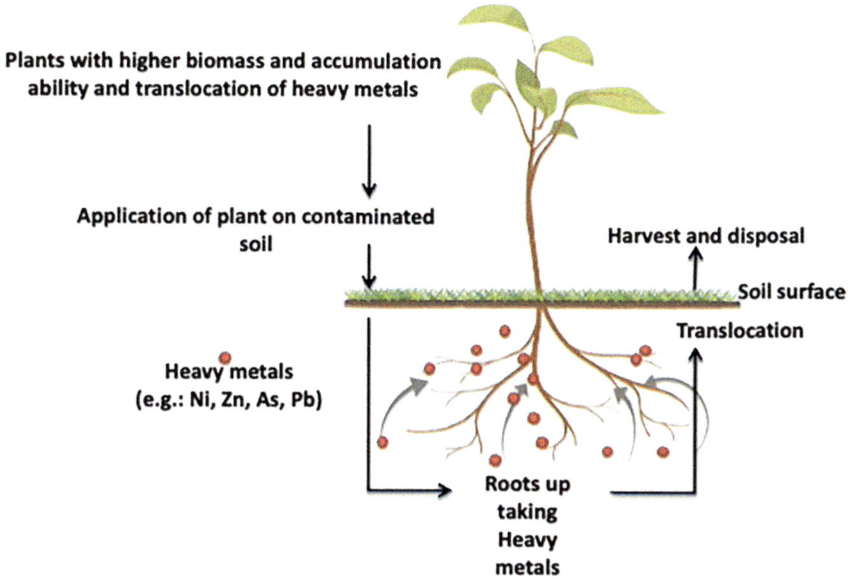

Fig. 2 The mechanism of heavy metal absorption and translocation through plants with high biomass and accumulation capacity.

Table 1 Steps and mechanisms for accumulation of heavy metals by plants.

Steps	Plant part(s)	Involved functional mechanism(s)
Uptake of metals	Root and root exudates	Chelation, mobilization, precipitation, and chemical modification
	Cell walls	Adsorption, chemisorption, ion exchange, and microprecipitation
	Plasma membrane	Transporters involved in metal uptake and efflux
Root-to-shoot transfer of metals	Vascular tissues	Uploading to xylem sap via various biological moieties
Detoxification and sequestration	Trichomes	Metal distribution in nonmetabolic cells
	Vacuoles	Tonoplast-specific transporters accumulate the substances within the vacuoles
	Biomolecules	Numerous functional groups in biomolecules allow them to bind to metal ions
	Specially synthesized biomolecules	Phytochelatins and metallothioneins
	Heavy metal-induced oxidative stress and biomolecules	Enzymes, Glutathione, Heat shock proteins, carotenoids, proline, salicylic acid, nitric oxide polyamines

4.3 Phytoextraction of metals and soil microorganisms

Microorganisms in the soil have been demonstrated to have a variety of ways for modifying metal bioavailability for absorption by the roots. Roots of plants are capable of absorbing metals absorbed by microbes through redox processes that are accelerated by microbes. Furthermore, root mycorrhizal associations have been demonstrated to influence metal absorption rates (Sarwar et al., 2017; Ponting et al., 2021).

4.3.1 Uptake of heavy metals through roots

Most metals are taken up by the roots from an aqueous phase. A significant amount is rendered by strong adhesion to soil particles and/or insoluble soil metal fraction precipitation and is hence inaccessible for absorption by plants. A major issue limiting the possibility for phytoextraction of large metal pollutants such as lead (Pb) is low soil bioavailability. The primary goal of contemporary phytoremediation research is to improve the tendency of lead absorption by plant roots by inducing Pb desorption from the soil into solution (Raikova et al., 2019; Egendorf et al., 2020).

4.3.2 Mechanisms based on plants

Metal hyperaccumulator species identification shows that plants have the genetic ability to eliminate harmful metals from contaminated soils. The importance of plant-based restorative processes cannot be overstated. This explanation, for instance, may provide suggestions for improving efficiency through proper agronomic practices. Furthermore, identifying and biochemically characterizing the restorative processes require preparatory stages to isolate plant genes responsible for the development of the remediating phenotype. The discovery and isolation of these genes may pave the way for biotechnology to improve plants for environmental cleaning (Maiti et al., 2004; Macnair et al., 2020).

4.4 Improving plants for phytoremediation

Plants, in general, exhibit an imperfect set of remediating characteristics. For example, the majority of metal hyperaccumulators are small and sluggish to expand. A biotechnological approach, such as transferring desirable features between hyperaccumulators and high biomass nonaccumulators, is employed to overcome these deficiencies (Khan, 2005). Suitable plants for phytoextraction should have a number of characteristics, including the capacity to grow outside of their collecting region, rapid growth, large biomass, ease of harvesting, and a variety of heavy metal accumulation in harvestable portions (Pajević et al., 2016; Paz-Alberto and Sigua, 2013). There is currently not a single plant that meets all of these requirements. A quickly growing nonaccumulator plant, on the other hand, might be changed and/or designed to attain the majority of the aforementioned characteristics. Significant progress has been made in establishing the molecular and biochemical basis for metal buildup, providing us with a solid scientific basis for defining some approaches for attaining this purpose (Ma et al., 2016).

4.5 Heavy metal accumulator plants

Plants selectively absorb metals, with some metals being more readily absorbed than others. Numerous heavy metals are critical micronutrients like Cu, Zn, Ni, and manganese (Mn), which plants require for developing and completing their life cycle. Plants growing in metalliferous soil are differentiated into three major classes:

(i) Excluders are plants in which metal concentrations in the shoot remain below a critical level over a wide range of soil concentrations. Toxic metals are prevented from entering the root cells by these devices. The use of excluders can help stabilize soil and prevent further contamination from spreading (Lasat, 2002; Fernández et al., 2017).

(ii) Metal concentrations in accumulators vary from lower to higher soil levels in the aboveground parts of the plant. Metals are prevented from being taken up by the roots, allowing for high levels of metal bioaccumulation (Ghosh and Singh, 2005; Martínez-Sánchez et al., 2012).

(iii) The indicators are points at which the internal concentration reflects the external concentration (McGrath et al., 2002).

4.6 High-capacity hyperaccumulators

Hyperaccumulator plant species discoveries have transformed phytoremediation technology as these have a higher inbuilt ability for metal absorption than do typical plants (Lasat, 1999). These accumulator plants collect significant quantities of metals in their leaves and are generally endemic to naturally mineralized soils (Baker et al., 2020). Metal hyperaccumulators can spontaneously collect heavy metals in the aboveground plant parts without emerging toxic symptoms. Plants that can concentrate metals to 0.1% for Co, Cu, Cr, Pb, Ni, and Al; 1% for Mn and Zn; and 0.1% for Se and Cd are metal hyperaccumulators (Baker and Brooks, 1989). Plants take time to lower the heavy metal quantity, and their biomass production is the concentration ratio of the metal in the soil to the shoot tissues (McGrath and Zhao, 2003). It is governed by the root's ability to store, absorb, collect, and detoxify metals while sustaining growth, biomass production, and metabolism (Shahid et al., 2017). Most plants, apart from hyperaccumulators, exhibit metal bioconcentration values of less than one, implying that it will take more time than human beings will to lower soil pollutants by half (Peuke and Rennenberg, 2005). Heavy metal-ion hyperaccumulation is a stunning phenomenon observed in around 2% of angiosperms (Rascio and Navari-Izzo, 2011; Baker and Whiting, 2002). More than 450 species of these hyperaccumulators are found in 45 angiosperm families (Table 2),

Table 2 Metal hyperaccumulator plants.

Metals	Plant species with specific accumulating metals	Family	References
Nickel (Ni)	*Alyssum crenulatum, Alyssum serpyllifolium, A. bertolonii, Bornmuellera kiyakii, Alyssum heldreichii,*	Brassicaceae	Perrier et al. (2004); Reeves et al. (2009); Becerra-Castro et al. (2009); Moradi et al. (2010); Barzanti et al. (2011)
	Walsura monophylla	Meliaceae	
	Sebertia acuminata	Sapotaceae	
	Berkheya coddii	Asteraceae	
Zinc (Zn)	*Brassica juncea, Thlaspi caerulescens Arabis gemmifera, A. paniculata, Arabidopsis halleri*	Brassicaceae	Zhao et al. (2000); Sun et al. (2005); Küpper and Kochian (2010);Du et al. (2011)
	Silene vulgaris	Caryophyllaceae	
	Sedum alfredii	Crassulaceae	
	Picris divaricata	Asteraceae	

Table 2 Metal hyperaccumulator plants—cont'd

Metals	Plant species with specific accumulating metals	Family	References
Lead (Pb)	*B. juncea*, *Thlaspi caerulescens*, *Arabis gemmifera*, *A. paniculata*, *Arabidopsis halleri*	Brassicaceae	Zhao et al. (2000); Kubota and Takenaka (2003); Sun et al. (2005); Küpper and Kochian (2010); Du et al. (2011)
	Euphorbia cheiradenia	Euphorbiaceae	
	Picris divaricata	Asteraceae	
	Sedum alfredii	Crassulaceae	
Cadmium (Cd)	*Thlaspi caerulescens*, *Arabidopsis halleri*,	Brassicaceae	Dahmani-Muller et al. (2000); Basic et al. (2006)
	Avena strigosa	Poaceae	
	Vallisneria spiralis	Hydrodictyaceae	
	Tagetes erecta	Asteraceae	
	Solanum photeinocarpum	Solanaceae	
	Azolla pinnata	Azollaceae	
Chromium (Cr)	*Pteris vittata*	Pteridaceae	Gardea-Torresday et al. (2005); Zhang et al. (2007)
	B. juncea	Brassicaceae	
	Leersia hexandra	Poaceae	
	Salsola kali	Amaranthaceae	
Copper (Cu)	*Aeolanthus biformifolius*, *Haumaniastrum robertii*	Lamiaceae	Küpper et al. (2009); Wang and Zhong (2011)
	Crassula helmsii	Crassulaceae	
	Larrea tridentata	Zygophyllaceae	
	Ipomea alpina	Convolvulaceae	
	Commelina communis	Commelinaceae	
Arsenic (As)	*Pteris quadriaurita*	Pteridaceae	Niazi et al. (2016); Saxena et al. (2019)
	Corrigiola telephiifolia	Caryophyllaceae	
Selenium (Se)	*Stanleya pinnata*, *B. juncea*	Brassicaceae	Galeas et al. (2007); Freeman et al. (2010); Hladun et al. (2011)
	Astragalus bisulcatus	Fabaceae	

Continued

Table 2 Metal hyperaccumulator plants—cont'd

Metals	Plant species with specific accumulating metals	Family	References
Aluminum (Al)	*Hordeum vulgare*	Poaceae	
Manganese (Mn)	*Gossia bidwillii, Austromyrtus bidwillii*	Myrtaceae	Bidwell et al. (2002); Fernando et al. (2007)
	Phytolacca americana	Phytolaccaceae	
	Maytenus founieri	Celastraceae	
	Virotia neurophylla	Proteaceae	
Magnesium (Mg)	*Schima superba, Camellia oleifera*	Theaceae	Yao et al. (2012); Yu et al. (2021)
Molybdenum (Mo)	*Thlaspi caerulescens*	Brassicaceae	Haque et al. (2008)
Silver (Ag)	*Brassica juncea*	Brassicaceae	Gardea-Torresdey et al. (2005)
Cerium (Ce)	*Helianthus annuus*	Asteraceae	Moreira et al. (2019)
Mercury (Hg)	*Pistia stratiotes*	Araceae	Kumar et al. (2019)
Thallium (Tl)	*Brassica oleracea, Iberis intermedia*	Brassicaceae	Leblanc et al. (1999); Al-Najar et al. (2005)
Cobalt (Co)	*Haumaniastrum robertii*	Lamiaceae	Kabeya et al. (2018)
	Crotalaria cobalticola	Fabaceae	Oven et al. (2002)

along with members of the *Caryophyllaceae, Brassicaceae, Violaceae, Asteraceae, Flacourtiaceae, Fabaceae, Cunoniaceae, Cyperaceae, Poaceae, Euphorbiaceae*, and *Lamiaceae* families (Padmavathiamma and Li, 2007). The angiosperm *Thlaspi caerulescens* is a well-known metal hyperaccumulator, capable of accumulating large amounts of Cd (1800 mg/kg) and Zn (39,600 mg/kg) (Basic et al., 2006; Rascio and Navari-Izzo, 2011). Self-pollinating diploid plants may readily grow in the laboratory, making it an ideal experimental organism for exploring the processes of metal intake, tolerance, and accumulation in connection to metal phytoextraction. *Brassica juncea*, in addition to *Thlaspi caerulescens*, has been utilized as a model system to study the biochemistry and physiology of metal accumulation in plants.

4.7 High biomass of crops

For effective and economically viable phytoextraction, plants with a biomass production of ten tonnes per hectare (t/ha) and a metal bioconcentration factor of 20 are required; alternatively, plants with a biomass production of 20 t/ha and a metal bioconcentration factor of 10 are required (Peuke and Rennenberg, 2005). The phytoextraction rate is directly proportional to a plant's growth rate, and the entire quantity of the phytoextracted metal is associated with plant biomass, making the phytoextraction process extremely steady (Shah and Nongkynrih, 2007). This demands the discovery of rapidly growing and highly metal-accumulating genotypes. Despite owning one-third of Zn content in its tissues, *Brassica juncea* is believed to be highly efficient in removing Zn from contaminated soils than is *T. caerulescens*, which is already considered a Zn hyperaccumulator (Ebbs et al., 1997). This benefit is because *B. juncea* produces 10 times higher biomass than does *T. caerulescens* (Suman et al., 2018).

Recently, there has been a sudden interest in utilizing crops with high biomass for metal phytoextraction (Pajević et al., 2016). Because of their vast root system, rates of high water intake, massive biomass, transpiration, quick growth, and ease of harvesting with subsequent resprouting without disturbing the land, trees that grow rapidly and are inexpensive are more suitable for phytoextraction (Peuke and Rennenberg, 2005). There has been considerable interest in phytoremediation of metal-contaminated soil by a variety of trees that exhibit promising heavy metal remediation properties (Bian et al., 2020).

4.8 Endophytes and phytoextraction

Endophytes are microorganisms that live within plant tissues without damaging the host (Kurissery et al., 2022). Many studies have been conducted on the role of endophytic bacteria in phytoextraction in recent years (Ullah et al., 2021; Ye et al., 2020). Endophytes can help plants grow faster and be more resistant to herbivores, infections, and drought (Decunta et al., 2021). To date, most studies have focused on the biodegradation of organic pollutants and the employment of endophytic bacteria for heavy metal phytoremediation. This has been sluggish because of the scarcity of desirable strains with heavy metal tolerance and detoxifying abilities (Luo et al., 2011). Major endophytic studies that support phytoremediation have focused on arbuscular mycorrhizal (AM) fungi; bacteria also play an important role in the absorption of metals through plants (Doty, 2008). Soil-borne, obligatory AM fungi biotrophs are essential to plant component lives with a mutual relationship of more than 80% with terrestrial plant roots (Smith and Read, 2008). These fungi have also been found in plant roots that grow in heavy metal-contaminated soils. They are considered to play an essential role in metal accumulation and tolerance (Karimi et al., 2011; Riaz et al., 2021). Treatment of contaminated soils with mycorrhizal plants requires the selection of appropriate AM species. The species chosen from heavy metal-contaminated regions are most effective because they have acquired the capacity to endure metal stress circumstances and, as a result, can function more effectively than other AM species (Miransari, 2011). Phytoextraction promoted by mycorrhizal fungi increases metal availability for plant absorption. Better phytoextraction after mycorrhization can be achieved by a variety of methods, including enhanced plant tolerance to metals, increased biomass output, improved plant growth, and higher metal concentration in plant tissues (Ashraf et al., 2019). The *Glomus* species predominate in the rhizospheres of growing plants in soils contaminated with heavy metals (Khade and Adholeya, 2009). However, the genotypic heterogeneity makes finding appropriate AM fungi for metal-contaminated soil regeneration challenging. As a result, ongoing attempts are necessary to find mycorrhizal strains with heavy metal tolerance, which are eventually used to treat soils polluted with metals (Javaid, 2011).

4.9 The biochemical mechanism of accumulators

Plants that demonstrate resistance to hazardous metals have various molecular and cellular pathways that may work in detoxification, tolerance of stress caused by heavy metals, and general homeostasis (Hall, 2002). Metal absorption by roots, complexation with chelating molecules, metal transport from roots to shoots, and compartmentalization into vacuoles are all believed to be important for accumulation (Hall, 2002; Pasricha et al., 2021). Metals should not be solitarily picked up quickly but instead should be transferred from roots to aerial plant regions for successful phytoextraction. Despite the presence of metals in soils at high levels, metal absorption is mostly affected by their proportion of bioavailability (Vamerali et al., 2010). Despite their abundance in nature, certain metals' bioavailability is restricted because of poor solubility in oxygenated water along with high soil particle-binding capacity. The availability of metals and their movement in the rhizospheres are also affected by rhizospheric bacteria and root exudates. With the exception of Fe, nothing is known regarding active trace element mobilization by plant roots. Rhizosphere acidification, carboxylate exudation, and processes that help in phosphorus acquisition all lead to increased bioavailability of specific micronutrients (Zhang et al., 2010). It is anticipated that the majority of the metals deposited are attached to ligands such as peptides, proteins, amino acids, and organic acids. There is evidence that ligands detoxify metals, but we lack a comprehensive picture of the many chelators involved in metal transit in plants and deposition in accumulators. However, there is no compelling evidence that hyperaccumulators release particular chelators in the rhizosphere to boost metal absorption (Gove et al., 2002). Numerous plants use highly specialized methods for increasing the bioavailability of metals in the rhizosphere to absorb metals. Siderophores are small-molecular-mass organic compounds generated by microbes and the Poaceae family members, which improve iron availability for absorption by plant roots (Scavino and Pedraza, 2013). Siderophores are iron (III) ligands that form a stable complex with metals such as Pb, Ni, Cd, Zn, and Cu (Nair et al., 2007; Alchoubassi et al., 2021). Metal chelation can be extremely useful in cleaning soil with a high metal concentration.

Transgenic plants might be produced to exude selective metal ligands into the rhizosphere, precisely solubilizing elements of interest (Yadav et al., 2018). Grass species that produce growing siderophores in association with plant accumulators can improve the efficiency of phytoextraction. Although this strategy has potential, phytosiderophores gain their specificity not from Fe chelation in soils but rather through the absorption of Fe phytosiderophores by a membrane carrier (Kumar and Chandra, 2019). Efforts have been undertaken to find ligands that were believed to be released by hyperaccumulator roots to improve the Ni release rate from soil and absorption by roots (Ameen et al., 2019). There are more ligands in the *Wheat* rhizosphere solution than in the *Thlaspi* rhizosphere solution (Zhao et al., 2001). Several studies indicate that plants can attain hyperaccumulation by regulating or constituting an element transporter in their plasma membranes. While secretion of a ligand and other typical root activities may not let the plant attain hyperaccumulation, they can affect the speciation of elements in the rhizosphere and influence elemental phytoavailability. Although the idea of plants secreting chelating ligands into the rhizosphere, and, subsequently, absorbing ligand-metal complexes into their roots, remains viable for phytoextraction, there is no indication that these strategies could be turned into a phytoextraction practical system (Chaney et al., 2007).

Aside from these natural methods, it is known that the use of synthetic chelates stimulates the metals released into the soil solution, hence increasing their capacity for absorption by the roots (Lasat, 2002). There have been numerous chelating agents developed, including diethylenetriaminepentaacetic acid (DTPA), citric acid, ethylenediaminetetraacetic acid (EDTA), nitrilotriacetic acid (NTA), ethylenediamine di-o-hydroxyphenyl acetic acid (EDDHA), ethylene glycol-O, tetraacetic acid (EGTA). For their ability to mobilize metals and boost their accumulation in plants, all of these compounds have been considered. It is EDTA that is mostly employed in phytoextraction studies and has been effectively used for heavy metal treatment in polluted soils (Chen et al., 2004). The amount of solubilization of heavy metals by chelation with organic complexing agents is defined by the order of stability constants in aqueous solutions using a metal to chelating agent ratio of 1:1. Nonetheless, despite the successful technology, concerns have been expressed about the mobility of the increased metals in soil along with the possible threat of leaking into groundwater. Aside from the potential leaching of metals, the usage of chelators is also constrained by their higher cost (Dhaliwal et al., 2020). Chelating agents that are quickly destroyed by soil microorganisms, and yet effectively chelate metals, are costlier and do not significantly reduce metal discharge from treated soil (Meers et al., 2005).

Following mobilization, metals are primarily attached to the cell walls, which are ion exchangers with poor selectivity. Secondary transporters, including H-coupled carrier proteins and/or channel proteins, subsequently moderate and promote uptake throughout the plasma membrane via transport networks and intracellular high-affinity binding sites (Chaney et al., 2007). In recent years, some cation transporters have been identified, the majority of which are members of the YSL (yellow stripe-like transporter), NRAMP (natural resistance-associated macrophage protein), ZIP (Fe-regulated transporter-like protein, Zn-regulated transporter), NAS (nicotinamine synthase), CDF (cation diffusion facilitator), FER (ferritin Fe (III) binding) SAMS (S-adenosyl methionine synthase), IREG (iron-regulated transporter), and HMA (heavy metal ATPase) families (Sharma et al., 2021). Once inside the plant, both root pressure and transpirational pull govern the passage of metals, transporting sap from the roots to the aerial regions. Transfer of metals to the shoot is mostly accomplished via the xylem. Metal-ion translocation to the shoot involves radial symplastic transit with active loading into the xylem (Song et al., 2017). Plants accelerate the redox process and change the chemistry of metal ions to enable their accumulation in nontoxic forms because metal ions are toxic at high intracellular concentrations. Heavy metals such as Zn, Cd, and Pb, however, do not exist in distinct oxidation states. Intracellular metals are detoxified in certain situations via binding to low-molecular-mass organic molecules, binding to histidine, or localization in vacuoles as a metal-organic acid complex (Rascio and Navari-Izzo, 2011). In order to regulate the cytoplasmic metal concentration of some metals, such as Zn, several mechanisms have been proposed: sequestration in a subcellular organelle, complexation that lowers uptake throughout the plasma membrane, precipitation as insoluble salts, organic ligands with low molecular masses, and active extrusion into apoplasts (Sinclair and Krämer, 2012). Chelation and metal sequestration by specific ligands are fundamental ways by which plants deal with metals stress. In plant cells, the metal-binding ligands are metallothionein (MT) and phytochelatin (PC). These ligands are extensively dispersed in the plant. Several genetic and physiological studies have shown that MT and PC are important for accumulating and tolerating metals in plants (Cobbett and Goldsbrough, 2002). These two generate stable compounds with metals in the cytosol that can be sequestered into the vacuoles later (Goldsbrough, 2020). Overproduction of PC appears to be an inducible rather than a constitutive process, found most notably in metal-intolerant plants as part of their metal defense system (Freeman et al., 2005). Phytochelatins have also been linked to As detoxification (Singh et al., 2021). MTs are low-molecular-mass cysteine-rich proteins discovered in mammals as Cd-, Zn-, and Cu-binding proteins (Duan et al., 2018). MTs derived from plants and animals are being researched for use in plants to reduce metal buildup in shoots by trapping the metals in the roots. Plant tolerance to certain metals can be increased by overexpression of MTs. Vacuoles are often regarded as the primary storage place for metals within plant cells, particularly Zn and Cd. Metal compartmentalization in vacuoles is a major feature of various metal hyperaccumulator plant tolerance mechanisms. *Thlaspi goesingense* improves tolerance of Ni by compartmentalizing the majority of the internal leaf Ni into vacuoles (Srivastava, 2017). *Thlaspi goesingense* has an elevated level of metal-ion transporters, which is believed to be the major reason for its improved capacity to collect metal ions within shoot vacuoles (Persans et al., 2001).

4.10 Conclusions

The metal phytoextraction process has proven beneficial and desired in restoring a polluted environment. Metals interfere with plant processes and act as an external trigger, activating the defense system through multiple physiological and metabolic pathways. There are several phytoremediation technologies that provide cost-effective, environmentally benign methods for in situ metal cleanup. Metal hyperaccumulators have major physiological properties that can be used for the extraction, transformation, and/or stabilization of metals. The efficacy and combination of phytoremediation technologies are crucial in the meantime. Methodologies such as microorganism's potential, amendments, and genetic engineering can be utilized to improve phytoextraction and boost plant yield in soils with low levels of metals. Furthermore, phytoremediation capability and enhanced metal tolerance could be considered the first step toward maximizing plant species' accumulation potential. Metal phytoextraction requires a multidisciplinary approach because of the complex interactions that occur in site-specific settings. Plant scientists, soil microbiologists, agronomists, and environmental engineers will need to use a comprehensive analytical tool to integrate their work. In addition, to better understand the molecular mechanism of hyperaccumulators, field experiments at different locations are needed since most of this research is conducted in the lab. The molecular mechanism of phytoremediation mediated by soil amendments is likewise unknown.

Acknowledgment

This research is supported by the National Research Foundation, Prime Minister's Office, Singapore, under its Campus for Research Excellence and Technological Enterprise (CREATE) program. We are grateful to the Science and Engineering Research Board (SERB), New Delhi, for the Start-up Research Grant (SRG/2019/001940).

Conflict of interest

The authors declare that they have no known competing financial or personal interests that could have appeared to influence the work reported in this chapter.

References

Alchoubassi, G., Kińska, K., Bierla, K., Lobinski, R., Szpunar, J., 2021. Speciation of essential nutrient trace elements in coconut water. Food Chem. 339, 127680.

Al-Najar, H., Kaschl, A., Schulz, R., Römheld, V., 2005. Effect of thallium fractions in the soil and pollution origins on Tl uptake by hyperaccumulator plants: a key factor for the assessment of phytoextraction. Int. J. Phytoremed. 7 (1), 55–67.

Ameen, N., Amjad, M., Murtaza, B., Abbas, G., Shahid, M., Imran, M., Naeem, M.A., Niazi, N.K., 2019. Biogeochemical behavior of nickel under different abiotic stresses:

toxicity and detoxification mechanisms in plants. Environ. Sci. Pollut. Res. 26 (11), 10496–10514.

Ashraf, S., Ali, Q., Zahir, Z.A., Ashraf, S., Asghar, H.N., 2019. Phytoremediation: environmentally sustainable way for reclamation of heavy metal polluted soils. Ecotoxicol. Environ. Saf. 174, 714–727.

Baker, A.J., Brooks, R., 1989. Terrestrial higher plants which hyperaccumulate metallic elements. A review of their distribution, ecology and phytochemistry. Biorecovery. 1 (2), 81–126.

Baker, A.J., Whiting, S.N., 2002. In search of the holy grail: a further step in understanding metal hyperaccumulation? New Phytol., 1–4.

Baker, A.J., McGrath, S.P., Reeves, R.D., Smith, J.A.C., 2020. Metal hyperaccumulator plants: a review of the ecology and physiology of a biological resource for phytoremediation of metal-polluted soils. In: Phytoremediation of Contaminated Soil and Water, pp. 85–107.

Bali, A.S., Sidhu, G.P.S., Kumar, V., 2020. Root exudates ameliorate cadmium tolerance in plants: a review. Environ. Chem. Lett. 18 (4), 1243–1275.

Barzanti, R., Colzi, I., Arnetoli, M., Gallo, A., Pignattelli, S., Gabbrielli, R., Gonnelli, C., 2011. Cadmium phytoextraction potential of different Alyssum species. J. Hazard. Mater. 196, 66–72.

Basic, N., Keller, C., Fontanillas, P., Vittoz, P., Besnard, G., Galland, N., 2006. Cadmium hyperaccumulation and reproductive traits in natural *Thlaspi caerulescens* populations. Plant Biol. 8, 64e72.

Becerra-Castro, C., Monterroso, C., Garcia-Leston, M., Prieto-Fernandez, A., Acea, M.J., Kidd, P.S., 2009. Rhizosphere microbial densities and trace metal tolerance of the nickel hyperaccumulator *Alyssum serpyllifolium* subsp. lusitanicum. Int. J. Phytoremed. 11 (6), 525–541.

Bian, F., Zhong, Z., Zhang, X., Yang, C., Gai, X., 2020. Bamboo–an untapped plant resource for the phytoremediation of heavy metal contaminated soils. Chemosphere 246, 125750.

Bidwell, S.D., Woodrow, I.E., Batianoff, G.N., Sommer-Knudsen, J., 2002. Hyperaccumulation of manganese in the rainforest tree *Austromyrtus bidwillii* (Myrtaceae) from Queensland, Australia. Funct. Plant Biol. 29, 899e905.

Chaney, R.L., Angle, J.S., Broadhurst, C.L., Peters, C.A., Tappero, R.V., Sparks, D.L., 2007. Improved understanding of hyperaccumulation yields commercial phytoextraction and phytomining technologies. J. Environ. Qual. 36 (5), 1429–1443.

Chen, Y., Li, X., Shen, Z., 2004. Leaching and uptake of heavy metals by ten different species of plants during an EDTA-assisted phytoextraction process. Chemosphere 57 (3), 187–196.

Cherian, S., Weyens, N., Lindberg, S., Vangronsveld, J., 2012. Phytoremediation of trace element–contaminated environments and the potential of endophytic bacteria for improving this process. Crit. Rev. Environ. Sci. Technol. 42 (21), 2215–2260.

Cobbett, C., Goldsbrough, P., 2002. Phytochelatins and metallothioneins: roles in heavy metal detoxification and homeostasis. Annu. Rev. Plant Biol. 53 (1), 159–182.

Compant, S., Clément, C., Sessitsch, A., 2010. Plant growth-promoting bacteria in the rhizo-and endosphere of plants: their role, colonization, mechanisms involved and prospects for utilization. Soil Biol. Biochem. 42 (5), 669–678.

Dahmani-Muller, H., van Oort, F., Gelie, B., Balabane, M., 2000. Strategies of heavy metal uptake by three plant species growing near a metal smelter. Environ. Poll. 109, 231e238.

Decunta, F.A., Pérez, L.I., Malinowski, D.P., Molina-Montenegro, M.A., Gundel, P.E., 2021. A systematic review on the effects of epichloë fungal endophytes on drought tolerance in cool-season grasses. Front. Plant Sci. 12, 380.

Dhaliwal, S.S., Singh, J., Taneja, P.K., Mandal, A., 2020. Remediation techniques for removal of heavy metals from the soil contaminated through different sources: a review. Environ. Sci. Pollut. Res. 27 (2), 1319–1333.

Doty, S.L., 2008. Enhancing phytoremediation through the use of transgenics and endophytes. New Phytol. 179 (2), 318–333.

Du, R.J., He, E.K., Tang, Y.T., Hu, P.J., Ying, R.R., Morel, J.L., Qiu, R.L., 2011. How phytohormone IAA and chelator EDTA affect lead uptake by Zn/Cd hyperaccumulator *Picris divaricata*. Int. J. Phytoremed. 13, 1024e1036.

Duan, L., Kong, J.J., Wang, T.Q., Sun, Y., 2018. Binding of Cd (II), Pb (II), and Zn (II) to a type 1 metallothionein from maize (*Zea mays*). Biometals 31 (4), 539–550.

Ebbs, S.D., Lasat, M.M., Brady, D.J., Cornish, J., Gordon, R., Kochian, L.V., 1997. Phytoextraction of Cadmium and Zinc from a Contaminated Soil. Vol. 26 American Society of Agronomy, Crop Science Society of America, and Soil Science Society of America, pp. 1424–1430.

Edelstein, M., Ben-Hur, M., 2018. Heavy metals and metalloids: sources, risks and strategies to reduce their accumulation in horticultural crops. Sci. Hortic. 234, 431–444.

Egendorf, S.P., Groffman, P., Moore, G., Cheng, Z., 2020. The limits of lead (Pb) phytoextraction and possibilities of phytostabilization in contaminated soil: a critical review. Int. J. Phytoremed. 22 (9), 916–930.

Favas, P.J., Pratas, J., Varun, M., D'Souza, R., Paul, M.S., 2014. Phytoremediation of soils contaminated with metals and metalloids at mining areas: potential of native flora. Environ. Risk Assess. Soil Contam. 3, 485–516.

Fernández, S., Poschenrieder, C., Marcenò, C., Gallego, J.R., Jiménez-Gámez, D., Bueno, A., Afif, E., 2017. Phytoremediation capability of native plant species living on Pb-Zn and Hg-As mining wastes in the Cantabrian range, north of Spain. J. Geochem. Explor. 174, 10–20.

Fernando, D.R., Woodrow, I.E., Bakkaus, E.J., Collins, R.N., Baker, A.J.M., Batainoff, G. N., 2007. Variability of Mn hyperaccumulation in the Australian rainforest tree *Gossia bidwillii* (Myrtaceae). Plant Soil 293, 145e152.

Freeman, J.L., Garcia, D., Kim, D., Hopf, A., Salt, D.E., 2005. Constitutively elevated salicylic acid signals glutathione-mediated nickel tolerance in Thlaspi nickel hyperaccumulators. Plant Physiol. 137 (3), 1082–1091.

Freeman, J.L., Tamaoki, M., Stushnoff, C., 2010. Molecular mechanisms of selenium tolerance and hyperaccumulation in *Stanleya pinnata*. Plant Physiol. 153, 1630e1652.

Galeas, M.L., Zhang, L.H., Freeman, J.L., Wegner, M., Pilon-Smits, E.A.H., 2007. Seasonal fluctuations of selenium and sulfur accumulation in selenium hyperaccumulators and related nonaccumulators. New Phytol. 173, 51.

Gardea-Torresday, J.L., De la Rosa, G., Peralta-Videa, J.R., Montes, M., Cruz-Jimenez, G., Cano-Aguilera, I., 2005. Differential uptake and transport of trivalent and hexavalent chromium by tumble wed (*Salsola kali*). Arch. Environ. Contam. Toxicol. 48, 225e232.

Gardea-Torresdey, J.L., Peralta-Videa, J.R., De La Rosa, G., Parsons, J.G., 2005. Phytoremediation of heavy metals and study of the metal coordination by X-ray absorption spectroscopy. Coord. Chem. Rev. 249 (17–18), 1797–1810.

Ghori, Z., Iftikhar, H., Bhatti, M.F., Sharma, I., Kazi, A.G., Ahmad, P., 2016. Phytoextraction: the use of plants to remove heavy metals from soil. In: Plant Metal Interaction. Elsevier, pp. 385–409.

Ghosh, M., Singh, S.P., 2005. A comparative study of cadmium phytoextraction by accumulator and weed species. Environ. Pollut. 133 (2), 365–371.

Goldsbrough, P., 2020. Metal tolerance in plants: the role of phytochelatins and metallothioneins. In: Phytoremediation of Contaminated Soil and Water. CRC Press, pp. 221–233.

Gove, B., Hutchinson, J.J., Young, S.D., Craigon, J., McGrath, S.P., 2002. Uptake of metals by plants sharing a rhizosphere with the hyperaccumulator *Thlaspi caerulescens*. Int. J. Phytoremed. 4 (4), 267–281.

Gul, I., Manzoor, M., Silvestre, J., Rizwan, M., Hina, K., Kallerhoff, J., Arshad, M., 2019. EDTA-assisted phytoextraction of lead and cadmium by Pelargonium cultivars grown on spiked soil. Int. J. Phytoremed. 21 (2), 101–110.

Hall, J.Á., 2002. Cellular mechanisms for heavy metal detoxification and tolerance. J. Exp. Bot. 53 (366), 1–11.

Hao, L., Zhang, B., Feng, C., Zhang, Z., Lei, Z., Shimizu, K., Cao, X., Liu, H., Liu, H., 2018. Microbial vanadium (V) reduction in groundwater with different soils from vanadium ore mining areas. Chemosphere 202, 272–279.

Haque, N., Peralta-Videa, J.R., Jones, G.L., Gill, T.E., Gardea-Torresdey, J.L., 2008. Screening the phytoremediation potential of desert broom (*Baccharis sarothroides* Gray) growing on mine tailings in Arizona, USA. Environ. Pollut. 153 (2), 362–368.

Hladun, K.R., Parker, D.R., Trumble, J.T., 2011. Selenium accumulation in the floral tissues of two Brassicaceae species and its impact on floral traits and plant performance. Environ. Exp. Bot. 74, 90e97.

Javaid, A., 2011. Importance of arbuscular mycorrhizal fungi in phytoremediation of heavy metal contaminated soils. In: Biomanagement of Metal-Contaminated Soils. Springer, Dordrecht, pp. 125–141.

Jeevanantham, S., Saravanan, A., Hemavathy, R.V., Kumar, P.S., Yaashikaa, P.R., Yuvaraj, D., 2019. Removal of toxic pollutants from water environment by phytoremediation: a survey on application and future prospects. Environ. Technol. Innov. 13, 264–276.

Kabeya, F.I., Pongrac, P., Lange, B., Faucon, M.P., Van Elteren, J.T., Šala, M., Šelih, V.S., Eeckhoudt, E.V., Verbruggen, N., 2018. Tolerance and accumulation of cobalt in three species of Haumaniastrum and the influence of copper. Environ. Exp. Bot. 149, 27–33.

Karimi, A., Khodaverdiloo, H., Sepehri, M., Sadaghiani, M.R., 2011. Arbuscular mycorrhizal fungi and heavy metal contaminated soils. Afr. J. Microbiol. Res. 5 (13), 1571–1576.

Khade, S.W., Adholeya, A., 2009. Arbuscular mycorrhizal association in plants growing on metal-contaminated and noncontaminated soils adjoining Kanpur tanneries, Uttar Pradesh, India. Water Air Soil Pollut. 202 (1), 45–56.

Khan, A.G., 2005. Role of soil microbes in the rhizospheres of plants growing on trace metal contaminated soils in phytoremediation. J. Trace Elem. Med. Biol. 18 (4), 355–364.

Kubota, H., Takenaka, C., 2003. Field note: *Arabis gemmifera* is a hyperaccumulator of Cd and Zn. Int. J. Phytoremed. 5 (3), 197–201.

Kumar, V., Chandra, R., 2019. Bacterial-assisted phytoextraction mechanism of heavy metals by native hyperaccumulator plants from distillery waste–contaminated site for eco-restoration. In: Microbes for Sustainable Development and Bioremediation. CRC Press, pp. 1–41.

Kumar, V., Chopra, A.K., 2018. Phytoremediation potential of water caltrop (*Trapa natans* L.) using municipal wastewater of the activated sludge process-based municipal wastewater treatment plant. Environ. Technol. 39 (1), 12–23.

Kumar, V., Singh, J., Saini, A., Kumar, P., 2019. Phytoremediation of copper, iron and mercury from aqueous solution by water lettuce (*Pistia stratiotes* L.). Environ. Sustain. 2 (1), 55–65.

Küpper, H., Kochian, L.V., 2010. Transcriptional regulation of metal transport genes and mineral nutrition during acclimatization to cadmium and zinc in the Cd/Zn hyperaccumulator, *Thlaspi caerulescens* (Ganges population). New Phytol. 185 (1), 114–129.

Küpper, H., Mijovilovich, A., Götz, B., Kroneck, P.M.H., Küpper, F.C., MeyerKlaucke, W., 2009. Complexation and toxicity of copper in higher plants (I): characterisation of copper accumulation, speciation and toxicity in *Crassula helmsii* as a new copper hyperaccumulator. Plant Physiol. 151, 702e714.

Kurissery, S., Shaw, L.K., Kanavillil, N., 2022. A historic perspective of endophytes in vascular plants and their role in environmental sustainability. In: Research Anthology on Recent Advancements in Ethnopharmacology and Nutraceuticals. IGI Global, pp. 76–100.

Lasat, M.M., 1999. Phytoextraction of metals from contaminated soil: a review of plant/soil/metal interaction and assessment of pertinent agronomic issues. J. Hazard. Substance Res. 2 (1), 5.

Lasat, M.M., 2002. Phytoextraction of toxic metals: a review of biological mechanisms. J. Environ. Qual. 31 (1), 109–120.

Leblanc, M., Petit, D., Deram, A., Robinson, B., Brooks, R.R., 1999. The phytomining and environmental significance of hyperaccumulation of thallium by Iberis intermedia from southern France. Econ. Geol. 94, 109e113.

Lee, B.X.Y., Hadibarata, T., Yuniarto, A., 2020. Phytoremediation mechanisms in air pollution control: a review. Water Air Soil Pollut. 231 (8), 1–13.

Luo, S., Wan, Y., Xiao, X., Guo, H., Chen, L., Xi, Q., Zeng, G., Liu, C., Chen, J., 2011. Isolation and characterization of endophytic bacterium LRE07 from cadmium hyperaccumulator *Solanum nigrum* L. and its potential for remediation. Appl. Microbiol. Biotechnol. 89 (5), 1637–1644.

Luo, J., Tao, Q., Jupa, R., Liu, Y., Wu, K., Song, Y., Li, J., Huang, Y., Zou, L., Liang, Y., Li, T., 2019. Role of vertical transmission of shoot endophytes in root-associated microbiome assembly and heavy metal hyperaccumulation in *Sedum alfredii*. Environ. Sci. Technol. 53 (12), 6954–6963.

Ma, Y., Oliveira, R.S., Freitas, H., Zhang, C., 2016. Biochemical and molecular mechanisms of plant-microbe-metal interactions: relevance for phytoremediation. Front. Plant Sci. 7, 918.

Macnair, M.R., Tilstone, G.H., Smith, S.E., 2020. The genetics of metal tolerance and accumulation in higher plants. In: Phytoremediation of Contaminated Soil and Water. CRC Press, pp. 235–250.

Maiti, R.K., Piñero, J.L.H., Oreja, J.A.G., 2004. Plant based bioremediation and mechanisms of heavy metal. Proc. Indian Natl. Sci. Acad. B70 (1), 1–12.

Manzoor, M., Gul, I., Silvestre, J., Kallerhoff, J., Arshad, M., 2018. Screening of indigenous ornamental species from different plant families for Pb accumulation potential exposed to metal gradient in spiked soils. Soil Sediment Contam. Int. J. 27 (5), 439–453.

Martínez-Sánchez, M.J., García-Lorenzo, M.L., Pérez-Sirvent, C., Bech, J., 2012. Trace element accumulation in plants from an aridic area affected by mining activities. J. Geochem. Explor. 123, 8–12.

McGrath, S.P., Zhao, F.J., 2003. Phytoextraction of metals and metalloids from contaminated soils. Curr. Opin. Biotechnol. 14 (3), 277–282.

McGrath, S.P., Zhao, J., Lombi, E., 2002. Phytoremediation of Metals, Metalloids, and Radionuclides.

Meers, E., Ruttens, A., Hopgood, M., Lesage, E., Tack, F.M.G., 2005. Potential of *Brassic rapa*, *Cannabis sativa*, *Helianthus annuus* and *Zea mays* for phytoextraction of heavy metals from calcareous dredged sediment derived soils. Chemosphere 61 (4), 561–572.

Miransari, M., 2011. Hyperaccumulators, arbuscular mycorrhizal fungi and stress of heavy metals. Biotechnol. Adv. 29 (6), 645–653.

Moradi, A.B., Swoboda, S., Robinson, B., Prohaska, T., Kaestner, A., Oswald, S.E., Wenzel, W.W., Schulin, R., 2010. Mapping of nickel in root cross-sections of the hyperaccumulator plant *Berkheya coddii* using laser ablation ICP-MS. Environ. Exp. Bot. 69 (1), 24–31.

Moreira, C.G., de Carvalho, T.S., de Oliveira, C., de Abreu, L.B., de Castro, A.C.S., Ribeiro, P.G., Bispo, F.H.A., Boutin, C., Guilherme, L.R.G., 2019. Ecological risk assessment of cerium for tropical agroecosystems. Chemosphere 221, 124–131.

Nair, A., Juwarkar, A.A., Singh, S.K., 2007. Production and characterization of siderophores and its application in arsenic removal from contaminated soil. Water Air Soil Pollut. 180 (1), 199–212.

Niazi, N.K., Bashir, S., Bibi, I., Murtaza, B., Shahid, M., Javed, M.T., Shakoor, M.B., Saqib, Z.A., Nawaz, M.F., Aslam, Z., Wang, H., 2016. Phytoremediation of arsenic-contaminated soils using arsenic hyperaccumulating ferns. In: Phytoremediation. Springer, Cham, pp. 521–545.

Oven, M., Grill, E., Golan-Goldhirsh, A., Kutchan, T.M., Zenk, M.H., 2002. Increase of free cysteine and citric acid in plant cells exposed to cobalt ions. Phytochemistry 60, 467e474.

Padmavathiamma, P.K., Li, L.Y., 2007. Phytoremediation technology: hyper-accumulation metals in plants. Water Air Soil Pollut. 184 (1), 105–126.

Pajević, S., Borišev, M., Nikolić, N., Arsenov, D.D., Orlović, S., Župunski, M., 2016. Phytoextraction of heavy metals by fast-growing trees: a review. Phytoremediation, 29–64.

Pandey, V.C., Bajpai, O., 2019. Phytoremediation: From theory toward practice. In: Phytomanagement of Polluted Sites. Elsevier, pp. 1–49.

Pasricha, S., Mathur, V., Garg, A., Lenka, S., Verma, K., Agarwal, S., 2021. Molecular mechanisms underlying heavy metal uptake, translocation and tolerance in hyperaccumulators-an analysis: heavy metal tolerance in hyperaccumulators. Environ. Challenges 4, 100197.

Paz-Alberto, A.M., Sigua, G.C., 2013. Phytoremediation: A Green Technology to Remove Environmental Pollutants.

Perrier, N., Colin, F., Jaffré, T., Ambrosi, J.P., Rose, J., Bottero, J.Y., 2004. Nickel speciation in *Sebertia acuminata*, a plant growing on a lateritic soil of New Caledonia. Compt. Rendus Geosci. 336 (6), 567–577.

Persans, M.W., Nieman, K., Salt, D.E., 2001. Functional activity and role of cation-efflux family members in Ni hyperaccumulation in *Thlaspi goesingense*. Proc. Natl. Acad. Sci. 98 (17), 9995–10000.

Peuke, A.D., Rennenberg, H., 2005. Phytoremediation: molecular biology, requirements for application, environmental protection, public attention and feasibility. EMBO Rep. 6 (6), 497–501.

Ponting, J., Kelly, T.J., Verhoef, A., Watts, M.J., Sizmur, T., 2021. The impact of increased flooding occurrence on the mobility of potentially toxic elements in floodplain soil – a review. Sci. Total Environ. 754, 142040.

Qiao, J., Sun, H., Luo, X., Zhang, W., Mathews, S., Yin, X., 2017. EDTA-assisted leaching of Pb and Cd from contaminated soil. Chemosphere 167, 422–428.

Raikova, S., Piccini, M., Surman, M.K., Allen, M.J., Chuck, C.J., 2019. Making light work of heavy metal contamination: the potential for coupling bioremediation with bioenergy production. J. Chem. Technol. Biotechnol. 94 (10), 3064–3072.

Rascio, N., Navari-Izzo, F., 2011. Heavy metal hyperaccumulating plants: how and why do they do it? And what makes them so interesting? Plant Sci. 180 (2), 169–181.

Reeves, R.D., Adiguezel, N., Baker, A.J., 2009. Nickel hyperaccumulation in Bornmuellera kiyakii Aytaç & Aksoy and associated plants of the Brassicaceae from Kızıldağ (Derebucak, Konya-Turkey). Turk. J. Bot. 33 (1), 33–40.

Riaz, M., Kamran, M., Fang, Y., Wang, Q., Cao, H., Yang, G., Deng, L., Wang, Y., Zhou, Y., Anastopoulos, I., Wang, X., 2021. Arbuscular mycorrhizal fungi-induced mitigation of heavy metal phytotoxicity in metal contaminated soils: a critical review. J. Hazard. Mater. 402, 123919.

Sarwar, N., Imran, M., Shaheen, M.R., Ishaque, W., Kamran, M.A., Matloob, A., Rehim, A., Hussain, S., 2017. Phytoremediation strategies for soils contaminated with heavy metals: modifications and future perspectives. Chemosphere 171, 710–721.

Saxena, G., Purchase, D., Mulla, S.I., Saratale, G.D., Bharagava, R.N., 2019. Phytoremediation of heavy metal-contaminated sites: eco-environmental concerns, field studies, sustainability issues, and future prospects. Rev. Environ. Contam. Toxicol. 249, 71–131.

Scavino, A.F., Pedraza, R.O., 2013. The role of siderophores in plant growth-promoting bacteria. In: Bacteria in Agrobiology: Crop Productivity. Springer, Berlin, Heidelberg, pp. 265–285.

Shah, K., Nongkynrih, J.M., 2007. Metal hyperaccumulation and bioremediation. Biol. Plant. 51 (4), 618–634.

Shahid, M., Dumat, C., Khalid, S., Schreck, E., Xiong, T., Niazi, N.K., 2017. Foliar heavy metal uptake, toxicity and detoxification in plants: a comparison of foliar and root metal uptake. J. Hazard. Mater. 325, 36–58.

Sharma, P., Ngo, H.H., Khanal, S., Larroche, C., Kim, S.H., Pandey, A., 2021. Efficiency of transporter genes and proteins in hyperaccumulator plants for metals tolerance in wastewater treatment: sustainable technique for metal detoxification. Environ. Technol. Innov. 23, 101725.

Shin, M.N., Shim, J., You, Y., Myung, H., Bang, K.S., Cho, M., Kamala-Kannan, S., Oh, B. T., 2012. Characterization of lead resistant endophytic Bacillus sp. MN3-4 and its potential for promoting lead accumulation in metal hyperaccumulator *Alnus firma*. J. Hazard. Mater. 199, 314–320.

Sinclair, S.A., Krämer, U., 2012. The zinc homeostasis network of land plants. Biochim. Biophys. Acta Mol. Cell Res. 1823 (9), 1553–1567.

Singh, R., Misra, A.N., Sharma, P., 2021. Differential responses of thiol metabolism and genes involved in arsenic detoxification in tolerant and sensitive genotypes of bioenergy crop *Ricinus communis*. Protoplasma 258 (2), 391–401.

Smith, S.E., Read, D.J., 2008. Mycorrhizal Symbiosis, third ed. Academic Press, San Diego, CA.

Song, Y., Jin, L., Wang, X., 2017. Cadmium absorption and transportation pathways in plants. Int. J. Phytoremed. 19 (2), 133–141.

Srivastava, N., 2017. Remediation of polluted soils using hyperaccumulator plants. In: Enhancing Cleanup of Environmental Pollutants. Springer, Cham, pp. 187–214.

Sturz, A.V., Nowak, J., 2000. Endophytic communities of rhizobacteria and the strategies required to create yield enhancing associations with crops. Appl. Soil Ecol. 15 (2), 183–190.

Suman, J., Uhlik, O., Viktorova, J., Macek, T., 2018. Phytoextraction of heavy metals: a promising tool for clean-up of polluted environment? Front. Plant Sci. 9, 1476.

Sun, Q., Ye, Z.H., Wang, X.R., Wong, M.H., 2005. Increase of glutathione in mine population of *Sedum alfredii*: a Zn hyperaccumulator and Pb accumulator. Phytochemistry 66 (21), 2549–2556.

Talebi, M., Tabatabaei, B.E.S., Akbarzadeh, H., 2019. Hyperaccumulation of Cu, Zn, Ni, and Cd in Azolla species inducing expression of methallothionein and phytochelatin synthase genes. Chemosphere 230, 488–497.

Teng, Y., Ni, S., Zhang, C., Wang, J., Lin, X., Huang, Y., 2006. Environmental geochemistry and ecological risk of vanadium pollution in Panzhihua mining and smelting area, Sichuan, China. Chin. J. Geochem. 25 (4), 379–385.

Ullah, I., Mateen, A., Ahmad, M.A., Munir, I., Iqbal, A., Al-Ghamdi, K.M., Al-Solami, H. M., Siddiqui, M.F., 2021. Heavy metal ATPase genes (HMAs) expression induced by endophytic bacteria, "AI001, and AI002" mediate cadmium translocation and phytoremediation. Environ. Pollut., 118508.

Vamerali, T., Bandiera, M., Mosca, G., 2010. Field crops for phytoremediation of metal-contaminated land. A review. Environ. Chem. Lett. 8 (1), 1–17.

Vangronsveld, J., Herzig, R., Weyens, N., Boulet, J., Adriaensen, K., Ruttens, A., Thewys, T., Vassilev, A., Meers, E., Nehnevajova, E., van der Lelie, D., 2009. Phytoremediation of contaminated soils and groundwater: lessons from the field. Environ. Sci. Pollut. Res. 16 (7), 765–794.

Visioli, G., Visioli, G., Vamerali, T., Mattarozzi, M., Dramis, L., Sanangelantoni, A., 2015. Combined endophytic inoculants enhance nickel phytoextraction from serpentine soil in the hyperaccumulator *Noccaea caerulescens*. Front. Plant Sci. 638 (6), 1–10.

Wang, H., Zhong, G., 2011. Effect of organic ligands on accumulation of copper in hyperaccumulator and nonaccumulator *Commelina communis*. Biol. Trace Elem. Res. 143, 489e499.

Wang, L., Lin, H., Dong, Y., He, Y., Liu, C., 2018. Isolation of vanadium-resistance endophytic bacterium PRE01 from *Pteris vittata* in stone coal smelting district and characterization for potential use in phytoremediation. J. Hazard. Mater. 341, 1–9.

Wang, Q., Ge, C., Xu, S.A., Wu, Y., Sahito, Z.A., Ma, L., Pan, F., Zhou, Q., Huang, L., Feng, Y., Yang, X., 2020a. The endophytic bacterium Sphingomonas SaMR12 alleviates Cd stress in oilseed rape through regulation of the GSH-AsA cycle and antioxidative enzymes. BMC Plant Biol. 20 (1), 1–14.

Wang, L., Lin, H., Dong, Y., Li, B., He, Y., 2020b. Effects of endophytes inoculation on rhizosphere and endosphere microecology of Indian mustard (*Brassica juncea*) grown in vanadium-contaminated soil and its enhancement on phytoremediation. Chemosphere 240, 124891.

Weyens, N., Thijs, S., Popek, R., Witters, N., Przybysz, A., Espenshade, J., et al., 2015. The role of plant–microbe interactions and their exploitation for phytoremediation of air pollutants. Int. J. Mol. Sci. 16 (10), 25576–25604.

Wood, J.L., Tang, C., Franks, A.E., 2016. Microbial associated plant growth and heavy metal accumulation to improve phytoextraction of contaminated soils. Soil Biol. Biochem. 103, 131–137.

Wu, Y., Ma, L., Zhang, X., Topalović, O., Liu, Q., Feng, Y., Yang, X., 2020. A hyperaccumulator plant *Sedum alfredii* recruits Cd/Zn-tolerant but not Pb-tolerant endospheric bacterial communities from its rhizospheric soil. Plant Soil 455 (1), 257–270.

Xiao, X.Y., Miao, Y.A.N.G., Guo, Z.H., Jiang, Z.C., Liu, Y.N., Xia, C.A.O., 2015. Soil vanadium pollution and microbial response characteristics from stone coal smelting district. Trans. Nonferrous Metals Soc. China 25 (4), 1271–1278.

Xiao, X., Jiang, Z., Guo, Z., Wang, M., Zhu, H., Han, X., 2017. Effect of simulated acid rain on leaching and transformation of vanadium in paddy soils from stone coal smelting area. Process Saf. Environ. Prot. 109, 697–703.

Xiao, E., Ning, Z., Xiao, T., Sun, W., Qiu, Y., Zhang, Y., Chen, J., Gou, Z., Chen, Y., 2019. Variation in rhizosphere microbiota correlates with edaphic factor in an abandoned antimony tailing dump. Environ. Pollut. 253, 141–151.

Yadav, K.K., Gupta, N., Kumar, A., Reece, L.M., Singh, N., Rezania, S., Khan, S.A., 2018. Mechanistic understanding and holistic approach of phytoremediation: a review on application and future prospects. Ecol. Eng. 120, 274–298.

Yao, Y.A., Mou, D., Xu, G., Lutts, S., Achal, V., Ma, J., 2012. Contrasting performance and different tolerance of chestnut rose and grape to excess manganese. J. Plant Growth Regul. 31 (3), 416–426.

Ye, D., Li, T., Liu, J., Yi, Y., Zou, L., Zhang, X., Xie, M., 2020. Characteristics of endophytic bacteria from *Polygonum hydropiper* and their use in enhancing P-phytoextraction. Plant Soil 448 (1), 647–663.

Yu, F., Xin, M., Yao, Y., Wang, X., Liu, K., Li, Y., 2021. Manganese accumulation and plant physiology behaviour of *Camellia oleifera* in response to different levels of phosphate fertilization. J. Soil Sci. Plant Nutr. 21 (3), 2562–2572.

Zhang, X.H., Liu, J., Huang, H.T., Chen, J., Zhu, Y.N., Wang, D.Q., 2007. Chromium accumulation by the hyperaccumulator plant *Leersia hexandra* Swartz. Chemosphere 67, 1138e1143.

Zhang, F., Shen, J., Zhang, J., Zuo, Y., Li, L., Chen, X., 2010. Rhizosphere processes and management for improving nutrient use efficiency and crop productivity: implications for China. Adv. Agron. 107, 1–32.

Zhao, F.J., Lombi, E., Breedon, T.M.S.P., 2000. Zinc hyperaccumulation and cellular distribution in *Arabidopsis halleri*. Plant Cell Environ. 23 (5), 507–514.

Zhao, F.J., Hamon, R.E., McLaughlin, M.J., 2001. Root exudates of the hyperaccumulator *Thlaspi caerulescens* do not enhance metal mobilization. New Phytol. 151 (3), 613–620.

Zou, Q., Gao, Y., Yi, S., Jiang, J., Aihemaiti, A., Li, D.A., Yang, M., 2019. Multi-step column leaching using low-molecular-weight organic acids for remediating vanadium-and chromium-contaminated soil. Environ. Sci. Pollut. Res. 26 (15), 15406–15413.

CHAPTER 5

Seed priming with microbial inoculants for enhanced crop yield

Debasis Mitra[a], Marika Pellegrini[b], Adeyemi Nurudeen Olatunbosun[c], Rittick Mondal[d], Maddalena Del Gallo[b], Sourav Chattaraj[a], Debolina Chakroborty[e], Ankita Priyadarshini[f], Bahman Khoshru[g], B.E. Guerra Sierra[h], Sergio de los Santos-Villalobos[i], Ansuman Senapati[f], Rihab Djebaili[b], Pradeep K. Das Mohapatra[a], and Periyasamy Panneerselvam[f]

[a]Department of Microbiology, Raiganj University, Raiganj, West Bengal, India
[b]Department of Life, Health and Environmental Sciences, University of L'Aquila, L'Aquila, Italy
[c]Department of Plant Physiology and Crop Production, Federal University of Agriculture, Abeokuta, Nigeria
[d]Department of Sericulture, Raiganj University, Raiganj, West Bengal, India
[e]Guru Nanak Institute of Pharmaceutical Science and Technology, Maulana Abul Kalam Azad University of Technology, Kolkata, West Bengal, India
[f]Crop Production Division, ICAR-National Rice Research Institute, Cuttack, Odisha, India
[g]Department of Soil Science, Faculty of Agriculture, University of Tabriz, Tabriz, Iran
[h]Universidad de Santander, Facultad de Ciencias Naturales, Grupo Microbiota, Bucaramanga, Colombia
[i]Instituto Tecnológico de Sonora, Ciudad Obregón, Sonora, México

5.1 Introduction

In modern agriculture, innovative technologies are utilized to maximize crop yield and improve agricultural output (De Clercq et al., 2018). Meeting expanding food demands, remaining internationally competitive, and providing high-quality agricultural goods are all factors that pose new challenges to a country (Saiz-Rubio and Rovira-Más, 2020). Seed priming is a physiological and presowing technique in which seeds are mildly moistened to the point where pregermination metabolic pathways are triggered, allowing for rapid germination, seedling growth, and eventual yield in both natural and stressed environments (Hussain et al., 2016; Lutts et al., 2016; Mitra et al., 2021b). When compared to unprimed seeds, primed seeds display faster and more uniform seed germination due to differences in enzyme activation, metabolic activities, biochemical processes of cell repair, synthesis of protein, and antioxidant defense system improvement (Marthandan et al., 2020; Waqas et al., 2019).

Microbial Inoculants
https://doi.org/10.1016/B978-0-323-99043-1.00016-5

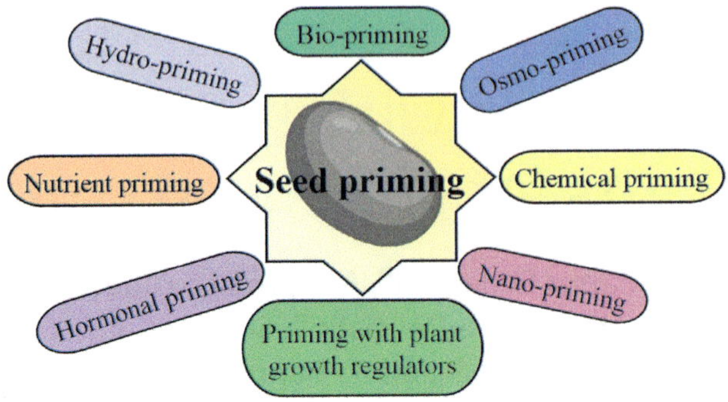

Fig. 1 Methods of seed priming.

Seed priming techniques are classified into two categories: conventional methods (biopriming, osmo-priming, chemical priming, hydro-priming, nutrient priming, polyamines (hormonal priming), and priming with plant growth regulators) and modern approaches (nano-priming and priming with physical agents) (Devika et al., 2021; Lemmens et al., 2019; Waqas et al., 2019) (Fig. 1). Biopriming is the use of a living microorganism's inoculum to apply plant growth-promoting microorganisms (PGPMs) to seeds (Rakshit et al., 2015). Over the last few decades, PGPM treatment has proven to be a viable and environmentally acceptable agricultural engineering strategy for boosting the food supply's durability while minimizing its ecological impact (Debasis et al., 2019; Mitra et al., 2021a; Prabhukarthikeyan et al., 2022). Most prominent biopriming agents, viz., *Pseudomonas* (Mitra et al., 2014), *Bacillus* (Mitra and Mukhopadhyay, 2014; Panneerselvam et al., 2019), *Paenibacillus*, *Azospirillum*, *Azotobacter*, *Enterobacter* (Panneerselvam et al., 2021), *Burkholderia*, *Serratia*, *Trichoderma*, and *Actinobacteria*, with plant growth-stimulating properties applied through seed inoculation help in faster germination and establishment of crops, and, as a result of their active metabolism, a beneficial impact is observed in the nutrient content and quality of the crops. Biopriming increases agricultural production by enhancing seed germination, viability, index vigor, growth promotion, and resistance to diseases (Debasis et al., 2020; Panneerselvam et al., 2019; Uniyal et al., 2018) (Tables 1 and 2). It is an environmentally acceptable method to enhance plant development, improve nutrient absorption, and protect plants from soil- and seed-borne diseases. This chapter focuses on biopriming approaches based on better PGPM inoculant solutions as an alternative to chemical treatments in current agriculture and as a more ecologically acceptable method of prospective farming.

Table 1 Beneficial importance of seed priming methods in agriculture.

Seed priming types	Plants/seeds/hosts	Effects	References
Hydro-priming	Rice	Seed germination increases	Nakao et al. (2020); Farooq et al. (2006)
	Wheat	Develops the seed germination rate and grain yield	Lemmens et al. (2019)
	Wheat	Enhances water-holding efficiency and grain yield	Meena et al. (2013)
	Mexican fir tree	Improves germination	Zulueta-Rodríguez et al. (2015)
Osmo-priming	Wheat	Develops the seed germination rate and grain yield	Lemmens et al. (2019)
	Caraway	Improves the germination rate, plumule dry and fresh weights, and length	Mirmazloum et al. (2020)
	Cowpea	Improves seed germination and seedling height	Singh et al. (2014)
Seed priming with hormones (gibberellins)	Wheat	Improves salt tolerance	Abido et al. (2019)
	Maize	Augments chlorophyll content and enhances drought tolerance	Nada and Hamza (2019)
Seed priming with hormones (cytokinins)	Wheat	Improves salt tolerance	Iqbal and Ashraf (2005)
	Soybean	Improves drought tolerance	Mangena (2020)
Seed priming with hormones (salicylic acid)	Wheat	Increases photosynthetic pigments	El-Shazoly et al. (2019)
	Maize	Enhances antioxidant activities	Farooq et al. (2008)
	Rice	Enhances antioxidant activities	Pouramir-Dashtmian et al. (2014)
Biopriming (*Bacillus subtilis*, *Pseudomonas putida*, and *P. fluorescens*)	Mexican fir tree	Improves germination	Zulueta-Rodríguez et al. (2015)

Table 2 Priming materials and their importance in seed priming.

Seed priming methods	Crops/plants/ hosts	Priming material	References
Hormonal priming	Tomato	Menadione sodium bisulfite	Jiménez-Arias et al. (2019)
Halo-priming	Tomato	NaCl	Demirkaya (2014)
Osmo-priming	Tomato	NaCl	İşeri et al. (2014)
Chemical priming	Maize	$ZnSO_4$	Imran et al. (2018)
Hydro-priming	Maize	Distilled water and leaf extracts of *Rosmarinus officinalis*	Panuccio et al. (2018)
Magneto-priming	Maize	Magnetic fields (MFs)/week (15 mT)	Karimi et al. (2017)
Hormonal priming	Maize	Salicylic acid	Kazemi et al. (2018)
Hormonal priming	Maize	Melatonin	Jiang et al. (2016)
Hormonal priming	Wheat	Salicylic acid	Karadağ and Yücel (2017)
Hormonal priming	Wheat	Salicylic acid	Yücel and Heybet (2016)
Halo-priming	Wheat	Silicon	Azeem et al. (2015)
Osmo-priming	Wheat	Ascorbic acid	Fercha et al. (2014)
Hormonal priming	*Vicia faba*	Melatonin	Dawood and El-Awadi (2015)
Hormonal priming	Rice	Gibberellic acid	Rhaman et al. (2021)
Hormonal priming	*O. sativa*	Salicylic acid	Sheteiwy et al. (2018)
Hormonal priming	*O. sativa*	Salicylic acid	Hussain et al. (2016)
Osmo-priming	*O. sativa*	Spermidine	Paul and Roychoudhury (2017)
Osmo-priming	*O. sativa*	PEG (10%), KNO_3, spermidine	Zheng et al. (2016)
Osmo-priming	*O. sativa*	Polyamines	Farooq et al. (2009)
Hormonal priming	*H. vulgaris*	ABA	Salama and Mansour (2015)

Table 2 Priming materials and their importance in seed priming—cont'd

Seed priming methods	Crops/plants/hosts	Priming material	References
Halo-priming	*Solanum lycopersicum*	NaCl	Nakaune et al. (2012)
Osmo-priming	*Sorghum bicolor*	PEG	Zhang et al. (2015)
Halo-priming	*Silybum marianum*	KNO_3	Zavariyan et al. (2015)
Hormonal priming	*Cucumis sativus*	Melatonin	Zhang et al. (2014)
Osmo-priming	*Cucurbita pepo*	Ascorbic acid	Fazlali et al. (2013)
Hormonal priming	*Helianthus annuus*	Jasmonic acid	Górnik and Lahuta (2017)
Osmo-priming	*Jatropha curcas*	Nitric oxide	Gadelha et al. (2017)
Osmo-priming	*Medicago sativa*	PEG	Yacoubi et al. (2013)
Osmo-priming	*M. sativa*	Mannitol	Amooaghaie (2011)
Osmo-priming	*Brassica napus*	Ascorbic acid	Razaji et al. (2014)
Halo-priming	*Brassica oleracea*	Methyl jasmonate	Hassini et al. (2017)
Hormone priming	*Cajanus cajan*	Jasmonic acid	Poonam et al. (2013)
Halo-priming	*Capsicum annuum*	KCl	Aloui et al. (2014)
Chemical priming	*C. annuum*	Thiourea	Patade et al. (2012)
Osmo-priming	*C. annuum*	Glycine betaine	Korkmaz and Sirikci (2011)
Halo-priming	*Cucumis melo*	NaCl	Farhoudi et al. (2011)
Halo-priming	*C. sativus*	$CaCl_2$	Joshi et al. (2013)

5.2 Role of seed priming methods in agricultural advances and farming practices

5.2.1 Biopriming

Biopriming is a technique of seed treatment for biotic and abiotic stress management, which combines seed inoculation using helpful microorganisms (biological aspect) with seed moisture adjustment (physiological aspect) (Mitra et al., 2021b) (Table 2). Biopriming also improves seed germination and seedling emergence as well as protects seeds from soil- and seed-borne infections (Mahmood et al., 2016). Beneficial microbes like *Azotobacter* sp. can be utilized in biopriming, which also have the ability to colonize and multiply in the rhizosphere, thus providing both direct and indirect assistance to plants (Prasad et al., 2016).

5.2.2 Osmo-priming

Osmo-priming is the process of soaking seeds in an osmotic solution. In this technique, before sowing the seeds, they are soaked for a while in sugar, polyethylene glycol (PEG), glycerol, sorbitol, or mannitol solution and are thereafter air-dried. Osmo-priming boosts seed germination and overall crop performance in both nonsaline and saline environments (Waqas et al., 2019).

5.2.3 Chemical priming

In stress responses, physiology, and crop stress management, chemical priming is a promising area. Recent research has suggested that plants can be primed by chemical compounds, viz., polyamines, sodium nitroprusside, β-aminobutyric acid, hydrogen peroxide, salicylic acid, gamma-aminobutyric acid, sodium hydrosulfide, melatonin, menadione sodium bisulfite, and selenium to better tolerate different abiotic stresses and crop production (Ashraf et al., 2019; Savvides et al., 2016). Treatment of plants with chemicals, prior to the occurrence of stress events, can alter their stress tolerance response. Plant growth, physiology, biochemistry, and molecular mechanisms are all influenced by chemical priming (Ashraf et al., 2019; Hussain et al., 2016).

5.2.4 Hydro-priming

Hydro-priming is a low-cost seed priming technique that comprises several events such as soaking seeds in water for a period of time, followed by a redrying stage before seed sowing (Lutts et al., 2016). Hydro-priming enhances the moisture content and other parameters of the seed under stressful conditions (McDonald, 2000). Different modes (such as drum priming, on-farm priming) of hydro-priming techniques are employed in modern-day agriculture (Lutts et al., 2016). In the case of drum priming, seeds are gently rotated in a drum and progressively hydrated by the addition of water in the form of vapor, and, in the case of on-farm priming, seeds are soaked with water and then their surfaces are dried (Sime and Aune, 2020). As per studies on drum priming and on-farm priming, primed seeds have better germination, flowering, and higher grain yield than do unprimed or controlled seeds (Carrillo-Reche et al., 2018; Sime and Aune, 2020; Warren and Bennett, 1997).

5.2.5 Nutrient priming

Nutrient seed priming is a method of increasing seed nutrient content in addition to the priming effect of water on improved seed quality for better germination and seedling establishment by soaking seeds in a nutrient solution instead of pure water (Imran et al., 2013; Nciizah et al., 2020). It is a simple and cost-effective strategy for farmers and farming systems with little resources. Priming seeds in macro- or micronutrient solutions has been shown to increase yield in plants, but priming at high nutrient concentrations has also been reported to damage seeds and inhibit germination (Ajouri et al., 2004; Imran, 2014). Nutrient priming of barley seeds with P and Zn improved the germination, seedling growth, and plant uptake of these elements (Ajouri et al., 2004). In soils with low or poor Zn availability, priming maize seeds with 1% $ZnSO_4$ solution enhanced plant development, ultimate grain production, and seed Zn content in harvested grains (Harris et al., 2007). Soybean seed priming using micronutrients such as zinc, manganese, and boron enhanced the germination rate and primed nutrient concentrations in young seedlings cultivated under salt stress conditions (Imran, 2014).

5.2.6 Hormonal priming

Hormonal priming, also known as hormopriming, is the treatment of seeds with various hormones to promote seedling growth and development (Lutts et al., 2016). Commonly used hormones are abscisic acid, auxins, gibberellins, kinetin, ethylene, polyamines, and salicylic acid (SA) (Rhaman et al., 2021).

5.2.7 Priming with plant growth regulators (PGRs)

Abiotic stresses, such as drought, salinity, heat, and cold, are a common occurrence in plants and are key constraints to agricultural productivity around the world (Aymen, 2018). PGR pretreatment is a typical method for improving seed germination under stressful conditions (Pallaoro et al., 2016). Under drought, salinity, metal, cold, and heat stress, it can be utilized to improve germination, seedling growth, and yield (Dezfuli et al., 2008; Pallaoro et al., 2016).

5.2.8 Nano-priming

Climate change, resource depletion, and biodiversity loss all pose a significant threat to agricultural crops. In order to boost crop production while maintaining food quality and safety in a sustainable manner, a new agricultural revolution is required (do Espirito Santo Pereira et al., 2021). Agriculture's long-term viability can be improved via nanotechnology. Seed nano-priming is a powerful technique that alters seed metabolism and signaling pathways, impacting not just germination of seed survival but also the plant's life cycle (do Espirito Santo Pereira et al., 2021). As an eco-friendly and sustainable nanotechnological technique, seed priming with silver nanoparticles (AgNPs) can improve seed germination, growth, and yield while retaining fruit quality (Acharya et al., 2020). Shah et al. (2021a) investigated the effects of TiO_2 nanoparticle seed priming on the overall germination and physiology of salinity-tolerant maize. Seeds of maize cultivars were soaked for 1 day in 40, 60, and 80 ppm aerated TiO_2 nanoparticle solutions and then exposed to salinity stress, i.e., 200 mM NaCl, on a paper culture system. The results showed that a 60-ppm TiO_2 nano-priming treatment improved the rate of maize seedling germination and growth under salinity stress.

5.3 Plant growth-promoting microorganisms (PGPMs)

Plant growth-promoting microorganisms (PGPMs) have gained recognition and importance in seed priming due to their dual benefits of stimulating plant growth and regulating plant pests and diseases, and they are widely utilized as microbial inoculants for improving agricultural production (Abhilash et al., 2016). Plant roots are surrounded by soil that contains microbes also known as the rhizosphere (Nihorimbere et al., 2011). *Rhizobacteria*, *Actinomycetes*, fungi, arbuscular mycorrhiza fungi, and endophytes in the rhizosphere can be either symbiotic (having a direct beneficial interaction with plants) or nonsymbiotic (having no direct beneficial connection with plants) (Debasis et al., 2019; Khoshru et al., 2020). The root system functions as a chemical factory, synthesizing and releasing phenolic chemicals to carry out other rhizospheric underground interactions (Rasmann and Turlings, 2016). These chemical substances and other ways attract a vast number of microbial communities or populations to the plant's roots (Huang et al., 2014). Both plants and microorganisms have chemical compositions that differ from one species to the next. The chemical composition is also influenced by the physiological state of plants.

Microorganisms like *Bacillus*, *Pseudomonas*, *Azospirillum*, *Rhizobium*, *Flavobacterium*, *Burkholderia*, *Clostridium*, *Klebsiella*, *Enterobacter*, *Beauveria*, *Frankia*, *Trichoderma*, *Serratia*, and *Streptomyces* and arbuscular mycorrhizal fungi (AMF) are used in the seed priming technique to increase agricultural productivity (Kumar et al., 2017; Panneerselvam et al., 2019, 2021; Parewa et al., 2021; Pellegrini et al., 2021b). Plant growth-promoting *Rhizobacteria* (PGPR)-inoculated crops account for a minor percentage of current agricultural practices around the world. The use of PGPM combinations in seed priming, as well as their integration with other methods such as host resistance and pesticides, has proven to be more effective in the control of a variety of disease issues. The use of PGPMs for growth promotion and resistance induction in various crops through seed priming has been demonstrated to be effective in treating a wide range of diseases (Ajee and Kaushik, 2021; Sattiraju et al., 2019). The practical application of PGPM-based seed priming solution is progressing, and many formulations are now accessible on a commercial basis, with more under development (Mahmood and Kataoka, 2018; Ma, 2019; Shah et al., 2021b). This chapter provides an overview of the scope and influence of PGPMs in plant growth promotion and crop disease management, with a focus on the details most relevant to the development and use of biological control techniques incorporating various microbial strains. Problems and opportunities for commercialization are also discussed as are the benefits and drawbacks of their use in seed priming and their potential for integrated management.

5.4 Application of PGPMs as inoculants in seed priming

Seed priming is one of the most promising techniques for PGPM inoculation, which assures cost- and time-saving and good seed colonization performances (Mahmood et al., 2016). Different PGPM methods are available for seed priming, using several seed pretreatments and different soaking and

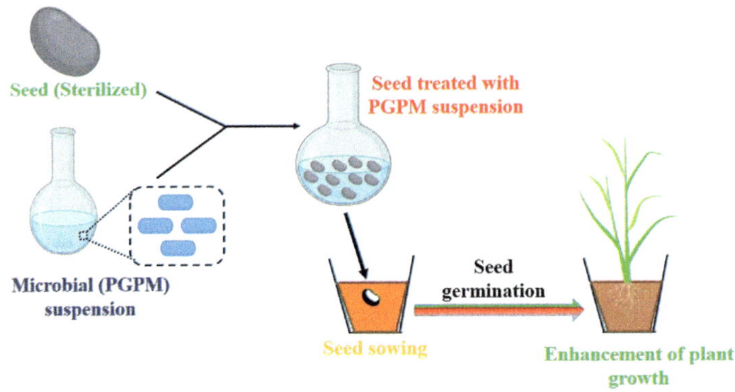

Fig. 2 Seed priming using PGPM inoculation.

temperature strategies (Singh et al., 2021). The effective association of microorganisms with seeds depends on these variables and on the type of microorganism, plant species, and cultivar of seeds used. Usually, the selection of indigenous microbes allows to achieve better performances and to avoid negative interactions with the endophytic microbiome (Sarkar et al., 2021). Seed priming can be carried out with single-species microorganisms or with a consortium of PGPMs. Using the PGPM consortia allows to improve the beneficial effects on plants. The combination of bacteria and fungi, for example, has shown to be effective in plant growth promotion, soil health, and fertility status improvement (Rashid et al., 2016) (Fig. 2). However, the compatibility of microorganisms with those in the consortium must be carefully evaluated (Woo and Pepe, 2018). Once biopriming and seed dehydration are carried out, the other variables that should be considered are storage conditions for optimal survival of microorganisms and shelf life of seeds. Microorganisms on seeds have the best survival and viability at low temperatures and moisture content and in the presence of polymers (Ma, 2019). Many authors have described seed priming with different PGPMs, including bacteria, *Actinobacteria*, and fungi. The following sections report the different applications described for these three major groups of PGPMs.

5.4.1 Bacteria

Different authors have described seed priming with bacteria for many vegetal species. The final densities applied are different and range from 10^6 to 10^{10} mainly depending on the types of bacteria used and the soaking duration. The bacterial densities of the suspensions are adjusted either spectrophotometrically or with the McFarland scale comparison, whereas the final bacterial density on seeds is determined by cultural methods (i.e., plate counting and most probable number). For spore-forming bacteria, suspensions can include spores instead of bacterial cells and are usually adjusted with the help of cell-counting chambers (Fichtel et al., 2007). Seed biopriming with bacteria can be carried out by diluting broth cultures or bacterial spores in water to obtain the final density (Pellegrini et al., 2021b). Bacterial-based solutions can also be enriched with gelling agents (e.g., alginate, guar gum, carboxymethyl cellulose) to help the bacterial adhesion to seed surface (Amruta et al., 2019). The soaking durations reported in the literature are extremely variable, with ranges from a few minutes up to more than 12 h. The soaking duration depends not only on bacterial density but also on the type of seed treated, its morphology, and physiology (Paparella et al., 2015). The bacterial suspensions used for biopriming can include bacterial consortia, including several strains, also belonging to different bacterial species (Deshmukh et al., 2020). The concurrent presence of PGPB strains enhances the effects on plants, creating synergisms among bacterial activities (Pellegrini et al., 2021a). The PGPB most used in biopriming belong to *Azotobacter*, *Arthrobacter*, *Agrobacterium*, *Azospirillum*, *Bacillus*, *Burkholderia*, *Enterobacter*, *Klebsiella*, *Pseudomonas*, and *Serratia* (Marthandan et al., 2020; Mitra et al., 2021b).

5.4.2 Actinobacteria

Within the PGPM group, we can also find plant growth-promoting *Actinobacteria* (PGPA), Gram-positive bacteria widely distributed in the environment with a strong contribution to plant–microbe interactions (Djebaili et al., 2020; Hasegawa et al., 2006). Even if *Actinobacteria* belong to the bacterial domain, they deserve to be discussed separately because of their unique features; they present elongated cells with a filamentous/hyphal structure and, owing to their metabolism, occupy many habitats, including those characterized by inhospitable environmental conditions (alkaline and acidic pH, high temperature, drought) (Pepper and Gentry, 2015). *Actinomycetes* with plant growth-promoting traits are many and mainly belong to the *Streptomyces* genus (Mitra et al., 2021a). Its filamentous growth and sporulation capabilities give this genus a major advantage over other microorganisms in terms of both survival and colonization (Viaene et al., 2016). Seed treatments with *Streptomyces* include the use of spore suspensions (Djebaili et al., 2020). Because of the wide range of molecules that *Streptomyces* and other

Actinobacteria produce, the cell-free supernatants (CFSs) of these bacteria are also valid agents for biopriming (Pellegrini et al., 2020) and can be used in association with the CFSs of other PGPMs (Manhas and Kaur, 2016).

5.4.3 Fungi

Plant growth-promoting fungi (PGPF) are other PGPMs used as biopriming agents. The inoculation of fungi via seed priming promotes colonization, association, and proliferation of PGPF on seeds (Mahmood and Kataoka, 2018). The most common and described PGPF is *Trichoderma harzianum* because of its wide biocontrol against phytopathogens and biostimulant activity on plants even under stress conditions (Nagaraju et al., 2012; Shoresh et al., 2010; Swain et al., 2021). The PGPF formulations used in biopriming are spore suspensions with a density of 10^8 spores/mL. The treatment usually takes 4–6 h, followed by a drying step of 12–18 h. Beyond *Trichoderma*, other PGPF used in biopriming belong to the genera *Acaulospora*, *Ambispora*, *Gigaspora*, *Glomus*, *Pacispora*, and *Paraglomus* (Deshmukh et al., 2020).

5.5 Enhancement of plant growth, disease control, and yield through seed priming techniques

Through seed biopriming, PGPMs establish an association with plants during the first stages of growth. The intimacy of this association and the beneficial effects on plant growth and development depend on many pedoclimatic variables. However, it is extensively reported that PGPM biopriming induces growth stimulation and protection against phytopathogens even under stress conditions (Qu et al., 2020) (Table 3). Growth stimulation has been described for several crops and includes enhancements that range from seed germination to crop yields and nutritional value. Plant growth parameters enhanced by PGPM priming include seed germination uniformity, sprouting, plant height, number of leaves, number of branches, fresh and dry matter of plant shoots and roots, stem diameter, chlorophyll contents, nodulation (in legumes), and primary and secondary metabolite accumulation (Sarkar et al., 2021) (Fig. 3). The mechanisms related to seed germination's uniformity and sprouting include enzyme activation and metabolism enhancement (Mahmood and Kataoka, 2018).

Plant growth and crop yield promotion and improved nutritional value are related to nutrient mobilization and uptake, hormonal homeostasis, and metabolism enhancement (Ma, 2019). Beyond improvement of crop yield

Table 3 Beneficial importance of microbial inoculants in seed priming.

Inoculants/PGPMs	Crop/plant/ host name	Effects on plant growth, defense, and yield	References
Enterobacter sp.	Rice	Growth promotion under saline conditions	Sarkar et al. (2018)
Streptomyces sp.	Rice	Enhanced growth and chlorophyll contents	Jaemsaeng et al. (2018)
Pseudomonas sp.	Rice	Stimulatory effects on growth and germination depending on strains	Nakbanpote et al. (2014)
Pseudomonas diminuta and *B. subtilis*	Rice	Reduced the effect of Xoo infection intensity of rice seeds	Palupi et al. (2017)
B. subtilis	Rice	Improved root and shoot growth, chlorophyll, and carotenoid concentrations in rice seedlings	Singh et al. (2016)
Pseudomonas sp.	Rice	Increased seedling length and dry weight	Sen and Chandrasekhar (2014)
Trichoderma harzianum	Rice	Alleviated salt tolerance	Rawat et al. (2012)
Pseudomonas sp. and *Bacillus* sp.	Soybean	Improved the overall growth and biomass improvement under saline conditions	Kumari et al. (2015)
P. fluorescens	Sunflower	Enhanced the ability of seedlings to grow uniformly and provided vigorous stand establishment	Moeinzadeh et al. (2010)

Continued

Table 3 Beneficial importance of microbial inoculants in seed priming—cont'd

Inoculants/PGPMs	Crop/plant/ host name	Effects on plant growth, defense, and yield	References
Chryseobacterium humi and *Ochrobactrum haematophilum*	Sunflower	Improved the N/K ratio, dry biomass, and enzymatic activity	Pereira et al. (2016)
P. fluorescens	Wheat	Improved the germination and growth parameters of seedlings	Safari et al. (2018)
Dietzia sp.	Wheat	Improved the overall growth and biomass	Bharti et al. (2016)
Bacillus sp. and *Staphylococcus* sp.	Wheat	Increased growth by 62%–78% under salinity stress	Orhan (2016)
Azospirillum brasilense and *Bacillus amyloliquefaciens*	Wheat	Induced drought tolerance in wheat by upregulating the expression of stress-related genes	Kasim et al. (2013)
Arthrobacter sp. and *Dietzia* sp.	Wheat	Enhanced seedling growth, dry weight, and transpiration	Barnawal et al. (2017)
Enterobacter cloacae	Wheat	Enhanced metabolic activities under salinity	Singh et al. (2017)
Bacillus, *Staphylococcus*, and *Arthrobacter*	*Ammodendron bifolium*	Creased seed germination and radicle elongation	Zhu and She (2018)
B. subtilis	Chickpea	Increased nodule number, root dry weight, and shoot dry weight	Egamberdieva et al. (2017)
Bacillus aquimaris	Maize	Induced salinity resistance	Li and Jiang (2017)
P. fluorescens	Maize	Enhanced root number, growth, and biomass	Zerrouk et al. (2016)
B. cereus and *B. subtilis*	Maize	Increased root length	Hassan (2017)

Table 3 Beneficial importance of microbial inoculants in seed priming—cont'd

Inoculants/PGPMs	Crop/plant/ host name	Effects on plant growth, defense, and yield	References
Pseudomonas sp. and *Burkholderia* sp.	*Echinacea purpurea* and *Lonicera japonica*	Increased root and shoot length, fresh and dry weight, and number of lateral roots	Gupta et al. (2016)
Bacillus endophyticus, *Bacillus tequilensis*, *Planococcus rifietoensis*, *Variovorax paradoxus*, and *Arthrobacter agilis*	*Salicornia europaea*	Increased germination and seedling growth	Zhao et al. (2016)
Pseudomonas sp.	*Brassica juncea*	Increased biomass	Ma et al. (2011)
Klebsiella sp.	Oat	Improved shoot and root growth	Sapre et al. (2018)
Pseudomonas sp.	Tomato	Enhanced ionic balance, physiological activity, growth, and total biomass	Win et al. (2018)
Bacillus fortis	Chili	Enhanced growth, biomass, and biochemical parameters in salt stress	Yasin et al. (2018)

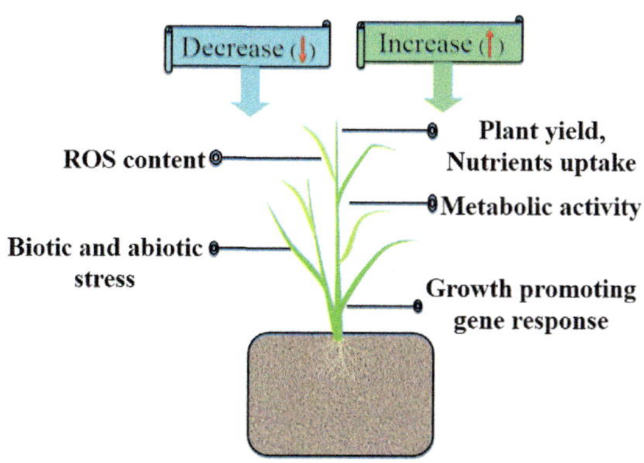

Fig. 3 Seed priming techniques modulating growth promotion, defense responses, and yield improvement in plants.

and growth, many authors have used seed priming as an inoculation method of PGPMs with biocontrol abilities against several biotic and abiotic stresses (Table 3). Most of the biotic stress studies investigated the biocontrol of fungal pathogens, including *Alternaria* spp., *Fusarium* spp., *Magnaporthe oryzae*, *Macrophomina phaseolina*, and *Rhizoctonia solani* (Amruta et al., 2019; Zulueta-Rodríguez et al., 2015). However, biopriming has been described as a valid tool to control the oomycete *Plasmopara halstedii* (Nagaraju et al., 2012), the insect pest *Spodoptera litura* (Song et al., 2017), and banana bunchy top virus (Harish et al., 2009). The mechanisms behind biotic and abiotic stress tolerance include induction of systemic resistance, phytohormone modulation, production of biocontrol metabolites, competition for nutrient and niche occupation, and biofilm formation (Rajendra Prasad et al., 2016).

5.6 Conclusions

Different stresses and environmental factors have a significant impact on plant growth and development. Seed priming is a useful technique for improving seed germination and plant growth, which leads to increased productivity under a variety of situations and challenges. Plants with efficient seed germination have a better chance of establishing themselves and developing a deep root system. Because of its favorable effects and unique physicochemical features, biopriming is the most successful of several seed priming methods. The impact of several priming methods on the germination and morphological and physiological features of crops to determine their establishment, growth, and production is addressed. In general, seed priming improves seed germination, seedling growth and development, vigor, seedling emergence rate, and subsequent plant performance. Future research topics of interest are briefly covered in this chapter.

Acknowledgments

The authors are thankful to Raiganj University, India; University of L'Aquila, Italy; Federal University of Agriculture, Nigeria; Maulana Abul Kalam Azad University of Technology, India; ICAR – National Rice Research Institute, India; Universidad de Santander, Colombia; and Instituto Tecnológico de Sonora, México. DM is grateful to the Government of West Bengal, India, for the Swami Vivekananda Merit Cum Means Ph.D. Scholarship (WBP191584588825). RM acknowledges DST, India, for the Inspire Fellowship (IF190457). The authors are thankful to the editors of *Microbial Inoculants: Recent Progress and Applications* for their invitation.

References

Abhilash, P.C., Dubey, R.K., Tripathi, V., Gupta, V.K., Singh, H.B., 2016. Plant growth-promoting microorganisms for environmental sustainability. Trends Biotechnol. 34 (11), 847–850.

Abido, W.A.E., Allem, A., Zsombik, L., Attila, N., 2019. Effect of gibberellic acid on germination of six wheat cultivars under salinity stress levels. Asian J. Biol. Sci. 12, 51–60.

Acharya, P., Jayaprakasha, G.K., Crosby, K.M., Jifon, J.L., Patil, B.S., 2020. Nanoparticle-mediated seed priming improves germination, growth, yield, and quality of watermelons (*Citrullus lanatus*) at multi-locations in Texas. Sci. Rep. 10 (1), 1–16.

Ajee, R.S., Kaushik, S., 2021. A promising approach of managing seed-borne pathogens through plant growth-promoting microbes. In: Antioxidants in Plant-Microbe Interaction. Springer, Singapore, pp. 315–338.

Ajouri, A., Asgedom, H., Becker, M., 2004. Seed priming enhances germination and seedling growth of barley under conditions of P and Zn deficiency. J. Plant Nutr. Soil Sci. 167 (5), 630–636.

Aloui, H., Souguir, M., Latique, S., Hannachi, C., 2014. Germination and growth in control and primed seeds of pepper as affected by salt stress. Cercetari agronomice în Moldova 47 (3), 83–95.

Amooaghaie, R., 2011. The effect of hydro and osmopriming on alfalfa seed germination and antioxidant defenses under salt stress. Afr. J. Biotechnol. 10 (33), 6269–6275.

Amruta, N., Prasanna Kumar, M.K., Kandikattu, H.K., Sarika, G., Puneeth, M.E., Ranjitha, H.P., Vishwanath, K., Manjunatha, C., Pramesh, D., Mahesh, H.B., Narayanaswamy, S., 2019. Bio-priming of rice seeds with novel bacterial strains, for management of seed-borne *Magnaporthe oryzae* L. Plant Physiol. Rep. 24, 507–520. https://doi.org/10.1007/s40502-019-00492-6.

Ashraf, M.A., Rasheed, R., Hussain, I., Iqbal, M., Riaz, M., Arif, M.S., 2019. Chemical priming for multiple stress tolerance. In: Priming and Pretreatment of Seeds and Seedlings. Springer, Singapore, pp. 385–415.

Aymen, E.M., 2018. Seed priming with plant growth regulators to improve crop abiotic stress tolerance. In: Advances in Seed Priming. Springer, Singapore, pp. 95–106.

Azeem, M., Iqbal, N., Kausar, S., Javed, M.T., Akram, M.S., Sajid, M.A., 2015. Efficacy of silicon priming and fertigation to modulate seedling's vigor and ion homeostasis of wheat (*Triticum aestivum* L.) under saline environment. Environ. Sci. Pollut. Res. 22 (18), 14367–14371.

Barnawal, D., Bharti, N., Pandey, S.S., Pandey, A., Chanotiya, C.S., Kalra, A., 2017. Plant growth-promoting rhizobacteria enhance wheat salt and drought stress tolerance by altering endogenous phytohormone levels and TaCTR1/TaDREB2 expression. Physiol. Plant. 161 (4), 502–514.

Bharti, N., Pandey, S.S., Barnawal, D., Patel, V.K., Kalra, A., 2016. Plant growth promoting rhizobacteria *Dietzia natronolimnaea* modulates the expression of stress responsive genes providing protection of wheat from salinity stress. Sci. Rep. 6 (1), 1–16.

Carrillo-Reche, J., Vallejo-Marín, M., Quilliam, R.S., 2018. Quantifying the potential of 'on-farm' seed priming to increase crop performance in developing countries. A meta-analysis. Agron. Sustain. Dev. 38 (6), 1–14.

Dawood, M.G., El-Awadi, M.E., 2015. Alleviation of salinity stress on *Vicia faba* L. plants via seed priming with melatonin. Acta Biol. Colomb. 20 (2), 223–235.

De Clercq, M., Vats, A., Biel, A., 2018. Agriculture 4.0: the future of farming technology. In: Proceedings of the World Government Summit, Dubai, UAE, pp. 11–13.

Debasis, M., Snežana, A., Panneerselvam, P., Manisha, S.A., Tanja, V., Ganeshamurthy, A. N., Devvret, V., Poonam, R.T., Divya, J., 2019. Plant growth promoting microorganisms (PGPMs) helping in sustainable agriculture: current perspective. Int. J. Agric. Sci. Vet. Med. 7 (2), 50–74.

Debasis, M., Ganeshamurthy, A.N., Komal, S., Radha, T.K., Rupa, T.R., 2020. Plant growth promotion and biocontrol activity of some typical harsh environment rhizomicrobes and their effects on *Amaranthus cruentus* plants. Res. J. Biotechnol., 1.

Demirkaya, M., 2014. Improvement in tolerance to salt stress during tomato cultivation. Turk. J. Biol. 38, 1–7.

Deshmukh, A.J., Jaiman, R.S., Bambharolia, R.P., Patil, V.A., 2020. Seed biopriming – a review. Int. J. Econ. Plants 7, 038–043. https://doi.org/10.23910/2/2020.0359.

Devika, O.S., Singh, S., Sarkar, D., Barnwal, P., Suman, J., Rakshit, A., 2021. Seed priming: a potential supplement in integrated resource management under fragile intensive ecosystems. Front. Sustain. Food Syst. 5, 209.

Dezfuli, P.M., Sharif-Zadeh, F., Janmohammadi, M., 2008. Influence of priming techniques on seed germination behavior of maize inbred lines (*Zea mays* L.). ARPN J. Agric. Biol. Sci. 3 (3), 22–25.

Djebaili, R., Pellegrini, M., Smati, M., Del Gallo, M., Kitouni, M., 2020. Actinomycete strains isolated from saline soils: plant-growth-promoting traits and inoculation effects on *Solanum lycopersicum*. Sustainability 12, 4617. https://doi.org/10.3390/su12114617.

do Espirito Santo Pereira, A., Caixeta Oliveira, H., Fernandes Fraceto, L., Santaella, C., 2021. Nanotechnology potential in seed priming for sustainable agriculture. Nano 11 (2), 267.

Egamberdieva, D., Wirth, S.J., Shurigin, V.V., Hashem, A., Abd Allah, E.F., 2017. Endophytic bacteria improve plant growth, symbiotic performance of chickpea (*Cicer arietinum* L.) and induce suppression of root rot caused by *Fusarium solani* under salt stress. Front. Microbiol. 8, 1887.

El-Shazoly, R.M., Metwally, A.A., Hamada, A.M., 2019. Salicylic acid or thiamin increases tolerance to boron toxicity stress in wheat. J. Plant Nutr. 42, 702–722.

Farhoudi, R., Saeedipour, S., Mohammadreza, D., 2011. The effect of NaCl seed priming on salt tolerance, antioxidant enzyme activity, proline and carbohydrate accumulation of Muskmelon (*Cucumis melo* L.) under saline condition. Afr. J. Agric. Res. 6 (6), 1363–1370.

Farooq, M., Basra, S.M.A., Afzal, I., Khaliq, A., 2006. Optimization of hydropriming techniques for rice seed invigoration. Seed Sci. Technol. 34 (2), 507–512.

Farooq, M., Aziz, T., Basra, S.M.A., Cheema, M.A., Rehman, H., 2008. Chilling tolerance in hybrid maize induced by seed priming with salicylic acid. J. Agron. Crop Sci. 194 (2), 161–168.

Farooq, M., Aziz, T., Wahid, A., Lee, D.J., Siddique, K.H., 2009. Chilling tolerance in maize: agronomic and physiological approaches. Crop Pasture Sci. 60 (6), 501–516.

Fazlali, R., Asli, D.E., Moradi, P., 2013. The effect of seed priming by ascorbic acid on bioactive compounds of naked seed pumpkin (*Cucurbita pepo* var. styriaca) under salinity stress. Int. J. Farm. Allied Sci. 2 (17), 587–590.

Fercha, A., Capriotti, A.L., Caruso, G., Cavaliere, C., Samperi, R., Stampachiacchiere, S., Laganà, A., 2014. Comparative analysis of metabolic proteome variation in ascorbate-primed and unprimed wheat seeds during germination under salt stress. J. Proteome 108, 238–257.

Fichtel, J., Köster, J., Rullkötter, J., Sass, H., 2007. Spore dipicolinic acid contents used for estimating the number of endospores in sediments. FEMS Microbiol. Ecol. 61 (3), 522–532.

Gadelha, C.G., de Souza Miranda, R., Alencar, N.L.M., Costa, J.H., Prisco, J.T., Gomes-Filho, E., 2017. Exogenous nitric oxide improves salt tolerance during establishment of *Jatropha curcas* seedlings by ameliorating oxidative damage and toxic ion accumulation. J. Plant Physiol. 212, 69–79.

Górnik, K., Lahuta, L.B., 2017. Application of phytohormones during seed hydropriming and heat shock treatment on sunflower (*Helianthus annuus* L.) chilling resistance and changes in soluble carbohydrates. Acta Physiol. Plant. 39 (5), 118.

Gupta, H., Saini, R.V., Pagadala, V., Kumar, N., Sharma, D.K., Saini, A.K., 2016. Analysis of plant growth promoting potential of endophytes isolated from *Echinacea purpurea* and *Lonicera japonica*. J. Soil Sci. Plant Nutr. 16 (3), 558–577.

Harish, S., Kavino, M., Kumar, N., Samiyappan, R., 2009. Biopriming banana with plant growth-promoting endophytic bacteria induces systemic resistance against banana bunchy top virus. Acta Hortic. 295–302. https://doi.org/10.17660/ActaHortic.2009.828.30.

Harris, D., Rashid, A., Miraj, G., Arif, M., Shah, H., 2007. 'On-farm' seed priming with zinc sulphate solution—a cost-effective way to increase the maize yields of resource-poor farmers. Field Crop Res. 102 (2), 119–127.

Hasegawa, S., Meguro, A., Shimizu, M., Nishimura, T., Kunoh, H., 2006. Endophytic actinomycetes and their interactions with host plants. Actinomycetologica 20, 72–81. https://doi.org/10.3209/saj.20.72.

Hassan, S.E.D., 2017. Plant growth-promoting activities for bacterial and fungal endophytes isolated from medicinal plant of *Teucrium polium* L. J. Adv. Res. 8 (6), 687–695.

Hassini, I., Martinez-Ballesta, M.C., Boughanmi, N., Moreno, D.A., Carvajal, M., 2017. Improvement of broccoli sprouts (*Brassica oleracea* L. var. italica) growth and quality by KCl seed priming and methyl jasmonate under salinity stress. Sci. Hortic. 226, 141–151.

Huang, X.F., Chaparro, J.M., Reardon, K.F., Zhang, R., Shen, Q., Vivanco, J.M., 2014. Rhizosphere interactions: root exudates, microbes, and microbial communities. Botany 92 (4), 267–275.

Hussain, S., Khan, F., Hussain, H.A., Nie, L., 2016. Physiological and biochemical mechanisms of seed priming-induced chilling tolerance in rice cultivars. Front. Plant Sci. 7, 116.

Imran, M., 2014. Nutrient seed priming improves abiotic stress tolerance in *Zea mays* L. and *Glycine max* L. Dr. Sc. Agr. /Ph.D. in Agricultural Sciences, Agricultural Sciences, University of Hohenheim.

Imran, M., Mahmood, A., Römheld, V., Neumann, G., 2013. Nutrient seed priming improves seedling development of maize exposed to low root zone temperatures during early growth. Eur. J. Agron. 49, 141–148.

Imran, M., Boelt, B., Mühling, K.H., 2018. Zinc seed priming improves salt resistance in maize. J. Agron. Crop Sci. 204 (4), 390–399.

Iqbal, M., Ashraf, M., 2005. Presowing seed treatment with cytokinins and its effect on growth, photosynthetic rate, ionic levels and yield of two wheat cultivars differing in salt tolerance. J. Integr. Plant Biol. 47 (11), 1315–1325.

İşeri, Ö.D., Sahin, F.I., Haberal, M., 2014. Sodium chloride priming improves salinity response of tomato at seedling stage. J. Plant Nutr. 37 (3), 374–392.

Jaemsaeng, R., Jantasuriyarat, C., Thamchaipenet, A., 2018. Molecular interaction of 1-aminocyclopropane-1-carboxylate deaminase (ACCD)-producing endophytic *Streptomyces* sp. GMKU 336 towards salt-stress resistance of *Oryza sativa* L. cv. KDML105. Sci. Rep. 8 (1), 1–15.

Jiang, X., Li, H., Song, X., 2016. Seed priming with melatonin effects on seed germination and seedling growth in maize under salinity stress. Pak. J. Bot. 48 (4), 1345–1352.

Jiménez-Arias, D., García-Machado, F.J., Morales-Sierra, S., Suárez, E., Pérez, J.A., Luis, J. C., Garrido-Orduña, C., Herrera, A.J., Valdés, F., Sandalio, L.M., Borges, A.A., 2019. Menadione sodium bisulphite (MSB): beyond seed-soaking. Root pretreatment with MSB primes salt stress tolerance in tomato plants. Environ. Exp. Bot. 157, 161–170.

Joshi, N., Jain, A., Arya, K., 2013. Alleviation of salt stress in Cucumis sativus L. through seed priming with calcium chloride. Indian J. Appl. Res. 3 (11), 22–25.

Karadağ, B., Yücel, N.C., 2017. Salicylic acid and fish flour pre-treatments affect wheat phenolic and flavonoid compounds, lipid peroxidation levels under salt stress. Cereal Res. Commun. 45 (2), 192–201.

Karimi, S., Eshghi, S., Karimi, S., Hasan-Nezhadian, S., 2017. Inducing salt tolerance in sweet corn by magnetic priming. Acta Agric. Sloven. 109 (1), 89–102.

Kasim, W.A., Osman, M.E., Omar, M.N., Abd El-Daim, I.A., Bejai, S., Meijer, J., 2013. Control of drought stress in wheat using plant-growth-promoting bacteria. J. Plant Growth Regul. 32 (1), 122–130.

Kazemi, J.S., Aboutalebian, M.A., Hamzei, J., Meskarbashee, M., 2018. Mycorrhiza and seed priming effect to improve the balance of sodium and potassium and some changes in antioxidants in the leaves of maize (*Zea mays* L.) under soil salinity. J. Agron. 17 (1), 18–27.

Khoshru, B., Mitra, D., Khoshmanzar, E., Myo, E.M., Uniyal, N., Mahakur, B., Mohapatra, P.K.D., Panneerselvam, P., Boutaj, H., Alizadeh, M., Cely, M.V.T., 2020. Current scenario and future prospects of plant growth-promoting rhizobacteria: an economic valuable resource for the agriculture revival under stressful conditions. J. Plant Nutr. 43 (20), 3062–3092.

Korkmaz, A., Sirikci, R., 2011. Improving salinity tolerance of germinating seeds by exogenous application of glycinebetaine in pepper. Seed Sci. Technol. 39, 377–388.

Kumar, D., Kumar, M., Verma, P., Shamim, M.D., 2017. Microbial biotechnology: role of microbes in sustainable agriculture. In: Plant Biotechnology. Apple Academic Press, pp. 415–460.

Kumari, S., Vaishnav, A., Jain, S., Varma, A., Choudhary, D.K., 2015. Bacterial-mediated induction of systemic tolerance to salinity with expression of stress alleviating enzymes in soybean (*Glycine max* L. Merrill). J. Plant Growth Regul. 34 (3), 558–573.

Lemmens, E., Deleu, L.J., De Brier, N., De Man, W.L., De Proft, M., Prinsen, E., Delcour, J.A., 2019. The impact of hydro-priming and Osmo-priming on seedling characteristics, plant hormone concentrations, activity of selected hydrolytic enzymes, and Cell Wall and Phytate hydrolysis in sprouted wheat (*Triticum aestivum* L.). ACS Omega 4 (26), 22089–22100.

Li, H.Q., Jiang, X.W., 2017. Inoculation with plant growth-promoting bacteria (PGPB) improves salt tolerance of maize seedling. Russ. J. Plant Physiol. 64 (2), 235–241.

Lutts, S., Benincasa, P., Wojtyla, L., Kubala, S., Pace, R., Lechowska, K., Quinet, M., Garnczarska, M., 2016. Seed priming: new comprehensive approaches for an old empirical technique. In: New Challenges in Seed Biology-Basic and Translational Research Driving Seed Technology, pp. 1–46.

Ma, Y., 2019. Seed coating with beneficial microorganisms for precision agriculture. Biotechnol. Adv. 37 (7), 107423.

Ma, Y., Rajkumar, M., Luo, Y., Freitas, H., 2011. Inoculation of endophytic bacteria on host and non-host plants—effects on plant growth and Ni uptake. J. Hazard. Mater. 195, 230–237.

Mahmood, A., Kataoka, R., 2018. Potential of biopriming in enhancing crop productivity and stress tolerance. In: Advances in Seed Priming. Springer Singapore, Singapore, pp. 127–145, https://doi.org/10.1007/978-981-13-0032-5_9.

Mahmood, A., Turgay, O.C., Farooq, M., Hayat, R., 2016. Seed biopriming with plant growth promoting rhizobacteria: a review. FEMS Microbiol. Ecol. 92 (8), fiw112.

Mangena, P., 2020. Effect of hormonal seed priming on germination, growth, yield and biomass accumulation in soybean grown under induced drought stress. Indian J. Agric. Res.

Manhas, R.K., Kaur, T., 2016. Biocontrol potential of *Streptomyces hydrogenans* strain DH16 toward *Alternaria brassicicola* to control damping off and black leaf spot of *Raphanus sativus*. Front. Plant Sci. 7. https://doi.org/10.3389/fpls.2016.01869.

Marthandan, V., Geetha, R., Kumutha, K., Renganathan, V.G., Karthikeyan, A., Ramalingam, J., 2020. Seed priming: a feasible strategy to enhance drought tolerance in crop plants. Int. J. Mol. Sci. 21, 8258. https://doi.org/10.3390/ijms21218258.

McDonald, M.B., 2000. Seed priming. In: Black, M., Bewley, J.D. (Eds.), Seed Technology and its Biological Basis. Sheffield Academic Press, Sheffield, pp. 287–325.

Meena, R.P., Sendhil, R., Tripathi, S.C., Chander, S., Chhokar, R.S., Sharma, R.K., 2013. Hydro-priming of seed improves the water use efficiency, grain yield and net economic return of wheat under different moisture regimes. SAARC J. Agric. 11 (2), 149–159.

Mirmazloum, I., Kiss, A., Erdélyi, É., Ladányi, M., Németh, É.Z., Radácsi, P., 2020. The effect of osmopriming on seed germination and early seedling characteristics of *Carum carvi* L. Agriculture 10 (4), 94.

Mitra, D., Mukhopadhyay, A.K.M.A., 2014. Enhance plant health by using of phosphate solubilizing *Bacillus* sp. as ecofriendly bio fertilizer. Int. J. Res. 1 (11), 1438–1448.

Mitra, D., Mondal, A.K., Mondol, M., Mukhopadhyay, A., 2014. Plant growth promoting traits and prospects of using phosphate solubilizing *Pseudomonas* sp. isolated from laterite soil. Annals. Food Sci. Technol. 15 (2), 309–314.

Mitra, D., Mondal, R., Khoshru, B., Senapati, A., Radha, T.K., Mahakur, B., Uniyal, N., Myo, E.M., Boutaj, H., Guerra Sierra, B.E., Panneerselvam, P., 2021a. Enhanced plant growth, nutrient acquisition, and crop protection by using Actinobacteria: advancements in soil, plant, and microbial multifactorial interactions. Pedosphere 2.

Mitra, D., Mondal, R., Khoshru, B., Shadangi, S., Das Mohapatra, P.K., Panneerselvam, P., 2021b. Rhizobacteria mediated seed bio-priming triggers the resistance and plant growth for sustainable crop production. Curr. Res. Microb. Sci. 2, 100071. https://doi.org/10.1016/j.crmicr.2021.100071.

Moeinzadeh, A., Sharif-Zadeh, F., Ahmadzadeh, M., Tajabadi, F., 2010. Biopriming of sunflower ('*Helianthus annuus*' L.) seed with '*Pseudomonas fluorescens*' for improvement of seed invigoration and seedling growth. Aust. J. Crop. Sci. 4 (7), 564–570.

Nada, H.S., Hamza, J.H., 2019. Priming of maize seed with gibberellin (GA_3) to tolerate drought stress. 2. Field emergence and its properties. Iraqi J. Des. Stud 9, 1–12.

Nagaraju, A., Sudisha, J., Murthy, S.M., Ito, S., 2012. Seed priming with Trichoderma harzianum isolates enhances plant growth and induces resistance against *Plasmopara halstedii*, an incitant of sunflower downy mildew disease. Australas. Plant Pathol. 41, 609–620. https://doi.org/10.1007/s13313-012-0165-z.

Nakao, Y., Sone, C., Sakagami, J.I., 2020. Genetic diversity of hydro priming effects on rice seed emergence and subsequent growth under different moisture conditions. Genes 11 (9), 994.

Nakaune, M., Hanada, A., Yin, Y.G., Matsukura, C., Yamaguchi, S., Ezura, H., 2012. Molecular and physiological dissection of enhanced seed germination using short-term low-concentration salt seed priming in tomato. Plant Physiol. Biochem. 52, 28–37.

Nakbanpote, W., Panitlurtumpai, N., Sangdee, A., Sakulpone, N., Sirisom, P., Pimthong, A., 2014. Salt-tolerant and plant growth-promoting bacteria isolated from Zn/Cd contaminated soil: identification and effect on rice under saline conditions. J. Plant Interact. 9 (1), 379–387.

Nciizah, A.D., Rapetsoa, M.C., Wakindiki, I.I., Zerizghy, M.G., 2020. Micronutrient seed priming improves maize (*Zea mays*) early seedling growth in a micronutrient deficient soil. Heliyon 6 (8), e04766.

Nihorimbere, V., Ongena, M., Smargiassi, M., Thonart, P., 2011. Beneficial effect of the rhizosphere microbial community for plant growth and health. Biotechnol. Agron. Soc. Environ. 15 (2), 327–337.

Orhan, F., 2016. Alleviation of salt stress by halotolerant and halophilic plant growth-promoting bacteria in wheat (*Triticum aestivum*). Braz. J. Microbiol. 47, 621–627.

Pallaoro, D.S., Avelino, A.C.D., Camili, E.C., Guimarães, S.C., Albuquerque, M.C.D.F., 2016. Priming corn seeds with plant growth regulator. J. Seed Sci. 38, 227–232.

Palupi, T., Ilyas, S., Machmud, M., Widajati, E., 2017. Effect of seed coating with biological agents on seed quality of rice. Biodiversitas J. Biol. Divers. 18 (2), 727–732.

Panneerselvam, P., Senapati, A., Kumar, U., Sharma, L., Lepcha, P., Prabhukarthikeyan, S. R., Jahan, A., Parameshwaran, C., Govindharaj, G.P.P., Lenka, S., Nayak, P.K., 2019.

Antagonistic and plant-growth promoting novel *Bacillus* species from long-term organic farming soils from Sikkim, India. 3 Biotech 9 (11), 1–12.

Panneerselvam, P., Senapati, A., Sharma, L., Nayak, A.K., Kumar, A., Kumar, U., Prabhukarthikeyan, S.R., Mitra, D., Sagarika, M.S., 2021. Understanding rice growth-promoting potential of *Enterobacter* spp. isolated from long-term organic farming soil in India through a supervised learning approach. Curr. Res. Microb. Sci. 2, 100035.

Panuccio, M.R., Chaabani, S., Roula, R., Muscolo, A., 2018. Bio-priming mitigates detrimental effects of salinity on maize improving antioxidant defense and preserving photosynthetic efficiency. Plant Physiol. Biochem. 132, 465–474.

Paparella, S., Araújo, S.S., Rossi, G., Wijayasinghe, M., Carbonera, D., Balestrazzi, A., 2015. Seed priming: state of the art and new perspectives. Plant Cell Rep. 34, 1281–1293. https://doi.org/10.1007/s00299-015-1784-y.

Parewa, H.P., Joshi, N., Meena, V.S., Joshi, S., Choudhary, A., Ram, M., Meena, S.C., Jain, L.K., 2021. Role of biofertilizers and biopesticides in organic farming. In: Advances in Organic Farming, pp. 133–159.

Patade, V.Y., Khatri, D., Manoj, K., Kumari, M., Ahmed, Z., 2012. Cold tolerance in thiourea primed capsicum seedlings is associated with transcript regulation of stress responsive genes. Mol. Biol. Rep. 39 (12), 10603–10613.

Paul, S., Roychoudhury, A., 2017. Effect of seed priming with spermine/spermidine on transcriptional regulation of stress-responsive genes in salt-stressed seedlings of an aromatic rice cultivar. Plant Gene 11, 133–142.

Pellegrini, M., Pagnani, G., Bernardi, M., Mattedi, A., Spera, D.M., Del Gallo, M., 2020. Cell-free supernatants of plant growth-promoting bacteria: a review of their use as biostimulant and microbial biocontrol agents in sustainable agriculture. Sustainability 12, 9917. https://doi.org/10.3390/su12239917.

Pellegrini, M., Pagnani, G., Rossi, M., D'Egidio, S., Del Gallo, M., Forni, C., 2021a. *Daucus carota* L. seed inoculation with a consortium of bacteria improves plant growth, soil fertility status and microbial community. Appl. Sci. 11, 3274. https://doi.org/10.3390/app11073274.

Pellegrini, M., Spera, D.M., Ercole, C., Del Gallo, M., 2021b. *Allium cepa* L. inoculation with a consortium of plant growth-promoting bacteria: effects on plants, soil, and the autochthonous microbial community. Microorganisms 9, 639. https://doi.org/10.3390/microorganisms9030639.

Pepper, I.L., Gentry, T.J., 2015. Earth environments. In: Environmental Microbiology. Elsevier, pp. 59–88, https://doi.org/10.1016/B978-0-12-394626-3.00004-1.

Pereira, S.I., Moreira, H., Argyras, K., Castro, P.M., Marques, A.P., 2016. Promotion of sunflower growth under saline water irrigation by the inoculation of beneficial microorganisms. Appl. Soil Ecol. 105, 36–47.

Poonam, S., Kaur, H., Geetika, S., 2013. Effect of jasmonic acid on photosynthetic pigments and stress markers in *Cajanus cajan* (L.) Millsp. seedlings under copper stress. Am. J. Plant Sci. 4, 817–823.

Pouramir-Dashtmian, F., Khajeh-Hosseini, M., Esfahani, M., 2014. Improving chilling tolerance of rice seedling by seed priming with salicylic acid. Arch. Agron. Soil Sci. 60 (9), 1291–1302.

Prabhukarthikeyan, S.R., Keerthana, U., Baite, M.S., Panneerselvam, P., Mitra, D., Kumar, R.N., Parameswaran, C., Cayalvizhi, B., Kumar, A.M., Harish, S., Rath, P.C., 2022. Bacillus rhizobacteria: a versatile biostimulant for sustainable agriculture. In: New and Future Developments in Microbial Biotechnology and Bioengineering. Elsevier, pp. 33–44.

Prasad, S.R., Kamble, U.R., Sripathy, K.V., Bhaskar, K.U., Singh, D.P., 2016. Seed biopriming for biotic and abiotic stress management. In: Microbial Inoculants in Sustainable Agricultural Productivity. Springer, New Delhi, pp. 211–228.

Qu, Z., Zhao, H., Zhang, H., Wang, Q., Yao, Y., Cheng, J., Lin, Y., Xie, J., Fu, Y., Jiang, D., 2020. Bio-priming with a hypovirulent phytopathogenic fungus enhances the connection and strength of microbial interaction network in rapeseed. npj Biofilms Microbiomes 6, 45. https://doi.org/10.1038/s41522-020-00157-5.

Rajendra Prasad, S., Kamble, U.R., Sripathy, K.V., Udaya Bhaskar, K., Singh, D.P., 2016. Seed bio-priming for biotic and abiotic stress management. In: Microbial Inoculants in Sustainable Agricultural Productivity. Springer India, New Delhi, pp. 211–228.

Rakshit, A., Sunita, K., Pal, S., Singh, A., Singh, H.B., 2015. Bio-priming mediated nutrient use efficiency of crop species. Nutrient use efficiency: from basics to advances, 181–191.

Rashid, M.I., Mujawar, L.H., Shahzad, T., Almeelbi, T., Ismail, I.M.I., Oves, M., 2016. Bacteria and fungi can contribute to nutrients bioavailability and aggregate formation in degraded soils. Microbiol. Res. 183, 26–41. https://doi.org/10.1016/j.micres.2015.11.007.

Rasmann, S., Turlings, T.C., 2016. Root signals that mediate mutualistic interactions in the rhizosphere. Curr. Opin. Plant Biol. 32, 62–68.

Rawat, L., Singh, Y., Shukla, N., Kumar, J., 2012. Seed biopriming with salinity tolerant isolates of *Trichoderma harzmnum* alleviates salt stress in rice: growth, physiological and biochemical characteristics. J. Plant Pathol., 353–365.

Razaji, A., Farzanian, M., Sayfzadeh, S., 2014. The effects of seed priming by ascorbic acid on some morphological and biochemical aspects of rapeseed (*Brassica napus* L.) under drought stress condition. Int. J. Biosci. 4 (1), 432–442.

Rhaman, M.S., Imran, S., Rauf, F., Khatun, M., Baskin, C.C., Murata, Y., Hasanuzzaman, M., 2021. Seed priming with phytohormones: an effective approach for the mitigation of abiotic stress. Plants 10 (1), 37.

Safari, D., Jamali, F., Nooryazdan, H.R., Bayat, F., 2018. Evaluation of ACC deaminase producing '*Pseudomonas fluorescens*' strains for their effects on seed germination and early growth of wheat under salt stress. Aust. J. Crop. Sci. 12 (3), 413–421.

Saiz-Rubio, V., Rovira-Más, F., 2020. From smart farming towards agriculture 5.0: a review on crop data management. Agronomy 10 (2), 207.

Salama, K.H., Mansour, M.M.F., 2015. Choline priming-induced plasma membrane lipid alterations contributed to improved wheat salt tolerance. Acta Physiol. Plant. 37 (8), 1–7.

Sapre, S., Gontia-Mishra, I., Tiwari, S., 2018. Klebsiella sp. confers enhanced tolerance to salinity and plant growth promotion in oat seedlings (*Avena sativa*). Microbiol. Res. 206, 25–32.

Sarkar, A., Ghosh, P.K., Pramanik, K., Mitra, S., Soren, T., Pandey, S., Mondal, M.H., Maiti, T.K., 2018. A halotolerant *Enterobacter* sp. displaying ACC deaminase activity promotes rice seedling growth under salt stress. Res. Microbiol. 169 (1), 20–32.

Sarkar, D., Singh, S., Parihar, M., Rakshit, A., 2021. Seed bio-priming with microbial inoculants: A tailored approach towards improved crop performance, nutritional security, and agricultural sustainability for smallholder farmers. Curr. Res. Environ. Sustain. 3, 100093. https://doi.org/10.1016/j.crsust.2021.100093.

Sattiraju, K.S., Kotiyal, S., Arora, A., Maheshwari, M., 2019. Plant growth-promoting microbes: contribution to stress management in plant hosts. In: Environmental Biotechnology: For Sustainable Future, pp. 199–236.

Savvides, A., Ali, S., Tester, M., Fotopoulos, V., 2016. Chemical priming of plants against multiple abiotic stresses: mission possible? Trends Plant Sci. 21 (4), 329–340.

Sen, S., Chandrasekhar, C.N., 2014. Effect of PGPR on growth promotion of rice (*Oryza sativa* L.) under salt stress. Asian J. Plant Sci. Res. 4, 62–67.

Shah, T., Latif, S., Saeed, F., Ali, I., Ullah, S., Alsahli, A.A., Jan, S., Ahmad, P., 2021a. Seed priming with titanium dioxide nanoparticles enhances seed vigor, leaf water status, and antioxidant enzyme activities in maize (*Zea mays* L.) under salinity stress. J. King Saud Univ. Sci. 33 (1), 101207.

Shah, A., Nazari, M., Antar, M., Msimbira, L.A., Naamala, J., Lyu, D., Rabileh, M., Zajonc, J., Smith, D.L., 2021b. PGPR in agriculture: a sustainable approach to increasing climate change resilience. Front. Sustain. Food Syst. 5, 667546. https://doi.org/10.3389/fsufs.

Sheteiwy, M.S., Gong, D., Gao, Y., Pan, R., Hu, J., Guan, Y., 2018. Priming with methyl jasmonate alleviates polyethylene glycol-induced osmotic stress in rice seeds by regulating the seed metabolic profile. Environ. Exp. Bot. 153, 236–248.

Shoresh, M., Harman, G.E., Mastouri, F., 2010. Induced systemic resistance and plant responses to fungal biocontrol agents. Annu. Rev. Phytopathol. 48, 21–43. https://doi.org/10.1146/annurev-phyto-073009-114450.

Sime, G., Aune, J.B., 2020. On-farm seed priming and fertilizer micro-dosing: Agronomic and economic responses of maize in semi-arid Ethiopia. Food Energy Secur. 9 (1), e190.

Singh, A., Dahiru, R., Musa, M., Sani Haliru, B., 2014. Effect of osmopriming duration on germination, emergence, and early growth of cowpea (*Vigna unguiculata* (L.) Walp.) in the Sudan savanna of Nigeria. Int. J. Agron. 2014.

Singh, V.P., Kumar, J., Singh, M., Singh, S., Prasad, S.M., Dwivedi, R., Singh, M.P.V.V.B., 2016. Role of salicylic acid-seed priming in the regulation of chromium (VI) and UV-B toxicity in maize seedlings. Plant Growth Regul. 78 (1), 79–91.

Singh, R.P., Runthala, A., Khan, S., Jha, P.N., 2017. Quantitative proteomics analysis reveals the tolerance of wheat to salt stress in response to *Enterobacter cloacae* SBP-8. PLoS One 12 (9), e0183513.

Singh, A.K., Kumar, S., Sinha, T., 2021. Antioxidants in plant–microbe interaction. In: Antioxidants in Plant-Microbe Interaction. Springer Singapore, Singapore, pp. 3–20, https://doi.org/10.1007/978-981-16-1350-0_1.

Song, G.C., Choi, H.K., Kim, Y.S., Choi, J.S., Ryu, C.-M., 2017. Seed defense biopriming with bacterial cyclodipeptides triggers immunity in cucumber and pepper. Sci. Rep. 7, 14209. https://doi.org/10.1038/s41598-017-14155-9.

Swain, H., Adak, T., Mukherjee, A.K., Sarangi, S., Samal, P., Khandual, A., Jena, R., Bhattacharyya, P., Naik, S.K., Mehetre, S.T., Baite, M.S., Kumar, M.S., Zaidi, N.W., 2021. Seed biopriming with Trichoderma strains isolated from tree bark improves plant growth, antioxidative defense system in rice and enhance straw degradation capacity. Front. Microbiol. 12. https://doi.org/10.3389/fmicb.2021.633881.

Uniyal, N., Chauhan, A., Sharma, K., Mitra, D., 2018. Plant growth promotion by bacterial inoculum with different conditions and estimation of enzyme activity from germinating seeds and bacterial isolates. Curr. Trends Biotechnol. Pharm. 12 (1), 53–64.

Viaene, T., Langendries, S., Beirinckx, S., Maes, M., Goormachtig, S., 2016. *Streptomyces* as a plant's best friend? FEMS Microbiol. Ecol. 92, fiw119. https://doi.org/10.1093/femsec/fiw119.

Waqas, M., Korres, N.E., Khan, M.D., Nizami, A.S., Deeba, F., Ali, I., Hussain, H., 2019. Advances in the concept and methods of seed priming. In: Priming and Pretreatment of Seeds and Seedlings. Springer, Singapore, pp. 11–41.

Warren, J.E., Bennett, M.A., 1997. Seed hydration using the drum priming system. HortScience 32 (7), 1220–1221.

Win, K.T., Tanaka, F., Okazaki, K., Ohwaki, Y., 2018. The ACC deaminase expressing endophyte *Pseudomonas* spp. enhances NaCl stress tolerance by reducing stress-related ethylene production, resulting in improved growth, photosynthetic performance, and ionic balance in tomato plants. Plant Physiol. Biochem. 127, 599–607.

Woo, S.L., Pepe, O., 2018. Microbial consortia: promising probiotics as plant biostimulants for sustainable agriculture. Front. Plant Sci. 9, 1801. https://doi.org/10.3389/fpls.2018.01801.

Yacoubi, R., Job, C., Belghazi, M., Chaibi, W., Job, D., 2013. Proteomic analysis of the enhancement of seed vigour in osmoprimed alfalfa seeds germinated under salinity stress. Seed Sci. Res. 23 (2), 99–110.

Yasin, N.A., Khan, W.U., Ahmad, S.R., Ali, A., Ahmad, A., Akram, W., 2018. Imperative roles of halotolerant plant growth-promoting rhizobacteria and kinetin in improving salt tolerance and growth of black gram (*Phaseolus mungo*). Environ. Sci. Pollut. Res. 25 (5), 4491–4505.

Yücel, N.C., Heybet, E., 2016. Salicylic acid and calcium treatments improves wheat vigor, lipids and phenolics under high salinity. Acta Chim. Slov. 63 (4), 738–746.

Zavariyan, A.M., Rad, M.Y., Asghari, M., 2015. Effect of seed priming by potassium nitrate on germination and biochemical indices in *Silybum marianum* L. under salinity stress. International journal of. Life Sci. 9 (1), 23–29.

Zerrouk, I.Z., Benchabane, M., Khelifi, L., Yokawa, K., Ludwig-Müller, J., Baluska, F., 2016. A *Pseudomonas* strain isolated from date-palm rhizospheres improves root growth and promotes root formation in maize exposed to salt and aluminum stress. J. Plant Physiol. 191, 111–119.

Zhang, H.J., Zhang, N.A., Yang, R.C., Wang, L., Sun, Q.Q., Li, D.B., Cao, Y.Y., Weeda, S., Zhao, B., Ren, S., Guo, Y.D., 2014. Melatonin promotes seed germination under high salinity by regulating antioxidant systems, ABA and GA 4 interaction in cucumber (*Cucumis sativus* L.). J. Pineal Res. 57 (3), 269–279.

Zhang, F., Yu, J., Johnston, C.R., Wang, Y., Zhu, K., Lu, F., Zhang, Z., Zou, J., 2015. Seed priming with polyethylene glycol induces physiological changes in sorghum (*Sorghum bicolor* L. Moench) seedlings under suboptimal soil moisture environments. PLoS One 10 (10), e0140620.

Zhao, S., Zhou, N., Zhao, Z.Y., Zhang, K., Wu, G.H., Tian, C.Y., 2016. Isolation of endophytic plant growth-promoting bacteria associated with the halophyte *Salicornia europaea* and evaluation of their promoting activity under salt stress. Curr. Microbiol. 73 (4), 574–581.

Zheng, M., Tao, Y., Hussain, S., Jiang, Q., Peng, S., Huang, J., Cui, K., Nie, L., 2016. Seed priming in dry direct-seeded rice: consequences for emergence, seedling growth and associated metabolic events under drought stress. Plant Growth Regul. 78 (2), 167–178.

Zhu, Y., She, X., 2018. Evaluation of the plant-growth-promoting abilities of endophytic bacteria from the psammophyte *Ammodendron bifolium*. Can. J. Microbiol. 64 (4), 253–264.

Zulueta-Rodríguez, R., Hernández-Montiel, L.G., Murillo-Amador, B., Rueda-Puente, E. O., Capistrán, L.L., Troyo-Diéguez, E., Córdoba-Matson, M.V., 2015. Effect of hydropriming and biopriming on seed germination and growth of two Mexican fir tree species in danger of extinction. Forests 6 (9), 3109–3122.

CHAPTER 6

Organic waste decomposition by microbial inoculants as an effective tool for environmental management

Ravikant Dubey, Deepak Rathore, and Amrita Dwivedi
Department of Humanistic Studies, Indian Institute of Technology (Banaras Hindu University), Varanasi, Uttar Pradesh, India

6.1 Introduction

Composting is a natural recycling process that turns organic matter into soil rich in nutrients. When organic waste decomposes in the presence of microorganisms, it is converted to black or brown mass similar to what humans term compost. The process of composting involves raw material like food waste, animal waste, waste from municipal sources, and some agricultural refuse. The requirements for composting are raw material, moisture, microorganisms, and worms with air rated soil (Zucconi et al., 1981). The compost is a black mass that is rich in humus content with many nutrients available for the plant to intake. Compost is a lighter soil structure that provides the plant with enough aeration to the root, as well as available nutrients for further growth (Walker and Willson, 1973). Composting is the natural decomposition of organic matter that can be carried out in piles or pits. Compost has specific qualities: for instance, it provides habitat for various microflora/beneficial microorganisms, it has the capacity to hold moisture, and has sufficient pore space for the plant roots for healthy growth. Hence compost serves as the best soil conditioner. Composting is a simple way of adding rich nutrients to the soil, like humus, which facilitates plant growth and restores the soil quality (Walker and Willson, 1973; Willson and Walker, 1973; Willson et al., 1980).

Compost supports the earthworm population and other soil microorganisms that unlock soil nutrients present in subsoil and transform it into an available form. However, the earthworms in the soil are regarded as the soil champions due to their nature as borers, which helps the soil by increasing

Microbial Inoculants
https://doi.org/10.1016/B978-0-323-99043-1.00010-4

the soil pore size significantly for root aeration. Increased aerated roots become more active for absorption of nutrients. The earthworm consumes organic matter along with its mucus and excretes it out in the form of worm casting, which is highly nutrition-rich with 5 times more nitrogen and 7–11 times more potash available than normal soil (Walker and Willson, 1973).

Composting is the process of converting organic matter into a valuable product. It can be achieved by simply rolling leaf litter and adding soil moisture that can be converted into compost. When any organic matter is added to the soil, it is broken down into simpler compounds by microorganisms. This unlocks the nutrients present in the organic compound, such as potassium (K), phosphorus (P), and nitrogen (N), which become more available in the soil (Rodale, 1943; Truman, 1949; Golueke, 1950). This black mass composed of spacious particles provides a dynamic for the plant roots and microbial interactions. Interactions occur in such a way that plants have more opportunity to access optimum resources with minimum effort. This not only benefits the roots but also the soil microbes because they inhabit and receive the benefit of the available nutrients that favor plant growth in many aspects (Walker and Willson, 1973; Willson et al., 1980). There are various procedures for composting: it can be done by piling the organic matter or accumulating it in a pit, where the microbes/worms interact to convert it into a humus-like substance that is compost.

Any organic matter after decomposition can be converted into compost (Bidlingmaier, 1996; de Bertoldi and Zucconi, 1986). The breakdown of organic matter has three basic types: aerobic, anaerobic, and vermicompost (Stentiford, 1993).

6.2 Aerobic composting

In aerobic composting, the breakdown of organic matter occurs in the presence of air, which facilitates the aerobic microorganism to act in the degradation of organic content (de Bertoldi et al., 1987, 1988; de Bertoldi, 1999; Stentiford, 1993). The organic content is rich in green and brown matter. Green matter is mostly rich in nitrogen (N), and brown matter (C) provides a source for carbon. These two elements support the microorganism in growth as well as development. The ratio of brown to green needs to be maintained by continuous addition of material in a fixed proportion. The composting procedure is aerated by a turning process, and there is also the need for maintenance of moisture content. The addition of water can be done through a water sprinkler that provides favorable conditions for microbes to act in moist conditions, slowly. The decomposition time depends on the aeration and temperature of the pile. Increased temperature shows that turning is required regularly to avoid overheating. An increase in temperature is also very important for speeding up bacterial growth; there is a continuous succession of different bacteria/microbes at different temperatures (Peters et al., 2000). The composition of the bacteria ranges from low- to high-temperature bacteria. Hence the microbial succession is helpful in continuous biological degradation of different complex compounds with the help of specific groups of bacterial strains (Bidlingmaier, 1996; de Bertoldi and Zucconi, 1986; de Bertoldi et al., 1988; Stentiford, 1993). Apart from this, increased temperature also kills or suppresses disease-causing microorganisms or any other harmful bacterial or pathogen growth. So care must be taken that the pile is continuously turned/aerated to avoid overheating or developing a foul odor. This occurs due to the development of anaerobic conditions during the decomposition process (Stentiford, 1993).

6.3 Anaerobic composting

In contrast, anaerobic composting does not need air or oxygen for degradation by anaerobic microbes (Bidlingmaier, 1996; de Bertoldi et al., 1988). Anaerobic composting does not need any turning/aeration or moisture maintenance. Therefore there is no need to add continuous moisture. The anaerobic process is a long-duration process that occurs in various landfills or in dumpsites or in leftover sites of organic waste. It is not very useful because during anaerobic conditions various harmful bacteria develop and overtake to spoil the organic waste. The conversion of organic waste sometimes results in a toxic compound. In other words, anaerobic composting can be termed an effortless process for composting. It is non-labor intensive, meaning no turning or moisture addition is required. It is the least preferred method because it generates methane (CH_4), which is again a greenhouse gas (GHG) that has a harmful impact on the environment (Amlinger et al., 2008). Due to its foul-smelling signature, it is mostly avoided.

6.4 Vermicomposting

Vermicomposting involves worms consuming organic matter and turning it into useful compost. Vermicompost is one of the best soil conditioners, in comparison to the previous two, because worms slowly break down the organic nutrients by partial digestion, converting them into vermicast.

Vermicast is a product of enriched, partially digested organic compounds blended with mucus, in such a way that the locked nutrients become easily available for plant intake (Brinton, 2000). As already discussed, the availability ratios of nitrogen to phosphorus and potassium increase at a very drastic rate. Favorable content for vermicomposting is mostly food waste, garden waste, uncooked food, some fruits, or vegetable peelings. Worms digest the waste matter mostly at neutral pH (Bhat et al., 2018). In nature, acidic media act as recalcitrants that oppose compost formation. Other things like oil or greasy content also reduce the worm action. The vermicomposting is easy to handle and can be done in the kitchen in any closed structure, like a bucket or other vessel. Vermicomposting does not require turning or continuous addition of moisture. Of course, moisture needs to be retained by covering the compost bed or keeping it in a dark or shady area (Bhat et al., 2018).

The presence of worms needs to be handled. The most preferred organism for vermicomposting is red worms/red wigglers, which are highly efficient in degrading and multiplying. The worms are very sensitive to bright light and high temperatures. Care needs to be taken to optimize the conditions for maximum output of vermicompost. The vermicompost is also termed black gold, as it has superior results after application to soil, i.e., high productivity and improvement in soil quality. Thus we can assume composting can be a sustainable process for treatment of organic waste into a valuable product without adversely impacting the environment (Brinton, 2000).

6.5 Conditions required for composting

Organic matter decomposes to form compost with the help of the following factors that favor decomposition (Finstein et al., 1983). The following factors regulate the time period for composting too.

- Fixed carbon:nitrogen ratio
- Aeration
- Moisture
- Surface area exposed
- Temperature.

6.5.1 Fixed carbon:nitrogen ratio

This is the most important factor in the decomposition of organic waste, because the specific carbon (C) to nitrogen (N) ratio is a requirement of the microbial community for its growth and development (Golueke,

1977; Finstein and Morris, 1975). Carbon acts as a source of energy, while nitrogen acts as a source of building blocks. This helps in the microbe life cycle of physical development and repair. The optimum C:N ratio required for composting is 30:1, where it can generally range from 25:1 to 40:1 for a healthy decomposition. The limit above or below leads to halting/damping of the process and development of anomalies that impact the quality of the compost. The carbon source in nature is dry mass like wood, dry leaves, and cereal straw, while the source for nitrogen in organic waste is green matter like clippings of garden waste, vegetable/fruit peelings, and green leaves. Though the process of composting is a natural process and thus organic matter can degrade on its own, the major factor influencing the decomposition is the C:N ratio. If the C:N ratio of any component is within the range of approximately 30:1, it will degrade rapidly. If the C:N ratio is found to be higher, it means a high amount of carbon is present with respect to nitrogen availability. This leads to hampering of decomposition; to restore the decomposition, we need to add nitrogen sources like green leaves or grass cuttings that balance the nitrogen ratio. If the C:N ratio decreases, it means available carbon becomes a smaller amount in comparison to nitrogen. Thus excess of nitrogen may lead to the emission of toxic gases like ammonia, which again halts the decomposition process by reducing the efficiency of the microbial community. This condition can be restored by the addition of high carbon sources like wood chips, dry leaves, and agricultural waste. This signifies the importance of the C:N ratio for decomposition. It needs to be regulated from time to time for continuous composting (Finstein and Morris, 1975).

6.5.2 Aeration

Aeration is the important component for decomposition in the case of aerobic composting. Aeration leads to the circulation of oxygen into the waste particles, where microbes are working. The greater the oxygen supply, the more microbial action and rapid decomposition of organic matter into compost occurs. However, it is found that aerobic composting takes less time and has a greater tendency to degrade the particles efficiently (Wylie, 1957). With piles or pits, the material present on the outer surface is more exposed to the air than the inner or core portion. Therefore it is important to turn the pile in aerobic composting regularly. The aeration can be determined by particle size and porosity in the material. The porosity of the material is calculated by the volume of space between the particles divided by the total volume of the pile. If the material has a coarser size, this means it has spaces between particles that are called pores and the tendency to have pores is called porosity. This certainly helps the air to move between the particles and provides space for microbes to exist in the spaces for action. Thus greater pore size leads to more efficient decomposition due to sufficient aeration. In a pile, fine coarse particles have fewer spaces, leading to the compaction of material, which impacts the decomposition. It is seen that, as the process of composting moves further, it leads to the creation of fine particle size and more and more compaction, and hence in the later stages, turning becomes more significant. To expose the surface and flush air into the inner core for proper degradation of the complete pile/heap, various instruments can be used, such as pitchforks or shovels, for turning of piles (Wylie, 1957; NRAES, 1992).

6.5.3 Temperature

In the process of composting, the action of microbes on waste organic particles leads to release of energy in the form of heat, causing an increase in the temperature of the pile (Browne, 1933). The temperature in the pile during the composting process ranges from 30 to 60°C, or 65 to 144°F (Finstein et al., 1983). The temperature increase indicates the efficient decomposition of organic matter, but if the temperature rises above 65 to 144°F, this shuts down the composting, because the microorganisms cannot cope with high temperature (Stentiford, 1993). However, a record of pile temperature should be kept using a soil thermometer on a regular basis. To manage the optimum temperature for composting, the temperature can be reduced through mixing or turning of the heap. A reduction in the pile temperature leads to restoration of the composting process (Finstein and Hogan, 1993). However, the high temperature is not only helpful in increasing composting rates but also kills harmful microbes developing in the pile during the composting process.

6.5.4 Surface area/particle size

Surface area is an important factor because the decomposition of waste materials through microbes occurs at the interaction point of particle surface and available air. The surface area of a material can be increased by cutting the material into small pieces with the help of specific tools like choppers, pruners, and knives. The smaller pieces have more surface area and contact surface for bacterial/microbial action, so they are more likely to be acted upon at higher rates or at an increased pace. Thus waste that tends to possess uneven particle size can be chopped/shredded into smaller pieces. This leads to the development of a more uniform size and increases surface area. This causes acceleration in rate and efficient decomposition of waste with the development of more heat. The surface area plays a very significant role in the case of limited spaces (Finstein and Hogan, 1993).

6.5.5 Moisture

Moisture is a necessary component for the degradation of organic matter by the microbial community because of the thin film of water formed on particles in piles (Finstein and Morris, 1975). The optimum range for percentage of water should be 40%–60% of total organic mass. This amount ensures the biodegradability of that organic component. Exceeding or reducing moisture levels below the optimum range can hamper the process of composting (Schulze, 1964). For instance, if the moisture content is more than 60%, it will limit aeration because the soil will become saturated due to the closing of the pores of the particle. Due to high saturation, the pile may start to compact and degrade anaerobically, producing a foul odor (Schlegelmilch et al., 2005), while moisture below 40% will lead to a reduction in the growth of bacteria that results in, again, a damping/slowdown of the degradation process. The moisture of 60% is most favorable and suitable for efficient degradability of organic matter by microbes/bacteria (Rolle and Orsanic, 1964). The moisture content in compost should remain in a balanced way such that by squeezing it, one or two drops of water are expelled. This condition represents the best sample compost (Table 1).

6.6 Different types of composting methods

6.6.1 Indore method of composting

The Indore method of composting was introduced by A. Howard and Y.D. Wad (at the Institute of Plant Industry, Indore). The Indore method of composting includes the aerobic method (Howard, 1935). In this method, piles of dimension 3.5 × 2 × 1 m are made. At the base, the first layer will be of a dry straw/chopped wood of 12 in. thickness, which contains dry matter acting as the foundation layer, followed by the addition of a 6-in. green layer, which may contain wet organic material like vegetable, food, and agricultural waste, and other wet waste from municipal sources. A further layer of

Table 1 Requirement and suitable conditions for standard compost.

Air	Air needed for efficient breakdown of organic matter in aerobic composting
Moisture	Moist conditions are needed for the action of microorganisms on waste
Vegetable matter	Essential raw material for composting; the most preferred material
Worms	Worms digest or decompose the waste by forming worm casting and release of nutrients
Carbon:Nitrogen ratio (brown and green waste)	Required for proper and complete decomposition of organic waste
Microorganisms	Key component for decomposition of organic waste
Soldier flies	Support the further decomposition in compost piles by breaking down green compostables
Other beneficial bugs	Beetles, sowbugs, and bin mites are other useful bugs acting on rotten organic material to eat and excrete, which boost bacterial and fungal populations in compost

manure is added, which is formed by cow dung soaked with urine, and a soil layer is added with 5 cm thickness to cover the matter. The manure layer is rich in microflora and acts as the inoculum for composting. Delivery of material continues as mentioned previously, until the pile reaches a height of 5 ft. The pile is continuously turned with a frequency of every 21 days in a period for 3 months and left further for use as manure (Howard, 1935).

6.6.2 Bangalore method of composting

The Bangalore method of composting was introduced by Dr. C.N. Acharya at IISc Bangalore in 1939. This is an anaerobic method of composting that can be done in pits. The dimensions of the pit are usually 1 m deep, and the length and width can vary in size according to the availability of raw material. The raw material required is organic wet waste and nightsoil. A wet organic layer of 15 cm is added to the pit and another addition of nightsoil of 5 cm is made and covered. The procedure continues until the pit is full. After some time, the volume of the pit is reduced and so more material can be added. The process of addition of organic refuse material along with the nightsoil continues for 2 months and finally the pit is sealed with mud and left for 6–8 months for maturation. After this period, the compost is ready for use. The advantages of this type of composting include the lack of a requirement for labor for turning or regular moisture additions. Its method is not at all affected by any weather condition or moisture loss (Narayanasamy et al., 2009).

6.6.3 Coimbatore method of composting

This method was introduced by Manickam in 1967. This is an anaerobic method of composting that includes composting in pits. The size of the pits can be 1 m deep and 1.88 m wide with a varying length according to the availability of material. The materials required for this method are organic refuse, cow dung, and soil. Initially an area of 15–20 cm is added to the base of the pit and a further 5 cm of cow dung is laid upon it. Then it is sprinkled with water to attain a moisture of about 50%–60%. The layers are continued to be laid one after another until a height of 0.6 m is achieved. Then it is plastered with mud and left for a month. However, the decomposition takes place in an anaerobic condition in the sealed pit. Further the pit is opened, turned, and sprinkled with water to add moisture. The compost is ready within 4 months through this process. It is again a low-labor procedure with optimum maintenance cost (Narayanasamy et al., 2009).

6.6.4 NADEP composting method

This method was introduced by N.D. Pandharipande from Maharashtra. It is an aerobic method that can take place in a constructed rectangular tank. The requirement for this method is a rectangular tank of dimensions 10 × 5 × 3 ft. The tank can be constructed with bricks. For aeration the alternating bricks are removed, making a void space of 6 in. in the tank walls. The voids are created above 1 ft. from the ground surface. These voids ensure aeration and help in the aerobic decomposition of the organic matter. The raw material required for the NADEP method is an organic waste of agriculture or municipal origin, biogas slurry/cow dung, water, and fine soil. This method starts with the addition of 4–6 in. of straw dry matter as a matrix at the bottom of the tank. Then the organic waste of agriculture or municipal origin is added, up to 4–6 in., and a solution of cow dung/slurry is sprinkled on the waste for continued decomposition. The walls of the tank are also coated with a cow dung layer, which increases the contact surface of microbial active elements acting as inoculum for the degrading organic waste. The organic layer is covered by 2 in. of a fine soil layer, and the surface is moistened by sprinkling water. The further layering occurs vice versa to that described previously, till it reaches the surface of the tank. At last, it is sealed with mud for 1 month and later the mud layer is broken, and the composted material is mixed thoroughly and left again for 3–4 months to decompose properly. After 4 months, the tank is filled with compost ready for application (Acharya, 1950).

6.6.5 Large-scale composting

In this method of composting, the green matter is brought into an open area and chopped into a small particle size of 0.5–2 in. Then it is mixed to form a heap of 5 ft. The pile or heap is continuously aerated by turning at a frequency of 2–3 days up to a month. The turning leads to proper mixing of waste, which can be degraded by microbes. The moisture content needs to be continuously observed and water added using a sprinkler as per requirements. After 1 month to 1.5 months, the heap is ready for use. In the initial stage of composting, second to third day of the week usually the pile heats up, reaching a temperature of 160°F. The composting procedure continues with the help of a series of microbial succession. In the first week, if the pile does not sink or reduce in size, it means the carbon content is higher than the nitrogen content, which creates a pungent odor release of methane. To initiate the composting procedure and as per requirements, further green matter is added. The matter is added to balance the C:N ratio and moisture is added as per requirements. The abbreviation of the heap is necessary until the formation of a black humus-like mass with a neutral pH (Millner et al., 1977).

6.7 Microbes in composting

In today's world, urbanization is inevitable. With urbanization comes the problem of waste management. Although the rural areas also have waste generation, the scale of waste generated in urban spaces is much larger. The absence of proper disposal policies creates problems for the residents of that area. With the pace of urbanization going on in the world, even orthodox policies are falling short in the disposal of waste. That's because researchers globally are coming up with innovative ideas, such as using microbes involved in the decomposition of solid waste (Diaz et al., 1987; Savage et al., 1985). Composting of organic matter is a long-working technique for mankind to get rid of organic waste. Composting not only decomposes the organic waste, but it also converts it into a useful commodity.

6.8 Role of microorganisms in composting

These living organisms are the primary reason for the process of composting. With primary research on the composting process, the work has come a long way (Waksman et al., 1939). There are many aspects of composting, but we will focus here on bacteria that help in accelerating the process of composting. The diversity of microorganisms ensures a good composting process. During the process of composting, the effect of the physical environment and other factors should be kept in mind while dealing with the bacteria. The process of composting is started by the microorganisms and their presence changes the characteristics of the organic matter; being a living entity, they also need vital nutrients for their survival (Fuchs, 2010). Some bacteria such as cyanobacteria can make their own food, while others obtain it through the environment by parasitism or other ways. There are two types of processes: aerobic and anaerobic. In the aerobic process, by degradation of organic compounds microorganisms can produce carbon dioxide, water, and energy. This process takes place in the presence of oxygen, which is why it is called aerobic.

Natural waste for composting mainly consists of organic matter that is the direct substrate of microorganisms. When we talk about organic waste, it is generally of two types, or we can say that it is from two types of sources: one is a plant source and the second is an animal source. Globally, the main organic matter is of animal origin, but in Northern Europe it is of plant origin. Here, plant-source organic matter is used more, as compared to animal-source organic matter. In the case of an infectious disease outbreak, one of the most environmentally friendly and effective methods is to decompose the remains of human bodies, or to call it composting (Glanville et al., 2013). Likewise, slaughterhouse waste is a good candidate for composting; besides being nutrient-rich, it fulfills other criteria that help in composting, about which we will talk later (Franke-Whittle and Insam, 2013). One of the main problems with the plant-origin organic matter is that it is bulky as compared to animal-origin organic matter, so first it should be converted into smaller pieces so that the substrate comes in the vicinity of microorganisms, which will accentuate the process of composting. Shredded organic matter is easy to handle and mix. Hemicellulose, cellulose, protein, carbohydrates, and sugars are some of the compounds found in the feedstock. Microorganisms need food to produce energy for their survival, and these compounds act as a substrate for their energy production. The energy produced by these microorganisms is not fully utilized. Heat comes out as the byproduct of respiration, and this heat signature we cannot observe in natural conditions, because in an open environment it is dissipated (Ryckeboer et al., 2003). In the case of composting organic waste, a closed system like a pit or container is used, due to which heat cannot escape to the surroundings (Ryckeboer et al., 2003). There are many factors that affect the microbial composting of organic waste (Partanen et al., 2010).

6.9 Relation of C:N ratio and temperature

As we were discussing, microorganisms feed on organic matter for their carbon needs, but it is not the only thing they need; there are several other micronutrients that they need for their survival, such as sulfur, potassium, phosphorus, magnesium, and, above all, nitrogen, as it is vital for many metabolic pathways and processes like nucleotide and amino acid synthesis. Microorganisms use more nitrogen than plants. In some plants, the tissue C:N ratio is up to 100, while in bacteria it is around 10 (Diaz et al., 2011).

Low nitrogen supply slows down the composting process. This fact gives us the conclusion that for good composting an optimal C:N ratio is very important. Considering composting as microbial farming, there should be a balance of nutrients, as we balance in the case of plants the N:P:K ratio. The optimal scale of the C:N ratio ranges from 20 to 30. The best ratio for composting is considered to be 25 (Day and Shaw, 2001). Below 20, decomposition of the organic matter will be too fast, due to which the temperature will go up to the extent that it will cause ashing. The consequences of this will include the formation of ammonia, which will escape to the atmosphere. A higher value of the C:N ratio, above 35, will slow down the decomposition of organic matter to the extent that it will destabilize the final output or the compost. As a low C:N ratio will account for a significantly high temperature, the opposite situation also exists, where the temperature drops to a critical level such that it is not sufficient to destroy the plant and human pathogens.

In the process of composting, as a result of microbial respiration, carbon escapes in the form of CO_2. Consequently, the amount of carbon drops and C:N also drops, because the amount of nitrogen lost will be much lower than the loss of carbon (Ryckeboer et al., 2003). From 46% to 62% of carbon escapes, while only 19%–42% of nitrogen is lost during composting. So for a good composting process, a certain C:N level has to be maintained. As it keeps on changing, it has to be analyzed at regular intervals and should be balanced by supplying the required compound. Waste of plant origin usually has a high C:N ratio (40–45) so it has to be mixed with high nitrogen content material, such as poultry litter or food waste, to make it suitable for composting. Nitrogen is estimated as a whole by Kjeldahl N, while for estimation of carbon content the ash content is required (Golueke, 1973).

The equation for this calculation is

$$\%C = (100 - \%ash)/1.8.$$

While the C:N ratio is considered to be a benchmark for the composting process, there are various organic compounds whose biodegradability defy the set parameters of this ratio. If compost acquires a value of 15 in its final stages, then it is an indicator of compost maturity, as it is evident that the ratio will go down with the escape of carbon in the form of CO_2. A C: N ratio of 20 should be in the compost when it is applied to the soil, as it will aid in the release of nitrogen for the plants and prevent immobilization of nitrogen. As we have discussed earlier, that C:N ratio will go down eventually in the process of composting, so to achieve a ratio of 20 the initial ratio should be around the range of 30 (Maeda et al., 2011). Nitrogen is also lost in the process of composting in the form of NH_3. To retain nitrogen in the system, nitrifying bacteria are used; they convert NH_3 to NO_3, which is an intermediate between them. This process of oxidation needs oxygen (Maeda et al., 2011). This group of bacteria needs a moderate environment to flourish, so in the middle stages when the temperature is high these bacteria become fewer (Yang et al., 2020). There are many microorganisms that facilitate the process of composting, but they are mainly from two groups: bacteria and fungi.

6.10 Bacteria

In the early days of composting research, the bacteria were neglected as a topic of interest because fungi were more easily visible as compared to the bacteria, although bacteria were more useful than fungi in some types of wastes such as sewage sludge. As discussed earlier, temperature is one of the most important factors when dealing with bacteria. The optimal temperature for the growth of bacteria ranges somewhere between 20 and 45°C in general, and for the genus *Bacillus* in particular. It is not the case that bacteria cannot thrive above this temperature (Strom, 1985). Some bacteria, like *Bacillus stearothermophilus*, often thrive at a temperature above 65°C.

In bacteria also there are many groups that we will discuss in this chapter, such as actinobacteria. This group of bacteria thrives in slightly alkaline or neutral pH; a special characteristic of this group of bacteria is that they can decompose complex substrates. Many of these bacteria can grow at relatively high temperatures (50–60°C), so they are also known as thermophilic bacteria. A good oxygen supply and humid conditions pave the way for sound actinobacterial growth. Since this group of bacteria is good at decomposing complex organic compounds, they come into the picture when simple organic compounds are degraded by other bacteria and also the temperature is more than 45°C, since they are thermophilic. The actinobacterial phase is critical for successful cultivation. If the temperature of the compost goes above 70°C, then ammonia will be released, and that is why the 48°C temperature is kept throughout the process by altering air supply and moisture content. Microbial reassimilation of ammonia is the main aim of the process (Insam and De Bertoldi, 2007).

Some bacteria have abnormally high growth temperature and they thrive on organic waste; these are the *Thermus*|*Deinococcus* group of bacteria. These bacteria have an optimal range of temperature starting from 40°C and going as high as 80°C. As already stated, temperature plays a vital role in composting, and it can go up to 60–70°C; at this temperature normal bacteria cease their activity. Here this group of bacteria come into the picture. Earlier, these bacteria were only reported in geothermal sites, but now they have well adapted to the high-temperature phase of the composting process. *Hydrogenobacter*, which was only reported at geothermal sites, resembles that now found in the composting process. This group of bacteria is nonsporing and shows good growth at a temperature ranging from 60 to 80°C with an optimal temperature of around 70–75°C. They use sulfur or hydrogen as their substrate for energy production and use CO_2 for their organic matter synthesis (Beffa et al., 1996). In some cases, archaebacteria are found in compost, but usually they are found only in hypothermal vents (Wintzingerode et al., 1997). As part of the composting process, methane also forms in the pit. So, if research goes on, then we will also see some methanogenic archaea in compost (Cabanas-Vargas and Stentiford, 2006). Archaea are relatively low in number in compost, because their doubling time is higher than bacteria, so they cannot adjust to the changing environment of the composting process.

Bacteria also need vectors to proliferate. Worms do this job for the bacteria, which is why there are an increased number of bacteria in vermicomposting. These bacteria, to survive and meet their energy demands, increase the value of the final product of vermicomposting. Earthworms are a very good candidate for vectors, as they disperse soil microbes very efficiently

(Madsen and Alexander, 1982). They can change the microbial community present in the soil by acting as reservoirs for the microbes (Pedersen and Hendriksen, 1993). This is the reason bacteria are more prominent in the vermicomposting pits than in simple composting. Also, microbial communities in the vicinity of earthworm habitats are distinctly different from the rest of the surroundings (Steward et al., 1996). Although the microbial community inside earthworms is underreported, still we have some clue that gram-negative bacteria are present inside. We are not very far into the successful segregation and classification of these bacteria, but some researchers have reported *Aeromonas hydrophila* and *Vibrio* sp. in the gut of earthworms (Toyota and Kimura, 2000).

For the classification of microbial communities present in soil or different composts, we use BIOLOG plates based on substrate utilization. Besides the biological method, we also use the chemical method for the identification of microbial communities, as they have their advantages, such as speedy results. Gas chromatography-mass spectrometry (GC-MS) and gas chromatography are very commonly used for information about microbial metabolites and other biochemical structural compounds present in the cells of microorganisms. There are different markers (chemical or biochemical) that are very precisely used to detect given microorganisms (Brondz and Olsen, 1986). To distinguish living microorganisms from dead ones, phospholipid fatty acids (PLFAs) act as an indicator. PLFA tests and carbon source utilization methods are tools for the investigation of soil microorganisms. They are also used to study compost in the making, to get a picture of microbial community succession (Carpenter-Boggs et al., 1998).

6.10.1 Microbial diversity in vermicompost

Vermicompost hosts many microorganisms, such as gram-negative bacteria, stalked bacteria, facultative anaerobes, and anaerobic bacteria. Sewage sludge vermicompost has more microbial diversity than cow manure compost. *Cytophaga*, *Acetobacter*, *Pseudomonas*, and *Caulobacter* are present in vermicompost. Some other microorganisms that are present are *Bacillus* and *Clostridium* (Yi et al., 2012).

Gram-positive bacteria having high G + C content, such as the *Streptomyces*, *Micromonospora*, *Nocardia*, and *Maddiromycetes* groups, are present in vermicompost in considerable numbers. As compared to cow manure compost, sewage sludge compost has higher microbial diversity. One reason for this can be substrate inhibition due to the presence of excess organic matter.

Eisenia foetida has *Vibrio*, *Aeromonas*, *Bacillus*, and *Pseudomonas* in its intestine and that's why they are also present in vermicompost. *Woloinella succinogens* are found in the rumen, but its presence in compost is unexpected. After the sewage sludge vermicomposting process, the numbers of *Streptomyces*, *Vibrio*, *Aeromonas*, *Sphingobacterium*, and *Acetobacter* spiked notably. Some of these microorganisms can fix nitrogen, while others contribute to nitrogen metabolism, thus increasing the plant productivity when supplied by vermicompost (Domínguez et al., 2021).

Matter transformation through biochemical mechanisms and nutrient availability decides the functional groups of microorganism interaction. A method developed by Nikitin and Kunc (1988) was able to identify microbial communities in the compost (Nikitin and Kunc, 1988). High hydrolyzing capacity-containing microbes had a better shot at survival. *Maduromycetes* group microbes having a mycelial property, like *Nocardia*, spore-forming (*Bacillus*), and cellulolytic (*Cytophaga*) are some of the microorganisms having high hydrolyzing properties. The product of these microorganisms will become a substrate for *Woloinella*, *Vibrio*, *Clostridium*, and *Bacteroides* (fermenting organisms). Organic material produced by hydrolysis is the source for *Sphingomonas*, *Sphingobacterium*, *Pseudomonas*, and *Acetobacter* (Verkhovtseva et al., 2002).

Based upon the number of microorganisms, some are considered dominant. The exact definition does not exist, but if some microorganism reaches 30% of the total number of microorganisms present in the community, then it can be said that the given microbe is dominant (Alekhina et al., 1997). In the case of vermicomposting, if the data during analysis show that some specific microorganism touches the 10% mark, then we label it as a dominant microorganism. *Bacteroides* (37%) and *Pseudomonas* (27%) are two dominant genera of microorganisms. In the primary organic matter, 64% of biomass comprises dominant microorganisms and they are the ones that form the basis of structural and functional interaction. Anaerobic fermenting microbes (*Bacteroides*) and aerobic microbes having high hydrolysis capacity (*Pseudomonas*) use sewage sludge organic matter firsthand, enhancing the presence of these bacteria in sewage sludge. Out of the two processes given here, the anaerobic process is dominant. Certain genera are dominant in vermicompost of sewage sludge. Four main groups have been identified: *Acetobacter* (20%), *Wolinella* (12%), *Pseudomonas* (11%), and *Bacteroides* (10%), constituting a whopping 61% in total. Out of these four genera, *Wolinella* and *Acetobacter* are known for their efficient nitrogen metabolism capabilities (Verkhovtseva et al., 2002). In comparison to sewage sludge vermicompost, cow manure vermicompost has only three genera that can be considered dominant: obligate and facultative anaerobes *Spirillum* and *Bdellovibrio* (17%), commonly found in wastewater. *Clostridium* (17%) and *Bacteroides* (15%) are commonly present in rumen. Due to the presence of these facultative and obligate anaerobes the process is predominantly anaerobic in cow manure vermicomposting (Verkhovtseva et al., 2002).

6.11 Fungi

The microbial community drives the conversion and decomposition process during composting, and especially fungi have a remarkable role. In the initial phase, bacteria compete with fungi for whatever substrate is easily available. Bacteria win this battle of survival because the maximum specific growth rate is of a higher degree in the case of bacteria by one order of magnitude. Fungi cannot keep pace with bacteria, so bacteria outnumber them easily. One more reason for fungi being outcompeted is the availability of oxygen. Fungi need an abundant supply of oxygen and even if we aerate the system, hypoxic conditions will occur, due to which fungi will find it difficult to survive (Dobrovol'skaya et al., 1997). Even higher temperatures also go against the survival of fungi, because they have a lower level of thermotolerance. In the thermophilic phase, fungi are wiped out. Fungi perform exceptionally well in lignin and cellulose-rich substrate mediums and emerge as one of the most important microorganisms. The water potential goes down in later stages, which acts as an advantage for fungi (Griffin, 1985). Griffin and other researchers have reported a 2:1 biomass ratio of fungi to bacteria in some cases (Wiegant, 1991).

Fungi can utilize several carbon sources, especially lignin and cellulose (Miller, 2020). Fungal application is not only limited to organic waste, but it is also now used for bioremediation of soil, because it also helps in the biodegradation of other complex and hazardous materials (Minussi et al., 2001). Fungi, on the one hand, help in the bioremediation of hazardous materials but, on the other hand, sometimes have proven hazardous to plants, humans, and animals (Summerbell et al., 1994). Thus it is very important to have an eye on fungal diversity. The succession of thermophilic fungi is well reported and documented in thermogenic composting. This information is valid for experimental compost received by the standardized and environmentally regulated processes. Industrial composting is a different ball game. In the case of vermicomposting, fungi have a different role, as discussed earlier (Masciandaro et al., 2000). Earthworms perform the very important task of removing pathogens, which includes bacteria and other microorganisms. Fungi lower the chance of plant disease and have a profound effect on soil fertility. They also have a profound effect on fungal diversity. Earthworms can affect the presence of fungal communities by interfering in spore formation and by providing microsites that might be useful for fungal growth. Since there is ambiguous information about the mesophilic phase of fungi, it should be properly studied to see its hygienic aspects. Changing source material and composting methods greatly affect the fungal diversity, so there should be constant monitoring for the fungal diversity in compost to decide its field of application and quality.

There are two main parameters that decide the usefulness of fungi. One is thermotolerance and the second is to degrade different types of organic waste. A lot of thermotolerant species have been identified and isolated from compost. Around 66 species have been identified that are commonly present in the composting material, which includes *Thermomyces*, *Penicillium*, *Malbranchea*, *Cladosporium*, *Aspergillus*, *and Acremonium*. A significant fungal load has been observed in vermicompost. This can be the consequence of thermophilic preconditioning, which in turn could give a boost to the proliferation and development of thermophilic species. These species were found well equipped to stand with high temperatures and can survive during the initial and final phases of the compost product (Fuentes, 2014). *Scedosporium* states of *Aspergillus fumigatus* and *Pseudallescheria boydii* are present in good amounts in both the composts. The presence of these two species is taken seriously because they are potent pathogens for humans and animals.

Besides these species, *Scopulariopsis* and *Chrysosporium* are found frequently in vermicompost. These two have keratinolytic activity, making them perfect subjects to parasitize and invade cornified tissues (Odds, 1991). This finding contradicts earlier research according to which earthworms can hamper the growth of *Chrysosporium* quantitatively and qualitatively. The degree of effect that earthworms have on pathogenic fungi can only be studied when the composting medium or matter is the same. Keratinolytic species thrive where there is an abundance of their substrate, i.e., keratin. Since our vermicompost has a sizable quantity of hair, nails, and animal skin, it might be a reason for the proliferation of keratinolytic species. All these studies show it is very important to monitor fungi in vermicompost and to keep track of pathogenicity caused by these fungi, so that we can manage and have suggestions for the users and workers using these composts (Marchisio, 2000).

Ascomycetes and *Basidiomycetes* were found abundant in vermicompost, although *Basidiomycetes* were a little fewer in number and variations. This result could also be the consequence of different matter compositions in two composts. One more possible reason for this can be the fast feeding of earthworms on rapidly growing fungus (such as mitosporic fungi and zygomycetes), making slow proliferating fungi (ascomycetes and basidiomycetes) less competitive, thus providing these fungi with an added advantage. *Corynascus sepedomium* and *Pseudallescheria boydii* are found in vermicompost in a perfect state, and earthworms are the main reason behind this. They protect these fungi from unfavorable microniches and stress, directly or indirectly. Sterile mycelia were prominent in vermicompost.

The relationship of dematiaceous fungi and earthworms is unpredictable. Some reports suggest earthworms have a preference for these fungi. But some reports suggest that intake of these fungi is restricted by some chemical protective layers, such as melanin present in hyphal walls. Species diversity of zygomycetes (particularly *Mortierella* spp.) was less prominent in vermicompost. One more thing to stress is that both composts have a lower number of potential phytopathogens (Turick et al., 2002). To support the preceding statement that phytotoxicity is absent in these composts, it is visible by the results of root elongation, seed germination, and vegetative tests (shoot height, root and shoot dry weight, and various other growth parameters) (Turick et al., 2002). These composts can be suggested as fertilizers for widespread use (Cuevas, 1997). Functional analysis is a viable option to determine applications and potentials of compost other than the characterization of fungal diversity in compost. Some initial studies have indicated that the presence of diverse fungal species reflects unconventional metabolic potential. Fungal strains found in compost are now a topic of interest for other bioremediation processes.

6.12 Conclusion

This chapter illustrates the entire specific requirements for degradation of organic waste by the microbial community. It discusses at length and separately the roles of bacteria and fungi in composting. It is also explained here that different strains of bacteria and fungi in various combinations can be a much better option as an inoculum for the composting of organic matter. It also summarizes various types of composting methods and their settings. It describes the optimum prevailing conditions for compost to provide useful output.

References

Acharya, C.N., 1950. Preparation of Compost Manure from Town Wastes. Indian Council of Agricultural Research, Ministry of Agriculture; Bulletin No. 60, Calcutta.

Alekhina, L.K., Nevskaya, D.V., Dobrovol'skaya, T.G., Zvyagintsev, D.G., 1997. Bacterial diversity in forest soils: a successional analysis. In: Microbiology-Aibs-C/C of Mikrobiologiia. vol. 66, pp. 466–469.

Amlinger, F., Peyr, S., Cuhls, C., 2008. Greenhouse gas emissions from composting and mechanical biological treatment. Waste Manag. Res. 26 (1), 47–60.

Beffa, T., Blanc, M., Lyon, P.F., Vogt, G., Marchiani, M., Fischer, J.L., Aragno, M., 1996. Isolation of thermus strains from hot composts (60 to 80 degrees C). Appl. Environ. Microbiol. 62 (5), 1723–1727.

Bhat, S.A., Singh, S., Singh, J., Kumar, S., Bhawana, V.A.P., 2018. Bioremediation and detoxification of industrial wastes by earthworms: vermicompost as powerful crop nutrient in sustainable agriculture. Bioresour. Technol. 252 (2018), 172–179.

Bidlingmaier, W., 1996. Odour emissions from composting plants. In: de Bertoldi, M., Sequi, P., Lemmes, B., Papi, T. (Eds.), The Science of Composting, Part 1. Blackie Academic and Professional, London, UK.

Brinton, W.F., 2000. Compost Quality Standards & Guidelines, first ed. Woods End Research Laboratory, New York.

Brondz, I., Olsen, I., 1986. Microbial chemotaxonomy: chromatography, electrophoresis and relevant profiling techniques. J. Chromatogr. B Biomed. Sci. Appl. 379, 367–411.

Browne, C.A., 1933. The spontaneous heating and ignition of hay and other agricultural products. Science 77, 223–229.

Cabanas-Vargas, D.D., Stentiford, E.I., 2006. Oxygen and CO2 profiles and methane formation during the maturation phase of composting. Compost. Sci. Util. 14 (2), 86–89.

Carpenter-Boggs, L., Kennedy, A.C., Reganold, J.P., 1998. Use of phospholipid fatty acids and carbon source utilization patterns to track microbial community succession in developing compost. Appl. Environ. Microbiol. 64 (10), 4062–4064.

Cuevas, V.C., 1997. Rapid Composting Technology in the Philippines: Its Role in Producing Good-Quality Organic Fertilizers. Food & Fertilizer Technology Center.

Day, M., Shaw, K., 2001. Biological, chemical and physical processes of composting. In: Compost Utilization in Horticultural Cropping Systems, pp. 17–50.

de Bertoldi, M., 1999. The control of the process and compost quality. In: Bidlingmaier, W., de Bertoldi, M., Diaz, L.F., Papadimitriou, E.K. (Eds.), Organic Recovery & Biological Treatment, ORBIT 99. Rombos Verlag, Berlin, Germany.

de Bertoldi, M., Zucconi, F., 1986. Composting of organic residues. In: Wise, D.L. (Ed.), Bioenvironmental Systems. vol. III. CRC Press, Boca Raton, FL, pp. 95–141.

de Bertoldi, M., Ferranti, M.P., L'Hermite, P., Zucconi, F., 1987. Compost Production, Quality and Use. Elsevier Applied Science, London and New York.

de Bertoldi, M., Rutili, A., Citterio, B., Civilini, M., 1988. Composting management: a new process control through O2 feedback. Waste Manage. Res. 6, 239–259.

Diaz, L.F., Savage, G.M., Golueke, C.G., 1987. An integrated resource recovery system. Biocycle 28 (10), 47–52.

Diaz, L.F., De Bertoldi, M., Bidlingmaier, W. (Eds.), 2011. Compost Science and Technology. Elsevier.

Dobrovol'skaya, T.G., Chernov, I.Y., Zvyagintsev, D.G., 1997. Characterizing the structure of bacterial communities. Microbiologia 66, 408–414.

Domínguez, J., Aira, M., Crandall, K.A., Pérez-Losada, M., 2021. Earthworms drastically change fungal and bacterial communities during vermicomposting of sewage sludge. Sci. Rep. 11 (1), 1–10.

Finstein, M.S., Hogan, J.A., 1993. Integration of composting process microbiology, facility structure and decision-making. In: Hoitink, H.A.J., Keener, H.M. (Eds.), Science and Engineering of Composting. The Ohio State University, USA.

Finstein, M.S., Morris, M.L., 1975. Microbiology of municipal solid waste composting. Adv. Appl. Microbiol. 19, 113–151.

Finstein, M.S., Miller, F.C., Strom, P.F., MacGregor, S.T., Psarianos, K.M., 1983. Composting ecosystem management for waste treatment. Biotechnology 1, 347–353.

Franke-Whittle, I.H., Insam, H., 2013. Treatment alternatives of slaughterhouse wastes, and their effect on the inactivation of different pathogens: a review. Crit. Rev. Microbiol. 39 (2), 139–151.

Fuchs, J.G., 2010. Interactions between beneficial and harmful microorganisms: from the composting process to compost application. In: Microbes at Work. Springer, Berlin, Heidelberg, pp. 213–229.

Fuentes, A.L., 2014. Fungal Communities in Composting Systems Revealed by DNA-Based Molecular Methods. The University of Manchester, United Kingdom.

Glanville, T.D., Ahn, H.K., Richard, T.L., Harmon, J.D., Reynolds, D.L., Akinc, S., 2013. Effect of envelope material on biosecurity during emergency bovine mortality composting. Bioresour. Technol. 130, 543–551.

Golueke, C.G., 1950. Composting for Disposal of Organic Refuse. University of California, Berkeley, CA, Sanitary Engineering Research Project. Technical Bulletin No. 1.

Golueke, C.G., 1973. Composting: A Study of the Process and Its Principles. Rodale Press.

Golueke, C.G., 1977. Biological Reclamation of Solid Waste. Rodale Press, Inc., Emmaus, PA.

Griffin, D.M., 1985. A comparison of the roles of bacteria and fungi. In: Bacteria in Nature. Springer, Boston, MA, pp. 221–255.

Howard, A., 1935. The manufacture of humus by the Indore process. J. Roy. Soc. Arts 84, 25.

Insam, H., De Bertoldi, M., 2007. Microbiology of the composting process. In: Waste Management Series. vol. 8. Elsevier, pp. 25–48.

Madsen, E.L., Alexander, M., 1982. Transport of Rhizobium and Pseudomonas through soil. Soil Sci. Soc. Am. J. 46 (3), 557–560.

Maeda, K., Hanajima, D., Toyoda, S., Yoshida, N., Morioka, R., Osada, T., 2011. Microbiology of nitrogen cycle in animal manure compost. Microb. Biotechnol. 4 (6), 700–709.

Marchisio, V.F., 2000. Keratinophilic fungi: their role in nature and degradation of keratinic substrates. In: Biology of Dermatophytes and Other Keratinophilic Fungi. vol. 17, pp. 86–92.

Masciandaro, G., Ceccanti, B., Garcia, C., 2000. "In situ" vermicomposting of biological sludges and impacts on soil quality. Soil Biol. Biochem. 32 (7), 1015–1024.

Miller, F.C., 2020. Composting of municipal solid waste and its components. In: Microbiology of Solid Waste. CRC Press, pp. 115–154.

Millner, P.D., Marsh, P.B., Snowden, R.B., Parr, J.F., 1977. Occurrence of Aspergillus fumigatus during composting of sewage sludge. Appl. Environ. Microbiol. 34 (6), 765–772. https://doi.org/10.1128/aem.34.6.765-772.1977. PMID: 339835; PMCID: PMC242745.

Minussi, R.C., De Moraes, S.G., Pastore, G.M., Duran, N., 2001. Biodecolorization screening of synthetic dyes by four white-rot fungi in a solid medium: possible role of siderophores. Lett. Appl. Microbiol. 33 (1), 21–25.

Narayanasamy, G., Arora, B.R., Biswas, D.R., Khanna, S.S., 2009. Fertilizers, manures and biofertilizers. In: Goswami, N.N., Rattan, R.K., Dev, G., Narayanasamy, G., Das, D.K., Sanyal, S.K., et al. (Eds.), Fundamentals of Soil Science. Indian Society of Soil Science, New Delhi, pp. 579–622.

Nikitin, D.I., Kunc, F.R.A.N.T.I.Š.E.K., 1988. Introduction to structure of microbial soil associations and some mechanisms of their autoregulation. In: Developments in Agricultural and Managed-Forest Ecology. vol. 17, pp. 159–190.

NRAES, 1992. On-farm composting. In: Rynk, R. (Ed.), Natural Resource, Agriculture, and Engineering Service, Cooperative Extension. NRAES, Ithaka, NY.

Odds, F.C., 1991. Potential for penetration of passive barriers to fungal invasion in humans. In: The Fungal Spore and Disease Initiation in Plants and Animals. Springer, Boston, MA, pp. 287–295.

Partanen, P., Hultman, J., Paulin, L., Auvinen, P., Romantschuk, M., 2010. Bacterial diversity at different stages of the composting process. BMC Microbiol. 10 (1), 1–11.

Pedersen, J.C., Hendriksen, N.B., 1993. Effect of passage through the intestinal tract of detritivore earthworms (Lumbricus spp.) on the number of selected gram-negative and total bacteria. Biol. Fertil. Soils 16 (3), 227–232.

Peters, S., Koschinsky, S., Schwieger, F., Tebbe, C.C., 2000. Succession of microbial communities during hot composting as detected by PCR-single-strand-conformation polymorphism-based genetic profiles of small-subunit rRNA genes. Appl. Environ. Microbiol. 66, 930–936.

Rodale, J.I. (Ed.), 1943. Compost – and How to Make It. Rodale Press, Inc., Emmaus, PA.

Rolle, G., Orsanic, E., 1964. A New Method of Determining Decomposable and Resistant Organic Matter in Refuse and Refuse Compost. International Research Group on Refuse Disposal, Bulletin 21.

Ryckeboer, J., Mergaert, J., Vaes, K., Klammer, S., De Clercq, D., Coosemans, J., Insam, H., Swings, J., 2003. A survey of bacteria and fungi occurring during composting and self-heating processes. Ann. Microbiol. 53 (4), 349–410.

Savage, G.M., Diaz, L.F., Golueke, C.G., 1985. Biological treatment of organic toxic wastes. Biocycle 26 (7), 30–34.

Schlegelmilch, M., Streese, J., Biedermann, W., Herold, T., Stegmann, R., 2005. Odour control at biowaste composting facilities. Waste Manag. 25 (9), 917–927.

Schulze, K.F., 1964. Relationship between moisture content and activity of finished compost. Compost. Sci. 2, 32.

Stentiford, E.I., 1993. Diversity of composting systems. In: Hoitink, H.A.J., Keener, H.M. (Eds.), Science and Engineering of Composting. The Ohio State University, USA.

Steward, C.C., Nold, S.C., Ringelberg, D.B., White, D.C., Lovell, C.R., 1996. Microbial biomass and community structures in the burrows of bromophenol producing and non-producing marine worms and surrounding sediments. Mar. Ecol. Prog. Ser. 133, 149–165.

Strom, P.F., 1985. Effect of temperature on bacterial species diversity in thermophilic solid-waste composting. Appl. Environ. Microbiol. 50 (4), 899–905.

Summerbell, R.C., Staib, F., Ahearn, D.G., Ando, M., Ajello, L., Crow, S.A., Fung, D., Gregor, T., Noble, J., Price, D.L., Simmons, R.B., 1994. Household hyphomycetes and other indoor fungi. J. Med. Vet. Mycol. 32 (sup 1), 277–286.

Toyota, K., Kimura, M., 2000. Microbial community indigenous to the earthworm Eisenia foetida. Biol. Fertil. Soils 31 (3), 187–190.

Truman, H.A., 1949. Disposal of Wastes—Composting. The New Zealand Institution of Engineers, Wellington, New Zealand.

Turick, C.E., Tisa, L.S., Caccavo Jr., F., 2002. Melanin production and use as a soluble electron shuttle for Fe (III) oxide reduction and as a terminal electron acceptor by Shewanella algae BrY. Appl. Environ. Microbiol. 68 (5), 2436–2444.

Verkhovtseva, N.V., Osipov, G.A., Bolysheva, T.N., Kasatikov, V.A., Kuzmina, N.V., Antsiferova, E.J., Alexeeva, A.S., 2002. Comparative investigation of vermicompost microbial communities. In: Microbiology of Composting. Springer, Berlin, Heidelberg, pp. 99–108.

Waksman, S.A., Cordon, T.C., Hulpoi, N.J.S.S., 1939. Influence of temperature upon the microbiological population and decomposition processes in composts of stable manure. Soil Sci. 47 (2), 83–114.

Walker, J.M., Willson, G.B., 1973. Composting sewage sludge: part I. Why? Compost. Sci. Land Util. 14 (4), 10–12.

Wiegant, W.M., 1991. A simple method to estimate the biomass of thermophilic fungi in composts. Biotechnol. Tech. 5 (6), 421–426.

Willson, G.B., Walker, J.M., 1973. Composting sewage sludge: part I. How? Compost. Sci. Land Util. 14 (5), 30–31.

Willson, G.B., Parr, J.F., Epstein, E., Marsh, P.B., Chaney, R.L., Colacicco, D., Burge, W. D., Sikora, L.J., Tester, C.F., Hornick, S., Ryan, J.A., 1980. Manual for Composting Sewage Sludge by the Beltsville Aerated-Pile Method. U.S. Environmental Protection Agency, Washington, DC. EPA/600/8-80/022.

Wintzingerode, F.V., Göbel, U.B., Stackebrandt, E., 1997. Determination of microbial diversity in environmental samples: pitfalls of PCR-based rRNA analysis. FEMS Microbiol. Rev. 21 (3), 213–229.

Wylie, J.S., 1957. Progress report on high-rate composting studies. Eng. Bull. Proc. (12th Industrial Waste Conf., Series No. 94).

Yang, Y., Awasthi, M.K., Wu, L., Yan, Y., Lv, J., 2020. Microbial driving mechanism of biochar and bean dregs on NH3 and N2O emissions during composting. Bioresour. Technol. 315, 123829.

Yi, J., Zheng, R., Li, F., Chao, Z., Deng, C.Y., Wu, J., 2012. Temporal and spatial distribution of Bacillus and Clostridium histolyticum in swine manure composting by fluorescent in situ hybridization (FISH). Appl. Microbiol. Biotechnol. 93 (6), 2625–2632.

Zucconi, F., Forte, M., Monaco, A., de Bertoldi, M., 1981. Biological evaluation of compost maturity. Biocycle 22 (4), 27–29.

CHAPTER 7

Microbial inoculants for the management of herbicide toxicity in plants

Adriana Heiss Siqueira and Rafaella Costa Bonugli Santos
Federal University of Latin American Integration (UNILA), Institute Latin American of Nature and Life Sciences (ILACNV), Interdisciplinary Center of Life Sciences (CICV), Foz do Iguaçu, PR, Brazil

7.1 Introduction

Herbicides, substances used to control undesired plants, are applied all over the world to protect plants from weeds. However, their application also causes toxicity to the target crop, which negatively affects the growth and development of plants. In addition, the intensification of agricultural practices, such as the use of herbicides, has been increasing soil degradation through the loss of organic matter. A study by Nie et al. (2019) showed that soil erosion caused by land preparation for agriculture leads to variations in carbon and nitrogen stocks. This change alters soil microbial composition and disrupts soil fertility and other physicochemical soil properties. Consequently, it results in weed competition. Henceforth, to control weeds and, subsequently, to augment crop production, herbicides are regularly applied, resulting in an increase in the amount of herbicides used each time (Shahid et al., 2018). This is a continuous closed circle: degradation resulting from agricultural practices that cause soil alteration, which results in increased negative effects for the plants, which results in the application of herbicides, which results in herbicide resistance, which results in further soil degradation and herbicide toxicity in the plants (Fig. 1). The effects caused by the excessive and disorderly use of pesticides are harmful to the entire ecosystem, as they cause toxicity not only in the soil but also in water and air.

7.2 Herbicide toxicity in plants

The uptake of herbicides by plants occurs through the xylem, reaching the internal parts and generating poisonous effects on the soil and the plant, such as high production of reactive oxygen species (ROS), stress oxidation, DNA

Microbial Inoculants
https://doi.org/10.1016/B978-0-323-99043-1.00011-6

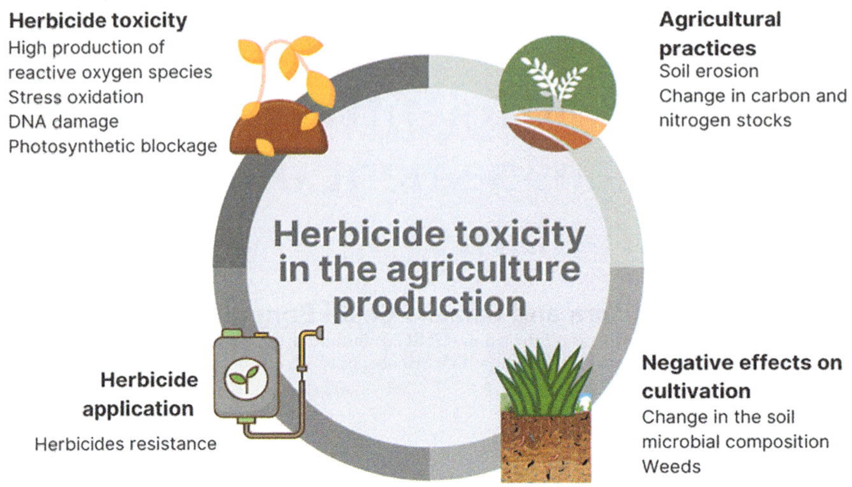

Fig. 1 A continuous closed circle of herbicide toxicity in agricultural production.

damage, photosynthetic blockage, and necrosis until the leaves twist, thus causing plant death (Alengebawy et al., 2021). When a herbicide is used, a toxic effect on the unwanted plant is expected, but the target crop is also affected. According to Sharma et al. (2019a), toxicity depends upon many factors such as use of pesticides, rate of application, spraying technique, climate conditions, organization of flora, humidity, and soil properties such as moisture, temperature, pH, texture, and microbial activity. In 2018, the same research group reviewed the undesired toxic effects of pesticides on crops, which are chlorosis, necrosis, and vein discoloration, leading to retarded growth and development by reducing photosynthetic efficiency and nitrogen and carbon metabolism; inhibiting photosynthesis by negatively affecting the plant photosystems; reducing root length, shoot length, fresh weight, and protein content; degrading the chlorophyll pigment; and enhancing levels of H_2O_2 among other effects, especially in the process of photosynthesis (Sharma et al., 2018).

Biotechnology has been central to the acceleration of crop improvement in herbicide application. The commercial transgenic crop "Flavr Savr," a variety of tomato, is the first example of the introduction of genes for herbicide tolerance (HT) in 1996. HT crops were created to reduce production costs and the environmental impact of using herbicides to control weeds (Mall et al., 2019). On the other hand, according to Schütte et al. (2017), the analyses' data and experiences do not support statements that herbicide-resistant crops provide consistently better yields than do conventional crops or that they reduce herbicide amounts. They rather show that the adoption of herbicide-resistant crops impacts agronomy, agricultural practice, and weed management and contributes to biodiversity loss in several ways. The adoption of herbicide-resistant crops has reduced crop rotation and favored weed management that is solely based on the use of herbicides. For example, continuous herbicide resistance cropping and the intensive use of glyphosate over the last 20 years have led to the appearance of at least 34 glyphosate-resistant weed species worldwide. Glyphosate-based herbicide products are the most widely used broad-spectrum herbicides in the world for postemergent weed control. However, there is a relationship between extensive glyphosate application and adverse non-target effects in agroecosystems, and the most significant among these concerns are persistence in the environment, effects on crop health, and interactions with crop nutrition (Kanissery et al., 2019).

In a plant, resistance may occur naturally or may be induced by techniques such as genetic engineering or selection of variants produced by tissue culture or mutagenesis (Holt, 2013). Herbicide resistance represents an extreme shift in weed species composition caused by the selection of plants possessing a gene or genes for resistance to a particular herbicide within a species that was formerly susceptible (Powles and Holtum, 1994). To reduce the impact of resistance, crop rotation is recommended; however, it is often performed incorrectly or does not prevent resistance (Schütte et al., 2017). Green (2014) indicated that new herbicides may not be commercialized due to development costs and the challenge of finding substances that are effective against weeds and with environmental and toxicological safety.

In this sense, when we talk about herbicide toxicity in plants, we are not only dealing with the biological effect on the plant but also with the whole biotic and abiotic chain involved in the application of the pesticide. Schütte et al. (2017) separate this process into indirect effects, which include the application of a broad-spectrum herbicide during crop growth and its impacts on weed abundance and diversity, and direct impacts, related to the toxicity of the herbicide, residues, and breakdown products. Direct effects promote herbicide resistance development.

In the application of a broad-spectrum herbicide during crop growth and its impacts on weed abundance and diversity, it is important to note that alterations in community dynamics can be caused by the initial use of the herbicide, which need not include resistance evolution.

Agriculture relies on ecosystem functions and services and on biodiversity, including pollination, biological pest control, maintenance of soil structure and fertility, nutrient cycling, and hydrological services (Schütte et al., 2017), so it is important to strengthen the strategies of management of herbicide toxicity. There are some works related to management to prevent or reduce the effect of herbicide resistance; however, based on the earlier study, resistance is not the only negative effect of the herbicide. The research study by Perotti et al. (2020) integrates agronomic and molecular knowledge of herbicide resistance. It covers not only the genetic basis of the most relevant resistance mechanisms but also the strengths and weaknesses of traditional and forthcoming agricultural practices. Schütte et al. (2017) highlight that diversification practices, such as cover crops, mixed cropping, intercropping, and agroforestry, help retain soil and soil moisture better than does intensive cropping and improve resiliency to climate disasters and thus support the structures of the agroecosystem that provide ecosystem services.

However, commercial agricultural cultivation cannot instantly follow this demand. We need practical strategies that can be applied to the current agricultural system. Microbes and their products play a key role in complementing chemical fertilizers and plant protection chemicals by eliciting defense mechanisms in crop plants, thereby providing immunity and resistance against diverse stresses. Among the different environmental technologies used to mitigate climate change, several microbiological interventions are promising (Velmourougane et al., 2022).

7.3 Microbial inoculants

Over the past decade, the agricultural sectors around the world have shown renewed interest in the use of microbial inoculants to enhance soil and plant health. Inoculants are products that have in their composition living microorganisms capable of benefiting the development of different plant species (Santos et al., 2019).

There is an interchangeable and confusing use of terms such as "inoculants," "bioinoculants," "biofertilizers," and "bioalleviators." Owen et al. (2015) propose a hierarchical classification in which each is a separate and exclusive group under a generalized classification of "bioresources." Bioresources can be any organic material applied to the soil to improve soil quality, nutrient supply, and plant growth. In summary, bio-softening are raw organic additives directly recycled into the soil, e.g., activated charcoal, and may contain stimulating microorganisms that provide a secondary supply of nutrients; biofertilizers are products that can contain micro- and macronutrients for plants (supply of primary nutrients) or specific organic components that, directly or indirectly, stimulate microbial activity; and bioinoculants are individual strains or consortia of microbes added to accelerate microbial decomposition and nutrient release (Owen et al., 2015).

These microbial inoculants can be categorized as bacteria and fungi, followed by subcategories such as intracellular and extracellular for bacteria, root-associated fungi (RAF), ectomycorrhizas (ECMs), and arbuscular mycorrhizas (AMs) (Gray and Smith, 2005; Owen et al., 2015; Patil and Solanki, 2016). According to Patil and Solanki (2016), bioinoculums are an additional constituent of soil and crop rotation, organic amendments, tillage protection, utilization of crop waste, soil fertility restoration, and management of pathogens; these processes can be extensively used for maintaining the sustainability of crop productions and can, by different methods, assist in the management of herbicide toxicity.

7.4 Management of herbicide toxicity caused by microbial inoculants

To circumvent the problems of herbicide toxicity caused by microbial inoculants, some paths are proposed in the literature, but, specifically, we identify two possibilities: plant growth-promoting microbes (PGPMs) that include some growth-promoting *Rhizobacteria* (PGPR) and soil and microbiota recovery. For PGPMs, the prevalence of bacteria as microbiological inoculants is remarkable. On the other hand, the use of fungi is more usual in the expression or by the name "soil and microbiota recovery." In the end, regardless of the name or expression, the objective is to recover the microenvironment, promoting the improvement of plant development and thus the management of herbicide toxicity.

These bioinoculants are mainly targeted for their effect on the rhizosphere. Unfortunately, there is still an incorrect use in the literature that the rhizosphere refers to the bacterial microbiota, and with that, the expression "*Rhizobacteria.*" In the rhizosphere, many interactions between microorganisms and invertebrates occur, affecting the biogeochemical cycles and the growth and tolerance of plants against biotic or abiotic stress (Fuentes et al., 2020).

We need to reinforce that for application in the management of herbicide toxicity, both the bacterial and fungal communities of the rhizosphere are important, including the association between groups (consortia). Fuentes et al. (2020), after reviewing, detail that rhizospheric microorganisms can promote plant growth by both direct and indirect mechanisms. For direct mechanisms, the most relevant are induction of systemic acquired resistance (SAR) in the plant and production of phytohormones such as auxins and gibberellins. On the other hand, indirect mechanisms include production of siderophores, production of low-molecular-weight organic acids, mineralization of organic matter to make nutrients available for plants, and protection against phytopathogens, among others. Under drought stress, changes in the structure of the microbial communities associated with the roots have been observed, indicating a selective pressure determined by the plant on the rhizospheric microbiome.

Patil and Solanki (2016) listed PGPMs involved in rhizoremediation; for bacteria, the authors highlight the genera *Arthrobacter* sp., *Bacillus* sp., *Enterobacter* sp., *Micrococcus* sp., *Pseudomonas* sp., *Rhodococcus* sp., *Sphingobium* sp., *Sphingomonas* sp., and *Streptomyces* sp. For fungi, the genera *Trichoderma* sp., *Trametes* sp., *Phanerochaete* sp., *Aspergillus* sp., and *Bjerkandera* sp. are highlighted. These PGPMs were used especially in the management of pesticides, such as atrazine, dimethoate, pentachlorophenol, hexachlorocyclohexane, chlorpyrifos, some fungicides, polycyclic aromatic hydrocarbons, and for metal treatment. Today, many other microorganisms have shown potential for use as inoculants in pesticide management, including arbuscular mycorrhizal fungi, ectomycorrhizas, yeasts, and *Actinomycetes* (Naik et al., 2019; Lopes et al., 2021; Soumare et al., 2021).

Regarding the main effects of the use of PGPMs to remediate organophosphate pesticides, Sundari et al. (2019) highlight the increased availability of nutrients, biomolecules like plant growth regulators, siderophores, trehalose, exopolysaccharides, volatile organic compounds (VOCs), and antibiotics, resulting in the modification of soil dynamics. However, despite countless works about PGPM effects, only limited data available specific to the action on pesticide toxicity are available.

Pesticides are metabolized by plants through enzyme-mediated pathways. This detoxification system occurs by a three-phased process, and, in the end, less toxic pesticide conjugates are carried to vacuoles/apoplasts or bind to cell walls/lignins (Sharma et al., 2019b). Plant growth regulators, like those produced by an association with PGPMs, play a fundamental role in perceiving and signaling varied stressful conditions, including pesticide toxicity and activation of various stress signaling pathways in response to pesticide toxicity (Ali and Baek, 2020). The main plant growth regulators used in recovery of plants under the herbicide glyphosate are cytokinin and salicylic acid. Salicylic acid is also involved as a plant growth regulator for paraquat and napropamide herbicides. For polychlorinated biphenyl present in some herbicides and for terbutryn, the plant growth regulator identification is 24-epibrassinolide (EBL) (Jan et al., 2020).

7.5 Microbial inoculants and the indigenous soil microbial community for the management of herbicide toxicity

Despite countless works on PGPM effects, only limited data are available on pesticide soil biodegradation by PGPM strains and their effects on microbial growth and the indigenous soil microbial community (Gupta and Pandey, 2019; Nathiya et al., 2020). So, in the use of microbial inoculation for the management of herbicide toxicity, we propose the experimental design listed in Fig. 2.

The selection of microbial strains can start from potential microorganisms obtained from the microbial collection or by isolation and identification of PGPMs. The indigenous soil microbial community has gained

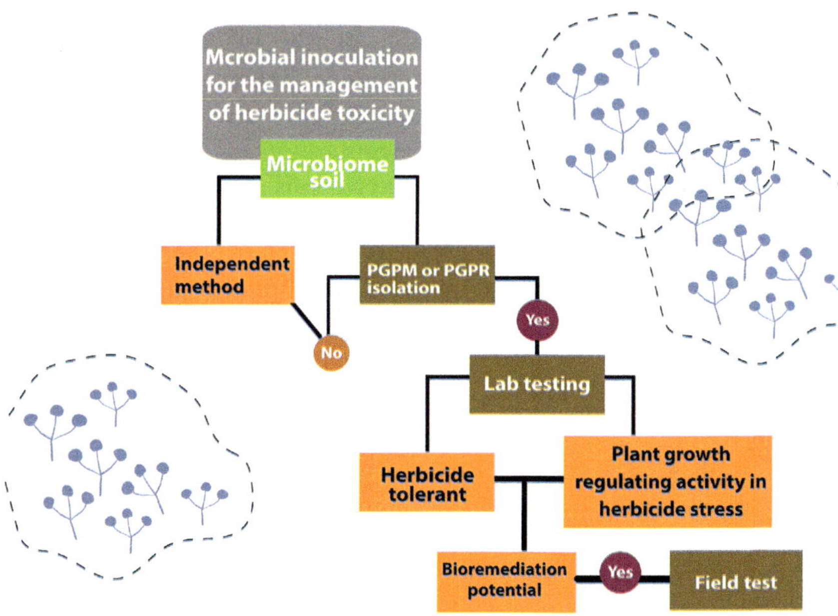

Fig. 2 Design for the study and use of PGPMs or PGPR (plant growth-promoting *Rhizobacteria*). Usually, the study starts with the isolation and/or selection of herbicide-tolerant microbial strains, the evaluation of plant growth-regulating activity in herbicide stress, and the bioremediation potential of herbicide-tolerant PGPM strains.

importance in recent years, and many studies have shown that selecting a consortium of indigenous microbes as inoculants is a potential approach to enhance the use and impacts of plant beneficial microorganisms on plant growth and possibly productivity (May et al., 2021; Nathiya et al., 2020; Shahid et al., 2018). When the microorganisms used as inoculants are isolated from a different environment, it makes them less apt to compete with indigenous microorganisms. The variable performance of microbial inoculants suggests that their success depends on interactions with the plant, with the environment, and with the community of microorganisms present in the soils (Gómez-Godínez et al., 2021).

To isolate and study PGPMs and their effects, we can choose dependent and independent methods of cultivation. For bacteria, there are different culture-dependent techniques used to quickly characterize them, and the majority are in Bergey's manual (Bergey and Holt, 2000). For fungi, as well as for many groups of bacteria, there are many articles available. Further knowledge about the ecology and biodiversity of communities that inhabit areas contaminated with herbicides is essential for the conduction of strategies to optimize the potential of these organisms in PGPMs. Environmental DNA (eDNA) investigation is a powerful technology for monitoring biodiversity in an ecosystem, and, through the use of culture-independent high-throughput sequencing as a descriptive technique, we can obtain the structure, diversity, and identification of the species. In addition, the omics sciences have also allowed the identification and characterization of genes involved in plant–microbiome interactions (Gómez-Godínez et al., 2021). However, although we cannot obtain isolated microorganisms for use, it is certainly complementary to dependent methods of cultivation and can help us in the correct definition of the culture medium, cultivation conditions, and the need for enrichment, depending on the study community.

After knowledge of the diversity and selection, the isolates are tested for herbicide-tolerant bacterial strains. These are simple cultivation tests aimed at evaluating growth in the presence of the herbicide. An important study in this process is to evaluate the impact of herbicides on active biomolecules' secreting ability of PGPMs. In this step, the expected compounds of each PGPM group are evaluated by different methodologies. The excretion of indole-3-acetic acid (IAA) can be determined by spectrophotometric methods using a calibration curve of pure IAA as a standard. Siderophores and hydrogen cyanide (HCN) by bacterial strains can be released by colorimetric methods (Bric et al., 1991; Nathiya et al., 2020;

Shahid et al., 2018). Exopolysaccharide (EPS) is a major polymeric compound, which protects bacteria from undesirable environments (Le and Stuckey, 2017), and it can also be a good strategy. Lopes et al. (2021) suggest that the ability of the microorganisms to produce catalase should also be considered in screening tests (it indicates the potential of the microorganisms to reduce oxidative stress in plants). In all these tests, different concentrations of the herbicide added to the culture medium must be evaluated. At the end of this phase, biodegradation of the herbicide must be evaluated.

Once the viability of the culture or consortium is confirmed, the inoculant formulation processes begin, and the formulation is usually related to the inoculation method. According to Lopes et al. (2021), besides the inoculum density and inoculation method, plant response to PGPM inoculation depends on root colonization, which varies with microorganisms' multiplication and distribution through the rhizosphere, microbial antagonism, soil humidity, pH, temperature, host, root exudates, and the plant's physiological state. For the inoculation of plants with beneficial microorganisms, different techniques are being employed, including seed, root, soil, and foliar inoculations. Foliar inoculation is the least used, whereas seed inoculation is the most used method (Lopes et al., 2021). In their review, the authors also highlight the use of nonsterile soil, test different inoculation methods and different abiotic stresses, and even select microorganisms from plants under abiotic stress.

7.6 Conclusions

Aiming to recover the microenvironment, promoting the improvement of plant development and thus the management of herbicide toxicity, PGPMs, which include many growth-promoting *Rhizobacteria*, are currently used. On the other hand, further studies are needed to assess the potential of other microorganisms involved in the rhizosphere or in the soil community. In this aspect, studies with environmental DNA that evaluate the microbial community and its effects in the presence of herbicides are essential to associate the impact of herbicides on the microbiota and on plant toxicity. In addition, it is known that many growth promoters help plants detoxify the herbicide, but few studies have shown the biodegradation effect of PGPMs on the soil and agricultural system as a biological community. Techniques already known from biodegradation studies should be incorporated to complete the management of herbicide toxicity in plants.

Acknowledgment

RCB S is thankful to the Federal University of Latin American Integration, Foz do Iguaçu, Brazil.

References

Alengebawy, A., Abdelkhalek, S.T., Qureshi, S.R., Wang, M.-Q., 2021. Heavy metals and pesticides toxicity in agricultural soil and plants: ecological risks and human health implications. Toxics 9, 42.

Ali, M.S., Baek, K.H., 2020. Jasmonic acid signaling pathway in response to abiotic stresses in plants. Int. J. Mol. Sci. 21, 621.

Bergey, D.H., Holt, J.G., 2000. Bergey's Manual of Determinative Bacteriology, ninth ed. Lippincott Williams & Wilkins, Philadelphia.

Bric, J.M., Bostock, R.M., Silverstone, S.E., 1991. Rapid in situ assay for indoleacetic acid production by bacteria immobilized on a nitrocellulose membrane. Appl. Environ. Microbiol. 57, 535–538.

Fuentes, A., Herrera, H., Charles, T.C., Arriagada, C., 2020. Fungal and bacterial microbiome associated with the rhizosphere of native plants from the Atacama desert. Microorganisms 8, 1–14.

Gómez-Godínez, L.J., Martínez-Romero, E., Banuelos, J., Arteaga-Garibay, R.I., 2021. Tools and challenges to exploit microbial communities in agriculture. Curr. Res. Microb. Sci. 2, 100062.

Gray, E.J., Smith, D.L., 2005. Intracellular and extracellular PGPR: commonalities and distinctions in the plant-bacterium signaling processes. Soil Biol. Biochem. 37, 395–412.

Green, J.M., 2014. Current state of herbicides in herbicide-resistant crops. Pest Manag. Sci. 70 (9), 1351–1357.

Gupta, S., Pandey, S., 2019. ACC deaminase producing bacteria with multifarious plant growth promoting traits alleviates salinity stress in French bean (*Phaseolus vulgaris*) plants. Front. Microbiol. 10, 1–17.

Holt, J.S., 2013. Herbicides. In: Encyclopedia of Biodiversity. Elsevier, pp. 87–95.

Jan, S., Singh, R., Bhardwaj, R., Ahmad, P., Kapoor, D., 2020. Plant growth regulators: a sustainable approach to combat pesticide toxicity. 3 Biotech 10, 1–11.

Kanissery, R., Gairhe, B., Kadyampakeni, D., Batuman, O., Alferez, F., 2019. Glyphosate: its environmental persistence and impact on crop health and nutrition. Plan. Theory 8, 1–11.

Le, C., Stuckey, D.C., 2017. Influence of feed composition on the monomeric structure of free bacterial extracellular polysaccharides in anaerobic digestion. Environ. Sci. Technol. 51, 7009–7017.

Lopes, M.J.S., Dias-Filho, M.B., Gurgel, E.S.C., 2021. Successful plant growth-promoting microbes: inoculation methods and abiotic factors. Front. Sustain. Food Syst. 5, 1–13.

Mall, T., Gupta, M., Dhadialla, T.S., Rodrigo, S., 2019. Overview of biotechnology-derived herbicide tolerance and insect resistance traits in plant agriculture. Methods Mol. Biol. 1864, 313–342.

May, A., Coelho, L.F., Pedrinho, A., Batista, B.D., Mendes, L.W., Mendes, R., Morandi, M.A.B., Barth, G., Viana, R.S., Vilela, E.S.D., 2021. The use of indigenous bacterial community as inoculant for plant growth promotion in soybean cultivation. Arch. Agron. Soil Sci., 1–16.

Naik, H.P., Mali, P.C., Kapse, V.D., Vaidya, K.P., Dhopavkar, R.V., Meshram, N.A., More, S.S., 2019. Response of F1 hybrids of cucumber (*Cucumis sativus* L.) to different levels of fertilizers under Konkan agro climatic condition. Pharma Innov. Int. J. 8 (12), 429–430.

Nathiya, S., Janani, R., Rajesh Kannan, V., 2020. Potential of plant growth promoting Rhizobacteria to overcome the exposure of pesticide in Trigonella foenum - graecum (fenugreek leaves). Biocatal. Agric. Biotechnol. 23, 101493.

Nie, X.J., Zhang, H.B., Su, Y.Y., 2019. Soil carbon and nitrogen fraction dynamics affected by tillage erosion. Sci. Rep. 9, 1–8.

Owen, D., Williams, A.P., Griffith, G.W., Withers, P.J.A., 2015. Use of commercial bio-inoculants to increase agricultural production through improved phosphrous acquisition. Appl. Soil Ecol. 86, 41–54.

Patil, H.J., Solanki, M.K., 2016. Microbial inoculant: modern era of fertilizers and pesticides. In: Microbial Inoculants in Sustainable Agricultural Productivity. Springer India, New Delhi, pp. 319–343.

Patil, H.J., Solanki, M.K., 2016. Microbial inoculant: modern era of fertilizers and pesticides. In: Singh, D.P., Singh, H.B., Prabha, R. (Eds.), Microbial Inoculants in Sustainable Agricultural Productivity: Vol. 1: Research Perspectives. Springer, New Delhi. https://doi.org/10.1007/978-81-322-2647-5_19.

Perotti, V.E., Larran, A.S., Palmieri, V.E., Martinatto, A.K., Permingeat, H.R., 2020. Herbicide resistant weeds: a call to integrate conventional agricultural practices, molecular biology knowledge and new technologies. Plant Sci. 290, 110255.

Powles, S.B., Holtum, J.A.M., 1994. Herbicide Resistance in Plants, first ed. Lewis Publishers, Boca Raton, FL, USA.

Santos, M.S., Nogueira, M.A., Hungria, M., 2019. Microbial inoculants: reviewing the past, discussing the present and previewing an outstanding future for the use of beneficial bacteria in agriculture. AMB Express 9, 205.

Schütte, G., Eckerstorfer, M., Rastelli, V., Reichenbecher, W., Restrepo-Vassalli, S., Ruohonen-Lehto, M., Saucy, A.G.W., Mertens, M., 2017. Herbicide resistance and biodiversity: agronomic and environmental aspects of genetically modified herbicide-resistant plants. Environ. Sci. Eur. 29, 5.

Shahid, M., Ahmed, B., Khan, M.S., 2018. Evaluation of microbiological management strategy of herbicide toxicity to greengram plants. Biocatal. Agric. Biotechnol. 14, 96–108.

Sharma, A., Kumar, V., Kumar, R., Shahzad, B., Thukral, A.K., Bhardwaj, R., 2018. Brassinosteroid-mediated pesticide detoxification in plants: a mini-review. Cogent Food Agric. 4, 1.

Sharma, A., Kumar, V., Thukral, A.K., Bhardwaj, R., 2019a. Responses of plants to pesticide toxicity: an overview. Planta Daninha 37.

Sharma, A., Yuan, H., Kumar, V., Ramakrishnan, M., Kohli, S.K., Kaur, R., Thukral, A.K., Bhardwaj, R., Zheng, B., 2019b. Castasterone attenuates insecticide induced phytotoxicity in mustard. Ecotoxicol. Environ. Saf. 179, 50–61.

Soumare, A., Diédhiou, A.G., Arora, N.K., Tawfeeq Al-Ani, L.K., Ngom, M., Fall, S., Hafidi, M., Ouhdouch, Y., Kouisni, L., Sy, M.O., 2021. Potential role and utilization of plant growth promoting microbes in plant tissue culture. Front. Microbiol. 12, 649878.

Sundari, S.K., Prakash, A., Yadav, P., Kumari, A., 2019. Plant Growth-Promoting Microbes as Front-Runners for On-site Remediation of Organophosphate Pesticide Residues in Agriculture Soils. Springer, Singapore, pp. 249–285.

Velmourougane, K., Thapa, S., Prasanna, R., 2022. Prospecting microbial biofilms as climate smart strategies for improving plant and soil health: a review. Pedosphere. https://doi.org/10.1016/j.pedsph.2022.06.037.

CHAPTER 8

Immobilization of microbial inoculants for improving soil nutrient bioavailability

Bruna Martins Dellagnezze[a] **and Isabel Natalia Sierra-Garcia**[b]
[a]Division of Microbial Resources, Research Center for Chemistry, Biology and Agriculture (CPQBA), Campinas University, Campinas, Brazil
[b]Department of Biology and CESAM, University of Aveiro, Aveiro, Portugal

8.1 Introduction

Soils play a crucial role in plant development. Besides offering physical support, water, and oxygen, soil sediments temporarily store nutrients essential for plant growth (Wiener et al., 2020). Therefore, soil composition is of the utmost importance for high yield and productivity (Mueller et al., 2010). The concept of soil health involves the broad aspect of the soil's capacity to maintain its natural biological process, sustainable crop productivity, and water quality without affecting human health (Lehmann et al., 2020). Soil nutrient bioavailability involves the capacity of the soil-plant system to supply/absorb essential nutrients through the root system (Comerford, 2005).

Anthropogenic or natural causes can contribute to several changes and disruptions in the nutrient availability of soil. Intensification of agricultural practices can lead to a scarcity of nutrients, thus decreasing soil quality and, consequently, plant growth. Soil degradation is a global problem, and, according to the UN's Food and Agriculture Organization, 25% of agricultural lands are highly degraded, mainly by salinization or erosions (DeLong et al., 2015). Moreover, despite their high efficiency, chemical compounds commonly applied as insecticides, fungicides, herbicides, and plant growth regulators have been demonstrated to be pernicious, especially in long-term use, representing another factor of soil chemical contamination. Nonetheless, these compounds have been strictly restrained in some parts of the world, especially in European Union countries (Nicolopoulou-Stamati et al., 2016).

Microbial Inoculants
https://doi.org/10.1016/B978-0-323-99043-1.00004-9

As a complex environment, the microbiota play a fundamental role as health/degradation indicators in soil system balance functioning depending on their intrinsic conditions (Trivedi et al., 2016). The soil microbiota are highly diverse (bacteria, fungi, protozoa) and act in synergy to carry out several biogeochemical functions such as nitrogen fixation, phosphorous solubilization, improvement of soil structure, and mitigation of toxins and pathogens. Microorganisms represent the core of good soil functioning, either directly or indirectly affecting crop productivity (Sammauria et al., 2020).

The search for sustainable alternatives to crop production, aiming to guarantee soil health and food safety, is currently a major concern; thus, the use of microbial inoculants has been highly considered. The successful application of microbial inoculants depends on their capability to thrive under environmental intrinsic conditions and competition with autochthonous microorganisms. Immobilization is a diverse process to attain better microbial resilience, enhancing its perpetuation in natural environments. This chapter will describe the application of microbial inoculants in crop production improvement by immobilization means.

8.2 Beneficial microorganisms

Plant growth-promoting bacteria (PGPB) is the designation given to a diverse group of bacteria capable of providing beneficial effects to plant growth. PGPB contribute to plant fitness by facilitating access to nutrients, releasing phytohormones, degrading or sequestering toxic pollutants, attenuating stress responses, stimulating immune defenses, and controlling phytopathogenic agents (Zhang et al., 2017). These bacteria belong to several genera such as *Azospirillum*, *Azotobacter*, *Pseudomonas*, *Bacillus*, *Burkholderia*, *Paenibacillus*, *Pantoea*, *Serratia*, *Streptomyces*, and *Rhizobia*, among others (Rocha et al., 2019). Specific PGPB denominated as plant growth-promoting *Rhizobacteria* (PGPR), present excellent root-colonizing ability and production of growth-promoting hormones such as indole-3-acetic acid (IAA) and gibberellic acid (GA3). They are classified into three main categories: (1) free-living bacteria, which can interact with plants under suitable conditions; (2) *Rhizobacteria*, which live in rhizospheric soil zones adjacent to roots or the phyllosphere zone (plant leaves); and (3) endophytic bacteria, which create associations with specific plant tissues (Khabbaz et al., 2019).

Fungi can also contribute to plant health and development. These fungi live in symbiotic associations, providing increased nutrient uptake and acquisition and controlling adverse effects of biotic and abiotic stresses. The most known are arbuscular mycorrhizal fungi (AMF) and *Trichoderma* (Rocha et al., 2019). AMF are capable of ameliorating soil aggregation, protecting against pathogens, and increasing water acquisition. *Trichoderma* spp. are free-living fungi that are well-known as a biocontrol agent against different phytopathogenic fungal and bacterial diseases in several crops. Moreover, these fungal genera are related to stimulatory function, enhancing nutritional quality and yield in extreme environments (Aamir et al., 2020; Rocha et al., 2019; Harman et al., 2004).

8.2.1 Microbial inoculants

Formulations of beneficial microorganisms are referred to as microbial inoculants. They are composed of a single strain or consortia of microorganisms, which are applied to plants or soils and can directly or indirectly stimulate biological activity and hence play an important role in sustainable agriculture (Alori and Babalola, 2018). Microbial inoculants have been applied for the last 15 years, aiming to enhance the resilience and capacity of plants in adverse conditions or to avoid a certain level of soil degradation (Vassilev et al., 2020). First, there is a broad "classification" defined as plant biostimulants (PBs), which are substances or microorganisms applied to plants to enhance their nutritional efficiency, abiotic stress tolerance, or quality traits (du Jardin, 2015). This category also includes a mixture of substances and microorganisms, such as chitosan, humic and fulvic acids, protein hydrolysates, phosphite, seaweed extracts, silicon, arbuscular mycorrhizal fungi, plant growth-promoting *Rhizobacteria* (PGPR), and *Trichoderma* spp. (Rouphael and Colla, 2020). Microorganisms used as inoculants are divided into three main categories, functioning as (i) biofertilizers (microbes or biological products that either fix atmospheric nitrogen or enhance the solubility of soil nutrients and have the potential to increase the yield of crops); (ii) biocontrol agents (an eco-friendly alternative to chemical pesticides against a variety of disease-causing agents, activating the host resistance mechanism and increasing biomass production and yield); and (iii) organic decomposers (some fungal and bacterial species capable of accelerating the decomposition of organic compounds, making them available to plants as nutrients (Bharti et al., 2017).

Some limiting factors are described as follows, regarding the effectiveness of microbial formulations in the field:

- Unavailability of a large number of viable cells that are required for the optimum functionality of inoculants (Bashan et al., 2014).
- High selectivity of PGPB, unlike chemical compounds, which are generally broad-spectrum products (Alori and Babalola, 2018).
- Viability maintenance of microorganisms during storage (O'Callaghan, 2016).

Without a proper carrier, the populations of most species of PGPB decay after inoculation into the soil, and, as such, difficulties in bacterial biomass growth can hinder the beneficial action of PGPB in the rhizosphere. A threshold number of cells, which differs among species, is essential to obtain the intended positive plant response (Nimnoi et al., 2017; Bashan et al., 2014).

The intrinsic properties of soil can contribute to nonthriving bioaugmented bacteria, which must compete with the autochthonous microbiota. Therefore, the formulation of inoculants is designed to provide an enhanced microenvironment, combined with physical protection, to avoid a rapid decline of bioaugmented microorganisms (Bashan et al., 2014).

8.3 Immobilization techniques

Microbial immobilization may be performed in different matrices and employed as carriers to enhance the effectiveness of microorganisms in natural environments to fulfill different metabolic functions (Siripattanakul and Khan, 2010). Overall, the immobilization process acts as a physical support, allowing biofilm formation and slow release of microbial cells into the external environment, resulting in an increased capacity to support stressful conditions. Besides microorganisms, this process has been applied to deliver enzymes and isolated molecules (Mohamad et al., 2015).

The use of microbial immobilization has broad application purposes, such as wastewater treatment (Bouabidi et al., 2018; Zhang et al., 2019), bioremediation of polluted environments (Dellagnezze et al., 2016; Catania et al., 2020), and crop production (Bashan et al., 2016). According to Bayat et al. (2015), immobilized cell systems are advantageous since they may provide high biomass, high resistance to toxic chemicals, pH and temperature variation, cell reuse, reduction of the cost of cell recovery and cell recycling, suitable microenvironmental conditions, and protection against damage.

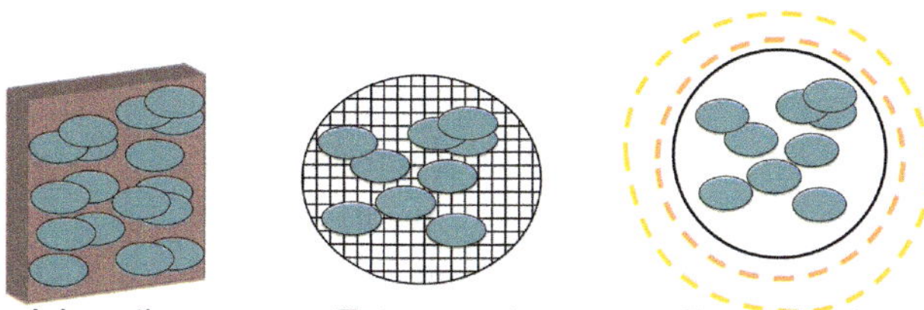

Fig. 1 Types of microbial immobilization processes. Microorganisms are represented in *green*.

There are two main methods of immobilization: chemical, such as adsorption, and physical, such as entrapment and encapsulation (Fig. 1).

8.3.1 Adsorption

This type of immobilization involves the adhesion of cells to the surface of the support, with subsequent colonization (Bouabidi et al., 2018). However, leakage from the matrix can occur due to weak and unstable interactions, which is the disadvantage of this method (Bayat et al., 2015).

8.3.2 Entrapment

This technique consists of enclosing microorganisms in a network, using different matrices such as agar, chitosan, alginate, celite, carrageenan, cellulose, collagen, gelatin, epoxy resin, photo-cross-linkable resins, polyester, polystyrene, polyurethane, and acrylic polymers (Bayat et al., 2015; Trelles and Rivero, 2013). The entrapment technique initially involves suspension of a microorganism in a solution of the selected matrix, followed by polymerization, thus forming spheres, film fibers, or cylinders. It has the advantage of being able to prevent cell release and diffusion of substrates into the external medium. Nonetheless, the exchange of nutrients may be limited, decreasing the activity of microorganisms (Trelles and Rivero, 2013; Dzionek et al., 2016).

8.3.3 Encapsulation

Encapsulation is a form of entrapment, in which a semipermeable membrane is used to immobilize cells (Dzionek et al., 2016). It is considered as a physiochemical or mechanical process of entrapping the encapsulated substance, which is called the core material, in a shell or coating, thus producing a capsule with a diameter of a few millimeters (Bouabidi et al., 2018).

8.4 Materials as carriers in crop production

The formulation of microbial inoculants involves the selection of suitable carriers and microorganisms (isolated or in consortia) to deliver the final stable and cost-effective product, regarding its performance in the field and resistance during production, storage, and shipping (John et al., 2011; Malusà et al., 2016). Carriers are based on several organic and inorganic materials that can have different forms, such as granules or powders (Vassilev et al., 2020; Stamenković et al., 2018). These support matrices should be biocompatible, providing a proper environment to cells to maintain the native conformation of enzymes and their biological activity. Several materials (matrices) of organic and inorganic nature have been proposed to be effective in cell immobilization (Fig. 2).

Polymers have been described as a more advantageous alternative to other organic carriers, such as peat, to immobilize microorganisms (Bashan et al., 2016; Stamenković et al., 2018). Carriers are divided into two types: natural carriers, such as alginate, carrageenan, agar, agarose, chitosan, and chitin, and synthetic carriers, such as acrylamide, polyurethane, polyvinyl, and resins. Another class of materials is inorganic carriers, which include zeolite, clay, porous glass, activated charcoal, and ceramics (Bayat et al., 2015; Bouabidi et al., 2018).

Polymers must have certain characteristics: (i) they should be nontoxic in nature (including for microorganisms and plants); (ii) be slowly degradable, so as to release bacteria uniformly; (iii) be able to provide physical protection from soil competitors and environmental stresses; and (iv) hold sufficient moisture for microorganisms' survival (Bashan et al., 2016). In addition,

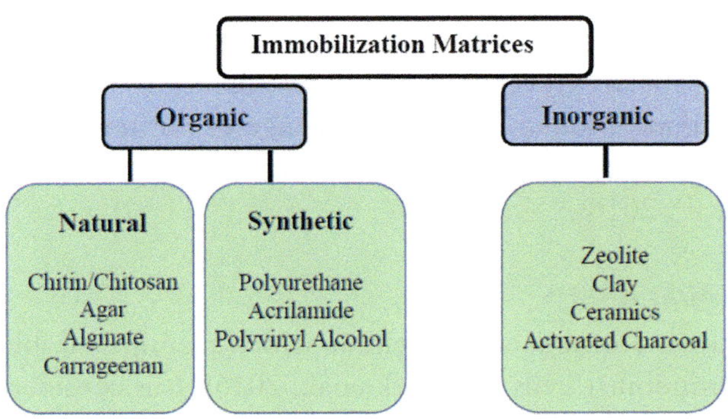

Fig. 2 Chemical nature of different materials used as carriers. *(Modified from Bouabidi, Z. B., El-Naas, M.H., Zhang, Z., 2018. Immobilization of microbial cells for the biotreatment of wastewater: a review. Environ. Chem. Lett. doi:10.1007/s10311-018-0795-7.)*

the ideal support properties should include hydrophilicity, resistance to microbial attacks, resistance to compression, and ready accessibility at a low cost (Bouabidi et al., 2018; Mohamad et al., 2015).

8.4.1 Organic carriers

In agriculture, the most used organic carriers are composed of raw materials such as peat soil, biochar (plant waste materials), and organic polymers, which have been used for microbial immobilization of PGPB or fungi (Stolarzewicz et al., 2011; John et al., 2011).

Peat is a complex organic soil material, considered as a traditional formulation. It is not available in all regions of the world, being more common in North and South America, Europe, and Australia (Bashan et al., 2014). Denton et al. (2009) demonstrated the efficacy of using peat slurry for rhizobial nodulation in legumes. Nonetheless, peat has shown some drawbacks such as decrease of inoculants due to possible contamination (Schoebitz and Belchi, 2016). Currently, it is considered a nonrenewable substrate, besides other impacts on its extraction and C emissions, and, therefore, a substitution with waste materials has been proposed (Ceglie et al., 2015).

Biochar can be formed from agricultural, animal, and municipal wastes (Taskin et al., 2019). It is a carbon-rich product obtained from the thermal decomposition of biomass and shows potential benefits for soil health, crop yield, carbon sequestration, and soil remediation (Egene et al., 2018). Biochar has also been applied as a support for plant-promoting bacteria such as *Pseudomonas* spp., *Bacillus* spp., *Rhizobium* spp., and *Azospirillum* spp., conferring benefits such as enhanced nutrient uptake and biological control against phytopathogenic fungi (Ajeng et al., 2020).

Organic polymeric formulations present more advantages than do organic carriers for industrial production. Agar, carrageenan, pectin, chitosan, bean gum, and alginate are all used in polymeric formulations for agriculture, using immobilization by encapsulation (Bashan et al., 2016; Chaudhary et al., 2020). Despite the several benefits of immobilized cell formulations described for these polymers, the major drawback is the high production cost of these polymeric compounds, which may hinder large-scale production (Vassilev et al., 2020).

Chitin and chitosan are other natural polymers, which are biodegradable and nontoxic. Chitin is a polysaccharide mostly found in the exoskeletons of arthropods (including insects, crustaceans, and arachnids), and chitosan is a derivative produced from the deacetylation process of chitin (Sharp, 2013).

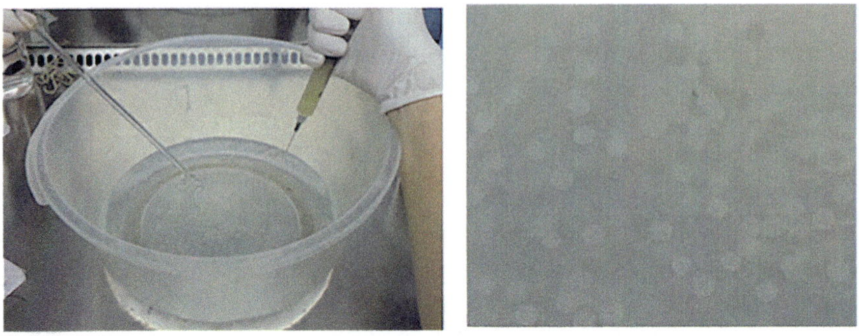

Fig. 3 Microbial immobilization process using chitosan on laboratory scale. (A) Dripping chitosan gel on tripolyphosphate solution and (B) formed beads. *(Dellagnezze, B.M., Vasconcellos, S.P., Angelim, A., Melo, V.M.M., Santisi, S., Cappello, S., Oliveira, V.M., 2016. Bioaugmentation strategy employing a microbial consortium immobilized in chitosan beads for oil degradation in mesocosm scale. Mar. Pollut. Bull. doi:10.1016/j.marpolbul.2016.04.011, images not shown in the paper.)*

Chitosan has been used as a polymeric matrix for microbial immobilization or coimmobilization, forming beads (Angelim et al., 2013) (Fig. 3). Both compounds are plant growth promoters and have been described as effective in plant defense systems under environmental stress conditions (Shahrajabian et al., 2021). They have been studied for soil amendment, showing efficient response against several pathogens such as fungi, bacteria, viruses, and nematodes and enhancing soil micronutrient delivery (Malerba and Cerana, 2016).

In tomato plants, *Bacillus* spp. immobilized in chitosan aiming to control cucumber mosaic virus (CMV) showed enhanced plant growth and protection against the virus (Murphy et al., 2003). In tomato cultivars, *Methylobacterium oryzae* CBMB20, an endophyte of rice plants immobilized in chitosan, improved plant growth and later also enhanced nutrient uptake (N, P, K, and Mg^{2+}) and photosynthetic efficiency under salt stress conditions (Chanratana et al., 2018, 2019). In cadmium-polluted soils, *Arthrobacter* sp. and *Micrococcus* sp. were coimmobilized, being effective in promoting the growth of *Chlorophytum laxum*, an ornamental plant capable of aiding in the phytoremediation process (Prapagdee and Wankumpha, 2017).

Alginate is another natural polymer found in different marine macroalgae. Sodium alginate/alginate is the most used polymer for the encapsulation of enzymes and microorganisms (Schoebitz and Belchi, 2016). Formulations prepared based on alginate generate beads, which are of two types, namely, microbeads and macrobeads with diameters of 2–3 mm and 50–200 μm, respectively, which can trap 10^9–10^{10} CFU g^{-1} (colony-forming unit per gram).

Microbial encapsulation in alginate is reported to benefit adverse field conditions, such as salinity and drought (Escalante et al., 2015). In general, salinity can induce water stress that can affect germinating seeds due to decreased water intake as a result of the osmotic effect; it can cause toxicity due to salt accumulation (El Semary et al., 2020). Encapsulated in alginate, the bacterium *Pantoea agglomerans* exhibited decreased sodium and increased calcium and potassium uptake in rice plants under stress salinity (Bhise and Dandge, 2019). He et al. (2016) evaluated the efficiency of encapsulated *Pseudomonas putida* Rs198 prepared with alginate sodium, bentonite, and starch on cotton roots under saline conditions. Under the same conditions, a *Klebsiella oxytoca* Rs-5 bacterial strain inhabiting saline soil in China was encapsulated in calcium alginate, which showed enhanced efficiency against free cells in cotton plant growth (Wu et al., 2011). Encapsulation in alginate beads (combined with additives in some cases) showed efficiency in yield and nutrient uptake using different crops with varied bacterial strains and conditions: *Azotobacter chroococcum* in wheat plants under water stress (Saad et al., 2020), *Azospirillum brasilense*, *A. chroococcum*, *Bacillus megaterium*, and *Trichoderma viride* in sunflower (Amalraj et al., 2014), *P. putida* in *Arabidopsis thaliana* (Liffourrena and Lucchesi, 2010), and *Pseudomonas fluorescens* as biocontrol (biopesticide) against *Fusarium solani* in potato (Pour et al., 2019).

A desirable characteristic of a microbial inoculant is the capacity to maintain metabolic viability after storage. Two strains, *Bradyrhizobium* sp. SEMIA6144 and *A. brasilense* Az39, immobilized in 2% of alginate beads showed an improvement in root length and biomass and an increase in nodulation, and the total plant N content in peanut plants increased after 12 months of storage (Cesari et al., 2017). PGPB *A. brasilense* and *Bacillus pumilus*, immobilized in alginate microbeads along with native arbuscular mycorrhizal fungi (AMF), were applied under a reforestation scenario in a desert area, where tested plants responded positively to bioaugmentation (Bashan et al., 2012). Agar and carrageenan are other natural polymers used in the microbial immobilization process for agriculture. El-Komy (2005) tested alginate and agar coimmobilization of *Azospirillum lipoferum* and *B. megaterium*, which enhanced the durability of the inoculum and increased phosphate solubilization in wheat plants. Still, aiming at phosphate solubilization, *Aspergillus awamori* S19, a phosphate-solubilizing filamentous fungus immobilized in Ca alginate and agar showed improved phosphate solubilization in mung bean plants (Jain et al., 2010).

The nitrogen-fixing strain *Ensifer fredii* (*Sinorhizobium*) immobilized in agar and added to kale plants showed metabolic activity after 50 days of introduction into the soil and improved plant growth and element content (plants and soils), including total nitrogen, phosphorus, and potassium (Pongsilp and Nimnoi, 2020).

8.4.2 Inorganic carriers

Inorganic inoculants are the oldest version of inoculants, and a few of them such as clays are experimentally used as a potential carrier for PGPB/PGPR (Bashan et al., 2016). Perlite and vermiculite are inert substances used as inoculant formulations. As an alternative to peat, inorganic carriers were tested which perlite was superior in better supporting the survival of several PGPB and enhancing the growth of soybeans (Albareda et al., 2008).

Clays can be applied in the form of granules, suspensions, or powders and are suitable to be used as desiccants, providing excellent storage for dried inoculants (Schoebitz et al., 2013b). They are reported to play a detoxifying role in heavy metal-contaminated fields, applied solely or in combination with other organic materials (Hamid et al., 2021; Zhang et al., 2019), and as plant growth promoters, as observed in rice (in the sepiolite form) (Li and Xu, 2018). Kadmiri et al. (2020) reported the formulation of microbial fertilizers, *P. fluorescens* Ms-01 (Pf) and *A. brasilense* DSM1690 (Ab), using two kinds of clay minerals combined with alginate. The application of the formulation to wheat plants significantly increased root and shoot biomass and nitrogen content in the roots. In a study to recover degraded soils in semiarid Mediterranean areas, coimmobilized consortia containing *A. brasilense* and *Pantoea dispersa* in clay were efficient in improving the plant biomass and nutrient content in soil (Schoebitz et al., 2014).

Zeolites are crystalline aluminosilicates, the most common minerals in sedimentary rocks, and are considered useful materials in agriculture to promote nutrient use efficiency (Ramesh and Reddy, 2011). Zeolite as a carrier for microbial immobilization is used in the bioremediation of polluted soils (Stelting et al., 2014), in phytoremediation, as a plant growth promoter, and for pest control (De Smedt et al., 2015). Saeed et al. (2019) report an application of a combination of zeolite and endophytic bacterium *Enterobacter* sp. MN17 on *Brassica napus* in cadmium (Cd)-contaminated soil. Despite bacteria not being entrapped with zeolite, the combined treatment (addition of *Enterobacter* sp. + zeolite) increased the structural (plant height, root length, dry biomass) and physiological attributes (photosynthetic rate, chlorophyll content, transpiration rate), decreasing Cd uptake from the soil compared to controls.

Studies reporting microbial formulations containing PGPB immobilized in organic and inorganic carriers are summarized in Table 1.

Microbial inoculants can vary widely according to the selected carrier, microorganism, and environmental condition. Their development for

Table 1 Compilation of immobilized bacteria and fungi in diverse organic and inorganic matrices for crop production improvement.

Immobilized microorganisms (PGPB)	Carrier	Crop	References
Rhizobacteria	Peat	Chickpea, fava bean, lentil, lupin, and pea	Denton et al. (2009)
Arthrobacter defluvii *Burkholderia cepacia* *Bacillus megaterium* *Pseudomonas frederiksbergensis* *Rhodanobacter* sp. *Streptomyces prasinopilosus* *Variovorax paradoxus*	Biochar	Rape (*Brassica napus*)	Zheng et al. (2019)
Methylobacterium oryzae CBMB20 *Bacillus* spp.	Chitosan	Tomato	Chanratana et al. (2018, 2019) and Murphy et al. (2003)
Pantoea agglomerans	Alginate	Rice	Bhise and Dandge (2019)
Pseudomonas putida *Klebsiella oxytoca*	Alginate	Cotton	He et al. (2016) and Wu et al. (2011)
P. putida	Alginate + perlite	*Arabidopsis thaliana*	Liffourrena and Lucchesi (2010)
Azospirillum brasilense, *Azotobacter chrococcum*, *B. megaterium*, and *Trichoderma viride*	Alginate	Sunflower	Amalraj et al. (2014)
Pseudomonas fluorescens	Alginate + gelatine	Potato	Pour et al. (2019)

Continued

Table 1 Compilation of immobilized bacteria and fungi in diverse organic and inorganic matrices for crop production improvement—cont'd

Immobilized microorganisms (PGPB)	Carrier	Crop	References
Bradyrhizobium sp. and *A. brasilense*	Alginate	Peanut	Cesari et al. (2017)
Azospirillum lipoferum and *B. megaterium*	Alginate and agar	Wheat	El-Komy (2005)
P. fluorescens *Serratia* sp.	Sodium alginate + potato starch	Wheat	Schoebitz et al. (2013a)
Aspergillus awamori	Alginate and agar	Mung bean	Jain et al. (2010)
A. brasilense *Bacillus pumilus*	Alginate microbeads	Leguminous trees	Bashan et al. (2012)
Ensifer fredii	Agar	Kale	Pongsilp and Nimnoi (2020)
P. fluorescens	Alginate	Sugar beet	Russo et al. (2001)
Glomus deserticola (AM mycorrhizae) *Yarrowia lipolytica* (yeast)	Alginate	Tomato	Vassilev et al. (2001)
Bacillus subtilis *Pseudomonas corrugata* *Serratia marcescens* CTM 50650 *and Enterobacter* sp. US468	Alginate	Maize	Trivedi et al. (2005) Farhat et al. (2014)
P. fluorescens and *A. brasilense*	Clay + alginate	Wheat	Kadmiri et al. (2020)
A. brasilense *Pantoea dispersa*	Clay	Shrub (*Cistus albidus*)	Schoebitz et al. (2014)
Enterobacter sp.	Zeolite	*B. napus*	Saeed et al. (2019)

commercial purposes must meet certain requirements, mainly, they should have the capacity to maintain viable and sufficient enough cells to accomplish their metabolic role in situ. According to Bashan et al. (2014, 2016), a formulation should have five main premises:

(i) *Chemical and physical characteristics*: The carrier should be nearly sterilized to avoid contamination of the inoculum, have high water-holding capacity (for wet carriers), and be suitable for different bacteria/strains of PGPB/PGPR. The raw material must be consistent and available for all carriers.

(ii) *Manufacturing:* A formulation should allow addition of nutrients, have an easily adjustable pH, and should be possibly made of low-cost raw materials with adequate supply and availability.

(iii) *Farm handling*: A formulation should be easy to handle by both producers and technology-resistant people, provide rapid and controlled release of bacteria into the soil, and be easily applied using standard equipment.

(iv) *Environmentally friendly*: A formulation should be nontoxic and biodegradable, leave no carbon footprint, and be nonpolluting; its application should minimize environmental risks.

(v) *Long storage quality*: The inoculant must support the viability of microorganisms after storage (shelf life) for proper application in the agricultural field.

8.5 Commercial inoculants

As previously described, microbial inoculants have several advantageous features over artificial chemicals, such as environmentally friendly renewable sources of nutrients and the capacity to activate soil biology and restore soil fertility. However, some research efforts to develop new and suitable bioformulations (biofertilizers and biopesticides) are based on the selective characteristics of PGPB/PGPR. It is an additional challenge when compared to conventional agrochemicals, which have broad-spectrum action on many different kinds of organisms (Timmusk et al., 2017).

Mehnaz (2016) reports, in a review, some globally and commercially available bioformulations (containing living microorganisms) and manufacturing companies in different parts of the world. Products from different companies, formulated from fungi and bacteria, isolated or in a consortia for several plant targets are listed in Table 2.

According to Research and Markets Report (2021), the global biofertilizer market was estimated at USD 2.6 billion in 2021 and is expected to reach a value of USD 4.5 billion by 2026, growing at a compound annual growth rate (CAGR) of 11.9% (Global Fertilizer Market Report, 2021). Some factors influencing the growth of the biofertilizer market include government agency initiatives, increase in the production and yield of crops, and increasing awareness about the need for sustainability in modern agriculture.

Table 2 List of some companies and commercial products developed based on microbial formulations for crop productivity enhancement.

Company	Country	Category	Product
Novozymes	Denmark	Biofertilizer	JumpStart Jumpstart LCO TagTeam TagTeamLCO Cell-Tech Optimize with LCO Promoter Technology Nitragin Gold Apron XL/Allegiance FL compatible RhizoMyco RhizoPlex RhizoMyx Glycimax Rhizomax Legumax
		Bioyield enhancers	RhizoBio RhizoMyx
		Biocontrol	Met52 Taegro
BioAgri	Sweden	Biopesticide	Cedomon Cerall Cedress
		Biofertilizer	Amase
Fertibio	Spain	Biopesticide	Bioscrop BT16
		Biofertilizer	Rhizosum N
Symborg	Spain	Biopesticide	MycoUp Attack
		Biofertilizer	MycoUp
Biagro	Argentina	Biofertilizer	Nodulest 10 Liquid PSA
Labiofam	Cuba	Biofertilizer	Nitrofix Bioenraiz
Flozyme	USA	Biofertilizer	Inogro
Mapleton Agri Biotec Pty Limited	Australia	Biofertilizer	Twin N CataPult Nitroguard

Table 2 List of some companies and commercial products developed based on microbial formulations for crop productivity enhancement—cont'd

Company	Country	Category	Product
AgriLife	India	Biofertilizer	Agrilife Nitrofix P Sol B Agri Life AgriVAM Fe Sol B K Sol B
		Biopesticide	Biotilis BioKuprum Downycare Ecosom TH Sheathguard
EM Pro-agriculture	Japan	Biofertilizer	EM

(Source: Mehnaz, S., 2016. An overview of globally available bioformulations. In: N.K. Arora, et al. (eds.), Bioformulations: For Sustainable Agriculture (Chapter 15). Springer India, pp. 267–281. doi:10.1007/978-81-322-2779-3_15.)

Specifically, for carrier-based biofertilizers, these formulations are predicted to grow at the highest CAGR between 2021 and 2026 in the Asia-Pacific region and in South America, mainly aiming at soil treatment.

8.6 Conclusions

Despite already commercially available and the ascension market, seeking more suitable microbial formulations is a challenge in a continuous changing scenario, to mitigate polluted soils or to adapt to climate changes, as high drought or colder seasons. Immobilization is a variable process, using multiple materials and techniques, and has been proven to be effective in aiding bioaugmented microorganisms to thrive under natural conditions, act as biofertilizers, improve nutrient uptake, and positively contribute to plant health and yield. Moreover, efforts from governmental and regulatory authorities should warrant these policies and formulations in countries where farmers have access to them at a large scale. In the face of anthropogenic interference in agricultural practices, leading to increasing degraded lands, the diffusion of green policies in crop management to ensure a sustainable production is fundamental.

References

Aamir, M., Rai, K.K., Zehra, A., Dubey, M.K., Kumar, S., Shukla, V., Upadhyay, R.S., 2020. Microbial bioformulation-based plant biostimulants: a plausible approach toward next generation of sustainable agriculture. In: Kumar, A., Radhakrishnan, E.K. (Eds.), Microbial Endophytes. Elsevier Inc, pp. 195–218, https://doi.org/10.1016/B978-0-12-819654-0.00008-9 (Chapter 8).

Ajeng, A.A., Abdullah, R., Ling, T.C., Ismail, S., Lau, B.F., Ong, H.C., Chew, K.W., Pau Show, P.L., Chang, J.S., 2020. Bioformulation of biochar as a potential inoculant carrier for sustainable agriculture. Environ. Technol. Innov. 20, 101168.

Albareda, M., Rodrıguez-Navarro, D.N., Camacho, M., Temprano, F.J., 2008. Alternatives to peat as a carrier for rhizobia inoculants: solid and liquid formulations. Soil Biol. Biochem. 40 (11), 2771–2779. https://doi.org/10.1016/j.soilbio.2008.07.021.

Alori, E.T., Babalola, O.O., 2018. Microbial inoculants for improving crop quality and human health in Africa. Front. Microbiol. 9, 2213. https://doi.org/10.3389/fmicb.2018.02213.

Amalraj, E.L.D., Mohanty, D., Kumar, G.P., Desai, S., Ahmed, S.K.M.H., Pradhan, R., Khan, S.S., 2014. Potential microbial consortium for plant growth promotion of sunflower (*Helianthus annuus* L.). Proc. Natl. Acad. Sci. India Sect. B Biol. Sci. https://doi.org/10.1007/s40011-014-0361-4.

Angelim, A.L., Costa, S.P., Farias, C.S., Freitas, F.L., Melo, M.V.M., 2013. An innovative bioremediation strategy using a bacterial consortium entrapped in chitosan beads. J. Environ. Manag. 127, 10–17. https://doi.org/10.1016/j.jenvman.2013.04.014.

Bashan, Y., Salazar, B.G., Moreno, M., Lopez, B.R., Linderman, R.G., 2012. Restoration of eroded soil in the Sonoran Desert with native leguminous trees using plant growth-promoting microorganisms and limited amounts of compost and water. J. Environ. Manag. 102, 26–36. https://doi.org/10.1016/j.jenvman.2011.12.032.

Bashan, Y., de-Bashan, L.E., Prabhu, S.R., Hernandez, J.P., 2014. Advances in plant growth-promoting bacterial inoculant technology: formulations and practical perspectives (1998–2013). Plant Soil 378, 1–33. https://doi.org/10.1007/s11104-013-1956-x.

Bashan, Y., de-Bashan, L.E., Prabhu, S.R., 2016. Superior polymeric formulations and emerging innovative products of bacterial inoculants for sustainable agriculture and the environment. In: Singh, et al. (Eds.), Agriculturally Important Microorganisms. Springer Science+Business Media, Singapore, pp. 15–39, https://doi.org/10.1007/978-981-10-2576-1_2 (Chapter 2).

Bayat, Z., Hassanshahian, M., Cappello, S., 2015. Immobilization of microbes for bioremediation of crude oil polluted environments: a mini review. Open Microbiol. J. 9, 48–54.

Bharti, N., Sharma, S.K., Saini, S., Verma, A., Nimonkar, Y., Prakash, O., 2017. Microbial plant probiotics: problems in application and formulation. In: Kumar, et al. (Eds.), Probiotics and Plant Health. Springer Nature Singapore Pte Ltd, pp. 317–335, https://doi.org/10.1007/978-981-10-3473-2_13 (Chapter 13).

Bhise, K.K., Dandge, P.B., 2019. Alleviation of salinity stress in rice plant by encapsulated salt tolerant plant growth promoting bacteria *Pantoea agglomerans* strain KL and its root colonization ability. Arch. Agron. Soil Sci. https://doi.org/10.1080/03650340.2019.1584395.

Bouabidi, Z.B., El-Naas, M.H., Zhang, Z., 2018. Immobilization of microbial cells for the biotreatment of wastewater: a review. Environ. Chem. Lett. https://doi.org/10.1007/s10311-018-0795-7.

Catania, V., Lopresti, F., Cappello, S., Scaffaro, R., Quatrinia, P., 2020. Innovative, eco-friendly biosorbent-biodegrading biofilms for bioremediation of oil-contaminated water. New Biotechnol. 58, 25–31. https://doi.org/10.1016/j.nbt.2020.04.001.

Ceglie, F.G., Bustamante, M.A., Ben Amara, M., Tittarelli, F., 2015. The challenge of peat substitution in organic seedling production: optimization of growing media formulation through mixture design and response surface analysis. PLoS One 10 (6), e0128600. https://doi.org/10.1371/journal.pone.0128600.

Cesari, A.B., Paulucci, N.S., Yslas, E.I., Dardanelli, M.S., 2017. Immobilization of *Bradyrhizobium* and *Azospirillum* in alginate matrix for long time of storage maintains cell viability and interaction with peanut. Appl. Microbiol. Biotechnol. https://doi.org/10.1007/s00253-020-10910-7.

Chanratana, M., Han, G.H., Joe, M.M., Choudhury, A.R., Sundaram, S., Halim, M.A., Sa, T., 2018. Evaluation of chitosan and alginate immobilized *Methylobacterium oryzae* CBMB20 on tomato plant growth. Arch. Agron. Soil Sci. 64 (11), 1489–1502. https://doi.org/10.1080/03650340.2018.1440390.

Chanratana, M., Joe, M.M., Choudhury, A.R., Anandham, R., Ramasamy Krishnamoorthy, R., Kim, K., Jeon, S., Choi, J., Choi, J., Sa, T., 2019. Physiological response of tomato plant to chitosan-immobilized aggregated *Methylobacterium oryzae* CBMB20 inoculation under salinity stress. 3Biotech 9, 397. https://doi.org/10.1007/s13205-019-1923-1.

Chaudhary, T., Dixit, M., Gera, R., Shukla, A.K., Prakash, A., Gupta, G., Shukla, P., 2020. Techniques for improving formulations of bioinoculants. 3Biotech 10, 199. https://doi.org/10.1007/s13205-020-02182-9.

Comerford, N.B., 2005. Soil factors affecting nutrient bioavailability. In: Bassiri Rad, H. (Ed.), Nutrient Acquisition by Plants, an Ecological Perspective, Ecological Studies. vol. 181. Springer-Verlag Berlin Heidelberg, pp. 1–14 (Chapter 1).

De Smedt, C., Someus, E., Spanoghe, P., 2015. Potential and actual uses of zeolites in crop protection. Pest Manag. Sci. 71, 1355–1367. https://doi.org/10.1002/ps.3999.

Dellagnezze, B.M., Vasconcellos, S.P., Angelim, A., Melo, V.M.M., Santisi, S., Cappello, S., Oliveira, V.M., 2016. Bioaugmentation strategy employing a microbial consortium immobilized in chitosan beads for oil degradation in mesocosm scale. Mar. Pollut. Bull. https://doi.org/10.1016/j.marpolbul.2016.04.011.

DeLong, C., Cruse, R., Wiener, J., 2015. The soil degradation paradox: compromising our resources when we need them the most. Sustainability 7, 866–879. https://doi.org/10.3390/su7010866.

Denton, M.D., Pearce, D.J., Ballard, R.A., Hannah, M.C., Mutch, L.A., Sorn Norng, S., Slattery, J.F., 2009. A multi-site field evaluation of granular inoculants for legume nodulation. Soil Biol. Biochem. 41, 2508–2516. https://doi.org/10.1016/j.soilbio.2009.09.009.

du Jardin, P., 2015. Plant biostimulants: definition, concept, main categories and regulation. Sci. Hortic. 196, 3–14. https://doi.org/10.1016/j.scienta.2015.09.021.

Dzionek, A., Wojcieszyńska, D., Guzik, U., 2016. Natural carriers in bioremediation: a review. Electron. J. Biotechnol. 23, 28–36. https://doi.org/10.1016/j.ejbt.2016.07.003.

Egene, C.E., Van Poucke, R., Ok, Y.S., Meers, E., Tacka, F.M.G., 2018. Impact of organic amendments (biochar, compost and peat) on cd and Zn mobility and solubility in contaminated soil of the Campine region after three years. Sci. Total Environ. 626, 195–202. https://doi.org/10.1016/j.scitotenv.2018.01.054.

El Semary, N.A.H., Alouane, M.H.H., Nasr, O., Aldayel, M.F., Alhaweti, F.H., Ahmed, F., 2020. Salinity stress mitigation using encapsulated biofertilizers for sustainable agriculture. Sustainability 12, 9218. https://doi.org/10.3390/su12219218.

El-Komy, H.M.A., 2005. Co-immobilization of *a. lipoferum* and *B. megaterium* for plant nutrition. Food Technol. Biotechnol. 43 (1), 19–27.

Escalante, F.M.E., Cortés-Jiménez, D., Tapia-Reyes, G., Suárez, R., 2015. Immobilized microalgae and bacteria improve salt tolerance of tomato seedlings grown hydroponically. J. Appl. Phycol. https://doi.org/10.1007/s10811-015-0651-0 (5th Congress of the International Society for Applied Phycology).

Farhat, M.B., Taktek, S., Chouayekh, H., 2014. Encapsulation in alginate enhanced the plant growth promoting activities of two phosphate solubilizing bacteria isolated from the phosphate mine of Gafsa. Neth. J. Agric. Sci. 2, 131–139.

Global Fertilizer Market Report, 2021. Global Biofertilizers Market by Form (Liquid, Carrier-based), Mode of Application (Soil Treatment, Seed Treatment), Type (Nitrogen-fixing, Phosphate Solubilizing & Mobilizing, Potash Solubilizing & Mobilizing), Crop Type, and Region. Forecast to 2026 Report. https://www.researchandmarkets.com.

Hamid, Y., Tang, L., Hussain, B., Usman, M., Liu, L., Ulhassan, Z., He, Z., Yang, X., 2021. Sepiolite clay: a review of its applications to immobilize toxic metals in contaminated soils and its implications in soil–plant system. Environ. Technol. Innov. 23, 101598. https://doi.org/10.1016/j.eti.2021.101598.

Harman, G.E., Howell, C.R., Viterbo, A., Chet, I., Lorito, M., 2004. *Trichoderma* species: opportunistic, avirulent plant symbionts. Nat. Rev. Microbiol. 2, 43.

He, Y., Wu, Z., Ye, B.C., Wang, J., Guan, X., Zhang, Z., 2016. Viability evaluation of alginate-encapsulated *Pseudomonas putida* Rs-198 under simulated salt-stress conditions and its effect on cotton growth. Eur. J. Soil Biol. 75, 135–141. https://doi.org/10.1016/j.ejsobi.2016.05.002.

Jain, R., Saxena, J., Sharma, V., 2010. The evaluation of free and encapsulated *Aspergillus awamori* for phosphate solubilization in fermentation and soil–plant system. Appl. Soil Ecol. 46 (2010), 90–94. https://doi.org/10.1016/j.apsoil.2010.06.008.

John, R.P., Tyagi, R.D., Brar, S.K., Surampalli, R.Y., Prévost, D., 2011. Bio-encapsulation of microbial cells for targeted agricultural delivery. Crit. Rev. Biotechnol. 31 (3), 211–226. https://doi.org/10.3109/07388551.2010.513327.

Kadmiri, I.M., Mernissi, N.E., Azaroual, S.E., Mekhzoum, M.E.M., Qaiss, A.E.K., Bouhfid, R., 2020. Bioformulation of microbial fertilizer based on clay and alginate encapsulation. Curr. Microbiol. 78 (1), 86–94. https://doi.org/10.1007/s00284-020-02262-2.

Khabbaz, S.E., Ladhalakshmi, D., Babu, M., Kandan, A., Ramamoorthy, V., Saravanakumar, D., et al., 2019. Plant growth promoting bacteria (PGPB)—a versatile tool for plant health management. Can. J. Pestic. Pest Manag. 1 (1), 1–25. https://doi.org/10.34195/can.j.ppm.2019.05.001.

Lehmann, J., Bossio, D.A., Knabner, I.K., Matthias, C., Rillig, M.C., 2020. The concept and future prospects of soil health. Nat. Rev. Earth Environ. https://doi.org/10.1038/s43017-020-0080-8.

Li, J., Xu, Y., 2018. Effects of clay combined with moisture management on Cd immobilization and fertility index of polluted rice field. Ecotoxicol. Environ. Saf. 158, 182–186. https://doi.org/10.1016/j.ecoenv.2018.04.031.

Liffourrena, A.S., Lucchesi, G.I., 2010. Alginate-perlite encapsulated *Pseudomonas putida* A (ATCC 12633) cells: preparation, characterization and potential use as plant inoculants. J. Biotechnol. 278, 28–33. https://doi.org/10.1016/j.jbiotec.2018.04.019.

Malerba, M., Cerana, R., 2016. Chitosan effects on plant systems. Int. J. Mol. Sci. 17, 996. https://doi.org/10.3390/ijms17070996.

Malusà, E., Pinzari, F., Canfora, L., 2016. Efficacy of biofertilizers: challenges to improve crop production. In: Singh, D.P., et al. (Eds.), Microbial Inoculants in Sustainable Agricultural Productivity. Springer India, pp. 17–40, https://doi.org/10.1007/978-81-322-2644-4_2 (Chapter 2).

Mehnaz, S., 2016. An overview of globally available bioformulations. In: Arora, N.K., et al. (Eds.), Bioformulations: For Sustainable Agriculture. Springer India, pp. 267–281, https://doi.org/10.1007/978-81-322-2779-3_15 (Chapter 15).

Mohamad, N.R., Marzuki, N.H.C., Buang, N.A., Huyop, F., Wahab, R.A., 2015. An overview of technologies for immobilization of enzymes and surface analysis techniques for immobilized enzymes. Biotechnol. Biotechnol. Equip. 29 (2), 205–220. https://doi.org/10.1080/13102818.2015.1008192.

Mueller, L., Schindler, U., Mirschel, W., Shepherd, T.G., Ball, B.C., Helming, K., et al., 2010. Assessing the productivity function of soils. A review. Agron. Sustain. Dev. 30 (3), 601–614. https://doi.org/10.1051/agro/2009057.

Murphy, J.F., Reddy, M.S., Ryu, C.M., Kloepper, J.W., Li, R., 2003. Rhizobacteria-mediated growth promotion of tomato leads to protection against cucumber mosaic virus. Phytopathology 93 (10), 1301–1307. https://doi.org/10.1094/PHYTO.2003.93.10.1301.

Nicolopoulou-Stamati, P., Maipas, S., Kotampasi, C., Stamatis, P., Hens, L., 2016. Chemical pesticides and human health: the urgent need for a new concept in agriculture. Front. Public Health 4, 148. https://doi.org/10.3389/fpubh.2016.00148.

Nimnoi, P., Pongsilp, N., Lumyong, S., 2017. The establishment and activity of bacterial inoculant immobilized in agar compared with those of alginate and perlite after being introduced into soil. Chiang Mai J. Sci. 44 (3), 751–767.

O'Callaghan, M., 2016. Microbial inoculation of seed for improved crop performance: issues and opportunities. Appl. Microbiol. Biotechnol. 100, 5729–5746. https://doi.org/10.1007/s00253-016-7590-9.

Pongsilp, N., Nimnoi, P., 2020. Inoculation of *Ensifer fredii* strain LP2/20 immobilized in agar results in growth promotion and alteration of bacterial community structure of Chinese kale planted soil. Sci. Rep. 10, 15857. https://doi.org/10.1038/s41598-020-72986-5.

Pour, M.M., Saberi-Riseh, R., Mohammadinejad, R., Hosseini, A., 2019. Investigating the formulation of alginate-gelatin encapsulated *Pseudomonas fluorescens* (VUPF5 and T17-4 strains) for controlling *fusarium solani* on potato. Int. J. Biol. Macromol. 133, 603–613. https://doi.org/10.1016/j.ijbiomac.2019.04.071.

Prapagdee, B., Wankumpha, J., 2017. Phytoremediation of cadmium-polluted soil by *Chlorophytum laxum* combined with chitosan-immobilized cadmium-resistant bacteria. Environ. Sci. Pollut. Res. https://doi.org/10.1007/s11356-017-9591-3.

Ramesh, K., Reddy, D.D., 2011. Zeolites and their potential uses in agriculture. In: Sparks, D.L. (Ed.), Advances in Agronomy. 113. Elsevier Inc, pp. 219–241, https://doi.org/10.1016/B978-0-12-386473-4.00004-X (Chapter 4).

Rocha, I., Ma, Y., Souza-Alonso, P., Vosátka, M., Freitas, H., Oliveira, R.S., 2019. Seed coating: a tool for delivering beneficial microbes to agricultural crops. Front. Plant Sci. 10, 1357. https://doi.org/10.3389/fpls.2019.01357.

Rouphael, Y., Colla, G., 2020. Editorial: biostimulants in agriculture. Front. Plant Sci. 11, 40. https://doi.org/10.3389/fpls.2020.00040.

Russo, A., Basaglia, M., Tola, E., Casella, S., 2001. Survival, root colonisation and biocontrol capacities of *Pseudomonas fluorescens* F113 LacZY in dry alginate microbeads. J. Ind. Microbiol. Biotechnol. 27 (6), 337–342. https://doi.org/10.1038/sj.jim.7000154.

Saad, M.M., Abo-Koura, H.A., El-Latif, K.M.A., Aly, M.M., 2020. Application of alginate bead encapsulated N-fixing bacteria is improving wheat yield under drought stress. Plant Arch. 20 (1), 1735–1747.

Saeed, Z., Naveed, M., Imran, M., Bashir, M.A., Sattar, A., Mustafa, A., et al., 2019. Combined use of *Enterobacter* sp. MN17 and zeolite reverts the adverse effects of cadmium on growth, physiology and antioxidant activity of *Brassica napus*. PLoS One 14 (3), e0213016. https://doi.org/10.1371/journal.pone.0213016.

Sammauria, R., Kumawat, S., Kumawat, P., Singh, S., Jatwa, T.K., 2020. Microbial inoculants: potential tool for sustainability of agricultural production systems. Arch. Microbiol. 202, 677–693. https://doi.org/10.1007/s00203-019-01795-w.

Schoebitz, M., Belchi, M.D.L., 2016. Encapsulation techniques for plant growth-promoting rhizobacteria. In: Arora, N.K., et al. (Eds.), Bioformulations: For Sustainable Agriculture. Springer India, pp. 251–265, https://doi.org/10.1007/978-81-322-2779-3_14 (Chapter 14).

Schoebitz, M., Ceballos, C., Ciampi, L., 2013a. Effect of immobilized phosphate solubilizing bacteria on wheat growth and phosphate uptake. J. Soil Sci. Plant Nutr. 13 (1), 1–10. https://doi.org/10.4067/S0718-95162013005000001.

Schoebitz, M., López, M.D., Roldán, A., 2013b. Bioencapsulation of microbial inoculants for better soil–plant fertilization. A review. Agron. Sustain. Dev. 33, 751–765. https://doi.org/10.1007/s13593-013-0142-0.

Schoebitz, M., Mengual, C., Roldán, A., 2014. Combined effects of clay immobilized *Azospirillum brasilense* and *Pantoea dispersa* and organic olive residue on plant performance and soil properties in the revegetation of a semiarid area. Sci. Total Environ. 466–467, 67–73. https://doi.org/10.1016/j.scitotenv.2013.07.012.

Shahrajabian, M.H., Chaski, C., Polyzos, N., Tzortzakis, N., Petropoulos, S.A., 2021. Sustainable agriculture systems in vegetable production using chitin and chitosan as plant biostimulants. Biomol. Ther. 11, 819. https://doi.org/10.3390/biom11060819.

Sharp, R.G., 2013. A review of the applications of chitin and its derivatives in agriculture to modify plant-microbial interactions and improve crop yields. Agronomy 3, 757–793. https://doi.org/10.3390/agronomy3040757.

Siripattanakul, S., Khan, E., 2010. Fundamentals and applications of entrapped cell bioaugmentation for contaminant removal. In: Shah, V. (Ed.), Emerging Environmental Technologies. vol. II. Springer, Dordrecht, pp. 147–169, https://doi.org/10.1007/978-90-481-3352-9_7 (Chapter 7).

Stamenković, S., Beškoski, V., Karabegović, I., Lazić, M., Nikolić, N., 2018. Microbial fertilizers: a comprehensive review of current findings and future perspectives. Span. J. Agric. Res. 16 (1), e09R01. https://doi.org/10.5424/sjar/2018161-12117.

Stelting, S.A., Burns, R.G., Sunna, A., Bunt, C.R., 2014. Survival in sterile soil and atrazine degradation of *Pseudomonas* sp. strain adp immobilized on zeolite. Bioremediation Journal 18 (4), 309–316. https://doi.org/10.1080/10889868.2014.938723.

Stolarzewicz, I., Biaecka-Florjañczyk, E., Majewska, E., Krzyczkowska, J., 2011. Immobilization of yeast on polymeric supports. Chem. Biochem. Eng. Q. 25 (1), 135–144.

Taskin, E., Bueno, C.C., Allegretta, I., Terzano, R., Rosa, A.H., Loffredo, E., 2019. Multi-analytical characterization of biochar and hydrochar produced from waste biomasses for environmental and agricultural applications. Chemosphere 233, 422–430. https://doi.org/10.1016/j.chemosphere.2019.05.204.

Timmusk, S., Behers, L., Muthoni, J., Muraya, A., Aronsson, A.C., 2017. Perspectives and challenges of microbial application for crop improvement. Front. Plant Sci. 8, 49. https://doi.org/10.3389/fpls.2017.00049.

Trelles, J.A., Rivero, C.W., 2013. Whole cell entrapment techniques. In: Guisan, J.M. (Ed.), Immobilization of Enzymes and Cells, third ed. vol. 1051. Springer Science+Business Media, New York, pp. 365–374, https://doi.org/10.1007/978-1-62703-550-7_24 (Chapter 24).

Trivedi, P., Pandey, A., Palni, L.M.S., 2005. Carrier-based preparations of plant growth-promoting bacterial inoculants suitable for use in cooler regions. World J. Microbiol. Biotechnol. 21, 941–945. https://doi.org/10.1007/s11274-004-6820-y.

Trivedi, P., Delgado-Baquerizo, M., Anderson, I.C., Singh, B.K., 2016. Response of soil properties and microbial communities to agriculture: implications for primary productivity and soil health indicators. Front. Plant Sci. 7, 990. https://doi.org/10.3389/fpls.2016.00990.

Vassilev, N., Vassileva, M., Azcon, R., Medina, A., 2001. Application of free and ca-alginate-entrapped *Glomus deserticola* and *Yarrowia lipolytica* in a soil-plant system. J. Biotechnol. 91, 237–242.

Vassilev, N., Vassileva, M., Martos, V., Garcia del Moral, L.F., Kowalska, J., Tylkowski, B., Malusá, E., 2020. Formulation of microbial inoculants by encapsulation in natural polysaccharides: focus on beneficial properties of carrier additives and derivatives. Front. Plant Sci. 11, 270. https://doi.org/10.3389/fpls.2020.00270.

Wiener, K.D., Simaika, J.P., Grenfell, S.E., Jacobs, S.M., 2020. Effects of invasive N2-fixing *Acacia mearnsii* on sediment nutrient concentrations in mountain streams: implications of sediment geochemistry for ecosystem recovery. Catena 195, 104786. https://doi.org/10.1016/j.catena.2020.104786.

Wu, Z., Zhao, Y., Kaleem, I., Li, C., 2011. Preparation of calcium alginate microcapsulated microbial fertilizer coating *Klebsiella oxytoca* Rs-5 and its performance under salinity stress. Eur. J. Soil Biol. 47, 152–159. https://doi.org/10.1016/j.ejsobi.2010.11.008.

Zhang, R., Vivanco, J.M., Shen, Q., 2017. The unseen rhizosphere root–soil–microbe interactions for crop production. Curr. Opin. Microbiol. 37, 8–14. https://doi.org/10.1016/j.mib.2017.03.008.

Zhang, W., Ren, X., He, J., Zhang, Q., Qiu, C., Fan, B., 2019. Application of natural mixed bacteria immobilized carriers to different kinds of organic wastewater treatment and microbial community comparison. J. Hazard. Mater. 377, 113–123. https://doi.org/10.1016/j.jhazmat.2019.05.068.

Zheng, B.X., Ding, K., Yang, X.R., Wadaan, M.A.M., Hozzein, W.N., Peñuelas, J., Zhu, Y.G., 2019. Straw biochar increases the abundance of inorganic phosphate solubilizing bacterial community for better rape (Brassica napus) growth and phosphate uptake. Sci. Total Environ. 647, 1113–1120. https://doi.org/10.1016/j.scitotenv.2018.07.454.

CHAPTER 9

Plant growth-promoting microbes and nanoparticles: Biotechnological potential in agrobiological systems

Amanda Carolina Prado de Moraes[a,b], Paulo Henrique Gomes Lisboa[a,b], Lucas da Silva Ribeiro[c], Emerson Rodrigues de Camargo[c], and Paulo Teixeira Lacava[a,b]

[a]Laboratory of Microbiology and Biomolecules—LaMiB, Department of Morphology and Pathology, Center for Biological and Health Sciences, Federal University of São Carlos, São Carlos, SP, Brazil
[b]Biotechnology Graduation Program, Federal University of São Carlos, São Carlos, SP, Brazil
[c]Interdisciplinary Laboratory of Electrochemistry and Ceramics, Department of Chemistry, Federal University of São Carlos, São Carlos, Brazil

9.1 Introduction

The application of microorganisms as inoculants in agriculture constitutes an economical and eco-friendly way to reduce the use of chemical fertilizers and pesticides. The involvement of these microbiological tools in the development of agricultural and biotechnological processes is evident since issues concerning practices that entirely dependent on synthetic fertilizers and pesticides may cause negative impacts on both the environment and human health (Brauer et al., 2019; Cullen et al., 2019; Omomowo and Babalola, 2019). Plant-associated microbes are well-known for their natural secondary metabolites and are also a vast and unexplored source of unique chemical structures of phytochemical compounds and growth promoters, which can be considered a sustainable approache to minimize the use of agrochemicals (Lacava and Azevedo, 2014; Quecine et al., 2014; Lisboa et al., 2021).

Nanotechnology has emerged as a new source of quality enrichment in the agricultural sector. The use of nanoparticles can be a successful method to meet the challenges regarding the effectiveness of biofertilizers in natural environments (Pramanik et al., 2020). Many nanomaterials have shown positive effects either on endophytes or on the growth of various plant species, thereby making these materials promising components of biofertilizer formulations (de Moraes et al., 2021). The approach of using nanomaterials

Microbial Inoculants
https://doi.org/10.1016/B978-0-323-99043-1.00014-1

in association with endophytes can increase microbial viability and resistance in the environment, improve their field stability, and, consequently, aid in plant growth promotion (Djaya et al., 2019; Rana et al., 2020). Given the novel sustainable strategies applied in agricultural systems, this chapter addresses the effects of nanoparticles on beneficial plant microbes for promotion of plant growth, management of plant diseases, and biocontrol of insect pests.

9.2 Plant growth-promoting microbes: Endophytes and Rhizobacteria

The importance of soil for plant development is well-known. Soil also serves as a habitat for several living beings, including microorganisms, and is the setting for complex interactions. Numerous microbes associated with plants are found to produce beneficial effects on plant development and can either colonize the rhizosphere or live intricately within plant tissues.

The rhizosphere is defined as the soil region around plant roots where they influence the biological and chemical characteristics due to the release of exudates and other substances in a process called "rhizodeposition" (Mendes et al., 2013). These modifications provide an ideal microenvironment for several species, thus making the rhizosphere a hot spot of microbial activity, facilitating the multiplication of bacteria beneficial to plant health and development. These bacteria are designated as plant growth-promoting rhizobacteria (PGPR) (Kloepper and Schroth, 1978). PGPR belong to several genera, including *Bacillus*, *Bradyrhizobium*, *Burkholderia*, *Enterobacter*, *Pseudomonas*, and *Serratia* (Gray and Smith, 2005). Among the microorganisms inhabiting the rhizosphere, some bacteria can penetrate plants through the stomata and microinjuries present in the roots and later reach stems, leaves, fruits, and other parts through the apoplasts in xylem vessels (Chi et al., 2005; Santoyo et al., 2016). In internal tissues, bacteria encounter environmental conditions in which variations in certain limiting growth factors such as temperature, osmotic potential, and ultraviolet radiation are less stressful (Santoyo et al., 2016; Lacava and Azevedo, 2013). These bacteria, termed "endophytes," are microorganisms that are nonphytopathogenic and live inside plant tissues for at least a part of their life cycle (Rosenblueth and Martínez-Romero, 2006).

Bacteria can promote plant growth through direct mechanisms, such as biological nitrogen fixation (BNF), inorganic phosphate solubilization, and production of phytohormones (e.g., indole-3-acetic acid (IAA)), and indirect mechanisms, such as biological control of phytopathogens (antagonism through the production of inhibitory metabolites, competition for nutrients, and induction of systemic resistance in the host plant) (Glick, 2012; Singh, 2018).

Nitrogen (N) is the primary element most demanded by plants and is required in almost all metabolic processes, essential for plant growth and productivity. Despite being abundant, atmospheric nitrogen is not accessible to plants, so it must be converted into an assimilable form (Ahemad and Kibret, 2014). BNF is a biological process involving the reduction of molecular nitrogen (N_2) to ammonia (NH_3) by nitrogen-fixing microorganisms, possessing the nitrogenase enzyme (Eady and Postgate, 1974; Burgess and Lowe, 1996). There are three known isoforms of nitrogenase, but the most commonly found is the molybdenum-dependent enzyme system formed by two metalloproteins: iron (Fe) protein and molybdenum-iron (MoFe) protein (Seefeldt et al., 2020). Iron (Fe) protein is the electron donor component and also the one that binds nucleotides, whereas molybdenum-iron (MoFe) protein contains the reducing site of the substrate (Burgess and Lowe, 1996). In this process, abundant H^+ ions in bacterial cells are incorporated into atmospheric nitrogen, leading to the formation of NH_4^+ ions, a form that can be assimilated by plants (Coutinho et al., 2000). Several genera of bacteria have been described as nitrogen fixers, such as *Azospirillum*, *Azotobacter*, *Bacillus*, *Burkholderia*, *Clostridia*, *Citrobacter*, *Enterobacter*, *Klebsiella*, *Mycobacterium*, *Pseudomonas*, *Rhizobium*, and *Serratia*. However, this list is constantly expanding due to research advances (Hayat et al., 2012).

Phytohormones are organic substances that play important roles in plant growth regulation. Both endophytic bacteria and PGPR are capable of producing phytohormones (Taiz et al., 2015; Gupta et al., 2021). Among phytohormones produced by different microorganisms, the most studied is indole-3-acetic acid, belonging to a class of auxins, which are compounds capable of stimulating plant growth by acting in cell expansion and elongation, cell division, rooting, and thus facilitating absorption of water and nutrients from the soil (Leinhos and Vacek, 1994). The production of IAA by bacteria is well-known and has been described in both endophytic bacteria and PGPR isolated from various crops such as soybean (Cattelan et al., 1999), sugarcane (Inui et al., 2012), jatropha (Jha et al., 2012), guarana (Batista et al., 2018), and strawberry (Dias et al., 2009), among others.

Phosphorus (P), after nitrogen, is the most essential element for plants. It plays an important role in all major plant metabolic processes and is available in soil in both organic and inorganic forms; however, plants only absorb it as soluble monobasic ($H_2PO_4^-$) and dibasic (HPO_4^{2-}) ions (Singh, 2018). The levels of soluble phosphorus in soil are typically low (in the order of 0.1 mg of P/L of soil), due to the low solubility of compounds and high adsorption capacity of soil particles, causing about 50% of phosphorus availability in soils in an insoluble form (Singh, 2018; Khan et al., 2009).

Thus, the use of phosphate-solubilizing bacteria (PSB) presents itself as an alternative that enables better use of phosphorus already existing in the soil. PSB solubilize inorganic phosphorus by releasing organic acids and/or mineralization of organic phosphorus through the synthesis of different phosphatases catalyzing the hydrolysis of phosphoric esters (Singh, 2018; Tao et al., 2008). Due to the characteristic potential of these bacteria, several studies have been carried out to assess the effects of PSB in cultures of economic interest. In general, PSB show excellent potential in field experiments, increasing the productivity of crops, such as tomato, cauliflower, pepper, apple, and apricot, among others (Walia et al., 2017).

The leading cause of productivity losses in agriculture is pests and diseases. Most importantly, fungi are the phytopathogenic agents of 85% of the diseases affecting crops of economic relevance (Baker et al., 2017). The conventional methods of controlling plant diseases and pests relying on agrochemicals are costly and described to exert a negative effect on environmental integrity. Therefore, the use of biological agents to manage considerable losses in agricultural productivity termed "biological control," is the need of the hour. In other words, biological control can be defined as a sustainable technique that reduces the population of a target species that are harmful to plants such as phytopathogenic fungi and insect pests (Stenberg et al., 2021). Bacteria inhabit the same niche as phytopathogens and need strategies to ensure their survival, such as antibiosis and induction of systemic resistance (Ahemad and Kibret, 2014). "Antibiosis" occurs via the production of antibiotics by bacteria with antagonistic potential. The term also applies to any metabolized compound capable of killing or inhibiting the growth or reproduction of phytopathogenic microorganisms capable of producing enzymes that degrade cell walls (Griffin, 2014). Mendes et al. (2007) reported endophytic bacteria of the genus *Burkholderia* isolated from sugarcane, which inhibited the growth of *Fusarium verticillioides*. In a study carried out by Abdalla et al. (2014), 27 Rhizobacteria belonging to the genus *Bacillus* were found to exhibit antagonistic activity against the phytopathogen *Alternaria alternata*.

Some bacteria can also systematically prepare the plant immune system and induce systemic resistance. The active penetration of these bacteria induces the synthesis of compounds acting on phytopathogens, avoids infection and colonization processes, and changes plant morphology (Santoyo et al., 2016; Compant et al., 2005). These morphological and physiological changes may include an increase in cell walls due to deposition of lignin, glucan, and increased cuticle thickness as well as the synthesis of phytoalexins and other phenolics, thus hindering the entry of pathogens and its further development in the host plant (Ahemad and Kibret, 2014). In a study by Benhamou et al. (2000), the endophytic bacterium *Serratia plymuthica* was reported to increase the levels of phenolic compounds in cucumber roots, thereby protecting the host plants from infections caused by *Pythium ultimum*. Zhang et al. (2004) reported the induction of systemic resistance by *Serratia marcescens* against tobacco blue mold (*Peronospora tabacina*).

Several studies have reported the potential of endophytic bacteria and PGPR in promoting plant growth and in the management of insect pests and plant diseases. Understanding the interactions between these bacteria and plants is essential for developing new strategies for effective and sustainable practices in agriculture.

9.3 Nanotechnology in agriculture

The rapid increase in population accompanied by greater food production has resulted in excessive application of fertilizers and pesticides to improve crop yields (Chhipa, 2017). However, despite allowing better food production, the excessive use of fertilizers has led to increased soil and groundwater pollution. Similarly, the careless use of pesticides has provoked the evolution of more resistant pests and pathogens (Tilman et al., 2002). Furthermore, there may be a substantial decrease in soil biodiversity and death of microbes beneficial to plant growth. Considering that the global consumption of pesticides alone is around 2 tons annually and that approximately 50%–70% of what is applied is lost through leaching, bioconversion, and mineralization, it is possible to see the major problems caused by the massive use of fertilizers and pesticides (De et al., 2014).

In an effort to reverse this situation, nanotechnology began to be applied in agriculture as well. Nano-based materials are well-known for their application in several areas such as medicine, biology, chemistry, electronics, physics, and energy. Nanoparticles have different optical, mechanical, and chemical properties than do respective bulk materials due to their small size and high surface area to volume ratio (Gutiérrez et al., 2011). Nanoparticles also have some advantageous properties such as greater reactivity, better bioactivity, and the possibility of being superficially modified, adding different properties to those already expected. In addition, nanoparticles allow the application of precision farming, the balanced use of techniques that enable an increase in crop yields considering the environmental situation (Chen and Yada, 2011). In this manner, formulations of nanofertilizers and nanopesticides can be produced, in which the nanoparticles act as a fertilizer, pesticide, or as a controlled release agent encapsulating the compound and releasing it in a controlled manner to a predetermined location, such as soil, roots, or leaves. This controlled release increases the effectiveness of fertilizers and pesticides, reduces the loss, and, consequently, lowers the amount of agrochemicals in crops (Duhan et al., 2017). Nanoparticles can be divided into three basic categories: metallic, oxide, and polymeric. In the case of metallic nanoparticles used in agriculture, silver (Ag), gold (Au), and copper (Cu) are well-known. These nanoparticles have a unique property of surface plasmon resonance, which allows their use as biosensors, in addition to being known for their antimicrobial properties (Kumar et al., 2018). The oxides of titanium (TiO_2), silicon (SiO_2), and zinc (ZnO) are popularly known (de Moraes et al., 2021). The metal oxides present an intrinsic charge separation capacity that differentiates them from the metal and permits their application as semiconductors. Unlike other nanoparticles, polymeric nanoparticles, especially biopolymers like chitosan and lignin, are usually exclusively used as encapsulating agents since they are not active either as a fertilizer or as a pesticide (Hasaneen et al., 2014). These nanoparticles can be obtained via cross-linking agents that form spheres of nanometric dimensions by capturing the compound to be released during their formation. Moreover, these nanoparticles may be of biodegradable nature, thus eliminating the process of their removal after use and eventually decreasing the chances of creating a toxic environment for plants.

The annual economic loss to crop productivity caused by microorganisms, nematodes, parasites, and insects is estimated to be around \$220 billion (Sharma et al., 2017). Generally, crop producers control the problem using pesticides. However, these substances may be harmful to the environment and may favor resistance development in phytopathogens in the long term. Therefore, nanopesticides are suitable alternatives to be explored in order to reduce huge economic loss while promoting greater sustainability (Pramanik et al., 2020). Silver is already known for its antimicrobial activity; however, other nanoformulations such as copper, zinc oxide, and titanium oxide have also shown antimicrobial activity against several phytopathogenic fungi and bacteria, such as *Aspergillus niger*, *A. alternata*, *Dickeya* sp.,

Fusarium oxysporum, *Pectobacterium* sp., and *Xanthomonas* sp. (Mondal and Mani, 2012; Kadri et al., 2013; Paret et al., 2013; Hafez et al., 2014; Kanhed et al., 2014). Furthermore, nanoformulations using nano-chitosan and poly (epsilon-caprolactone) also show herbicidal activity (Grillo et al., 2014; Pereira et al., 2014). Barik et al. (2012) observed the larvicidal and pupicidal activities of nanosilica against three mosquito strains. Nanosilica was also shown to be harmful to the reproduction of *Spodoptera littoralis*, leading to a considerable decrease in insect pest population and confirming its potential employment as a nano-insecticide (El-Bendary and El-Helaly, 2013).

Moreover, several studies have reported the promotion of plant growth by different types of nanoparticles and nanocomposites, such as nano-hydroxyapatite, nano-chitosan, silver nanoparticles, carbon nanotubes, gold nanoparticles, and others. Nanoparticles are reported to influence root and shoot elongation and increase germination chlorophyll content, and plant biomass (Tripathi et al., 2011; Arora et al., 2012; Sharma et al., 2012; Abdel-Aziz et al., 2016; Kottegoda et al., 2017).

9.4 Nanomaterials and plant growth-promoting bacteria

Microbial biofertilizers consist of live or latent microorganisms that increase nutrient availability in plants (Kour et al., 2020), such as plant growth-promoting bacteria (PGPB). Despite being eco-friendly and having several advantages over chemical fertilizers, biofertilizers can face limitations like stability in a given soil environment, causing decreased performance in plant processes (Glick, 2020). Such problems can be overcome using PGPB in conjunction with nanoparticles. Nanoparticle interactions with PGPB can facilitate bacterial colonization in plants, act as support for adherence, increase their resistance to extreme conditions, and enhance their beneficial properties.

Among the nanomaterials applied together with bacteria to promote plant growth, copper oxide (CuO), zinc oxide (ZnO), titanium dioxide (TiO_2), silicon dioxide (SiO_2), and metallic silver (AgNPs) and gold nanoparticles (AuNPs) are reported by different researchers (Dimkpa et al., 2012; Haris and Ahmad, 2017; Kibbey and Strevett, 2019; Tian et al., 2020; Ibrahim et al., 2019; Fernando et al., 2013). These nanoparticles have been mainly applied with *Pseudomonas fluorescens* and *Bacillus subtilis*. When used together, plants showed increased N_2 fixation, root and leaf weight, biomass, soil phosphorus, and IAA production by bacteria.

These nanoparticles can be commercially purchased, synthesized by wet chemical routes, or produced by endophytic bacteria. The latter has the advantage of negligible generation of toxic residues (Monowar et al., 2018). Examples of nanoparticles synthesized by bacteria are silver and copper nanoparticles. AgNPs can be produced using silver nitrate ($AgNO_3$) as a silver precursor and *Bacillus siamensis* as a reducing agent (Ibrahim et al., 2019). In the case of copper nanoparticles, the precursor used is copper sulfate pentahydrate ($CuSO_4 \cdot 5H_2O$) and the bacterium *Streptomyces capillispiralis* is responsible for reducing Cu^{2+} to Cu^0 (El-Din-Hassan et al., 2018). In both cases, a mild temperature close to 30°C is used for a period of at least 6h.

In the following section, there is a more detailed discussion on the combined application of NPs and PGPB, the mechanisms by which they affect plant growth, and the application of nanoparticles synthesized by endophytic bacteria (Fig. 1).

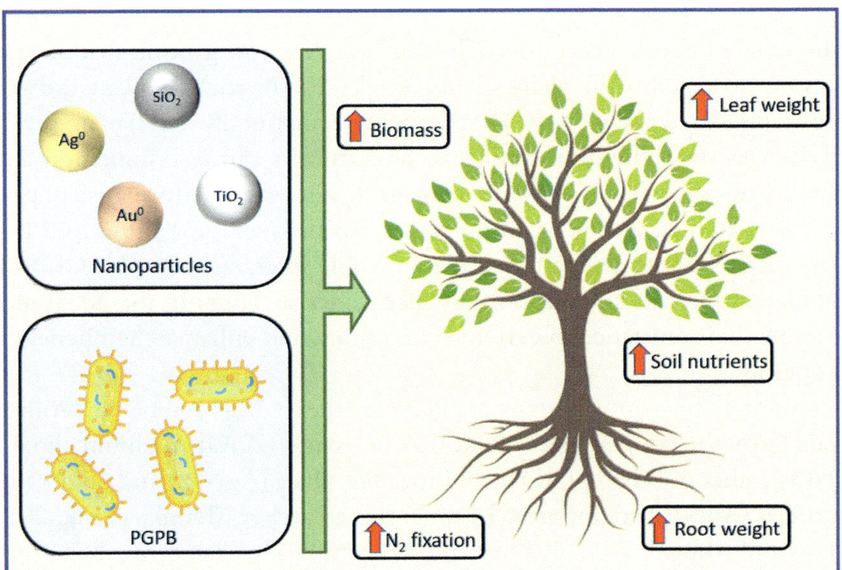

Fig. 1 The association of nanoparticles with plant growth-promoting bacteria (PGPB) can result in several gains for modern agriculture, such as improving the effectiveness of PGPB and the quality of crops (by authors).

9.4.1 The association between nanoparticles and beneficial microbes for plant growth promotion

Biofertilizer formulations containing nanoparticles designated as "nanobiofertilizers" are a promising and novel strategy to address limitations associated with biofertilizers such as shelf life, stability under environmental conditions, desiccation, and reproducibility (Mishra et al., 2017; El-Ghamry et al., 2018). These formulations can improve crop yield while reducing costs and environmental pollution by optimizing the beneficial traits of PGPB and improving soil fertility and plant growth-related molecules (Thirugnanasambandan, 2019; Kumari and Singh, 2020).

There are direct and indirect mechanisms by which NPs influence the plant-PGPB system. Directly, the NPs boost the synthesis of phytohormones and proteins involved in the plant growth machinery. Indirectly, nanomaterials stimulate PGPB, which results in enhanced beneficial activity via promoting plant growth even under adverse conditions such as abiotic stresses and pathogen attacks (Kalia and Kaur, 2019). The influence of NPs on the plant-PGPB system includes increased cell viability and bacterial growth rate, stimulation of specific beneficial microbial compounds, and improved bacterial attachment and colonization, owing to the unique properties of NP surfaces, which also protect these inoculants in the field (Ghalamboran and Ramsden, 2012; Karunakaran et al., 2013; Abdel Latef et al., 2018; Gudadhe et al., 2018).

Silica and gold NPs were found to induce growth of PGPB populations (Rangaraj et al., 2014; Shukla et al., 2015; Boroumand et al., 2020). They offer a significant advantage to microbial biofertilization as the bacterial population can drastically decrease after inoculation on plants in the field, thus reducing their beneficial effects on crops (Bashan et al., 2014). Silicates are the most abundant class of minerals in soils, which participate in soil solution regulation by forming chemical bonds with soluble substances (Luyckx et al., 2017) and by naturally interacting with plants and soil microorganisms in the environment. Therefore, silica NPs are promising and safer materials to be used with PGPB for plant growth promotion and protection (de Moraes et al., 2021). Inoculations of SiO_2 NPs with either *Pseudomonas stutzeri* or *Mesorhizobium* sp. enhanced N and P contents in the soil and augmented shoot and root dry weight of land cress (*Barbarea verna*) plants (Boroumand et al., 2020). Similar plant growth effects were obtained in maize (*Zea mays*) plants when treated with nano-silica and other PGPB (Karunakaran et al., 2013; Rangaraj et al., 2014; Pour et al., 2019). CuO

NPs at different concentrations were also found to promote positive effects on the beneficial traits of PGPB and plant growth. For example, 200 mg/L CuO NPs enhanced the IAA production by *Pseudomonas chlororaphis*, a plant growth-promoting bacterium (Haris and Ahmad, 2017). The inoculation of 50 mg/kg CuO NPs with PGPB increased N fixation and plant growth of wheat plants (*Triticum aestivum*) (Guan et al., 2020).

TiO_2 NPs can act as an instrument to deliver bacteria to specific tissues in plants (Palmqvist et al., 2015) due to their adhesion and bioluminescence properties as proved in a study by Park et al. (2008). Nano-titania also promotes bacterial stabilization and attachment to plant roots as shown in oilseed rape (*Brassica napus*) and wheat plants (Palmqvist et al., 2015; Timmusk et al., 2018). On the other hand, a higher concentration of TiO_2 NPs (200 μg/mL) reduced phosphate solubilization capacity and production of iron-binding siderophores by *Pseudomonas* spp. and *Bacillus amyloliquefaciens* (Kibbey and Strevett, 2019). The application of 1000 mg/kg TiO_2 and *P. fluorescens* inhibited the growth of *Trifolium repens* plants (Zand et al., 2020a), thus supporting the dose-dependent influence of nanomaterials on bacteria.

The application of nanomaterials associated with PGPB can also successfully address the negative impacts of abiotic stresses on crops. Under salt stress, formulations containing CuO NPs and PGPB (*Bacillus* spp. and *Lactobacillus casei*) amended DNA damage and DNA methylation in wheat plants, with the resultant improvement in tolerance and growth (Hosseinpour et al., 2021). Fetsiukh et al. (2021) found that silica NPs can improve bacterial development and performance in wheat plants under drought stress. The authors showed that although silica NPs did not significantly increase the PGPR population, they helped in bacterial attachment and colonization in plants, triggered PGPR exopolysaccharide (EPS) production, and increased the water-holding capacity and osmotic pressure of the biofilm matrix, leading to enhanced plant biomass in drought conditions (Fetsiukh et al., 2021).

Bioremediation is a low-priced and sustainable process to mitigate the contamination of heavy metals in water and soils (Singh and Gupta, 2016). Microorganisms can perform bioremediation using heavy metals as a source of their nutrition (Verma and Kuila, 2019). In this manner, the application of NPs and PGPB in plants has been shown to enhance the bacterial capacity for heavy metal uptake in contaminated soils. Khan and Bano (2015) reported that silver NPs (AgNPs) increased proline production and phytohormones levels and enhanced the PGPR bioremediation potential tested for Pb, Cd, and Ni in maize plants. Similarly, different concentrations of TiO_2 NPs used in combination with *P. fluorescens* increased the biomass of *T. repens* and enhanced its phytoremediation capacity for Cd (Zand et al., 2020a). Antimony (Sb), a metalloid harmful to the environment and human health, was shown to be more effectively removed from the soil by the coapplication of PGPR and 500 mg/kg zero-valent iron NPs than treatments exclusively composed of PGPR or zero-valent iron NPs only in *T. repens* plants, suggesting a helpful approach to the phytoremediation of Sb-contaminated soils (Zand et al., 2020b).

Nanomaterials can also improve the biocontrol potential of PGPB. NPs can increase the number of bacterial cells and provide an advantage to PGPB over phytopathogens in competition for niches and nutrients. In addition, the plant gets more resistant to phytopathogenic attacks as the NPs improve the beneficial traits of PGPB, increasing plant health (de Moraes et al., 2021). Djaya et al. (2019) reported an in vitro evaluation of a formulation containing PGPR, graphite, and silica NPs as a novel biocontrol delivery system (BDS) against *Ralstonia solanacearum* causing bacterial wilt on potatoes. The in vitro antagonism test consisted of a formulation containing the bacteria *Lysinibacillus* sp., *B. subtilis*, and *P. fluorescens* and graphite and silica NPs. Bacterial formulations with both NPs inhibited in vitro growth of *R. solanacearum*. In this perspective, nanobiopesticides can also be applied to replace or reduce chemical pesticides in current agricultural practices.

The interactions between nanoparticles and microorganisms seem to rely on the large surface area with a negative charge and different groups of the NPs regardless of the material. Some positively charged sites on the cell surface or solid particles are supposed to promote interactions between microbes and nanomaterials (Breznak et al., 2012; Kurdish, 2019). However, these interactions did not occur by electrostatic attraction only; the attachment also happened by chemical reactions on the bacterial surface such as those related to the phospholipid membrane (Palmqvist et al., 2015). Feng et al. (2015) reported that both the bacterial cell wall structure and the nanoparticle composition (charge and ligand type) determine the interaction between the NPs and the bacterium. Polyelectrolyte-wrapped AuNPs were most toxic to either Gram-negative or Gram-positive bacteria tested (Feng et al., 2015).

In summary, nanobiofertilizers have advantages over both regular nanofertilizers and biofertilizers in crop production. The combined action of nanomaterials and PGPB would require lower concentrations of nanoparticles than nanofertilizers alone to benefit plants, thereby diminishing the potential risks to the environment. Additionally, nanoparticles in combination with biofertilizers would work more effectively than biofertilizers in promoting plant growth under field conditions.

9.4.2 Nanoparticle biosynthesis using endophytes for nanobiopesticide formulations

Unquestionably, the excessive use of chemical fertilizers and pesticides over the past few years has resulted in the deterioration of soil quality, pollinator decline, and devastation of the natural habitats of friendly birds. Hence, crops have become more susceptible to diseases (Tilman et al., 2002). In the past few decades, the natural and biological control of insect pests and diseases to reduce pesticide use in agriculture has gained much attention (Lacava and Azevedo, 2014; Rana et al., 2020). Biocontrol has an extremely long association with agriculture and has played a constitutive role in integrated pest management programs worldwide (Orr and Lahiri, 2014; Damalas and Koutroubas, 2018). The interest in alternative control measures has further increased in recent years due to the growing population of pathogens resistant to chemical pesticides and consumer awareness of safer food (Kumar et al., 2013).

Many reports have described the feasibility of bacterial endophytic application in controlling phytopathogenic fungi (Berg et al., 2005; Sessitsch et al., 2004; Coombs et al., 2004), bacteria (Sessitsch et al., 2004), and insects (Lacava and Azevedo, 2014; Azevedo et al., 2000; Quecine et al., 2008).

Besides acting as biocontrol agents by their mechanisms, microbes have been proven to be a source for nanoparticle (NP) biosynthesis, another green strategy to control insect pests and diseases (El-Din-Hassan et al., 2018). The biosynthesis of NPs by endophytic microorganisms is considered clean, nonhazardous, and green nanotechnology (Rana et al., 2020). In this context, several studies have shown that NPs can play an important role in managing plant diseases and promoting plant growth (Kasprowicz et al., 2015; Pariona et al., 2019; Maluin et al., 2019). Some studies have demonstrated that metal NPs serve as potential fungicides (Chen et al., 2018; Abdallah et al., 2019). Silver nanoparticles (AgNPs) have been reported to significantly inhibit the growth of a variety of phytopathogenic fungi such as *A. alternata*, *Botrytis cinerea*, *Colletotrichum gloeosporioides*, *Fusarium oxysporum*, *F. solani*, *Monilinia fructicola*, and *Verticillium dahlia* (Malandrakis et al., 2019; Khalil et al., 2019). Therefore, the biosynthesis of AgNPs by microorganisms has been investigated by many authors (Masum et al., 2019; Ogunyemi et al., 2019; Hossain et al., 2019).

Fernando et al. (2013) reported the biosynthesis of AgNPs using a culture filtrate of the endophytic bacterium *B. siamensis* and their role in inhibiting two phytopathogenic bacteria in rice. Their study indicated that biosynthesized AgNPs showed a strong antibacterial effect against *Xanthomonas oryzae* pv. *oryzae* and *Acidovorax oryzae*, which cause rice bacterial leaf blight and bacterial brown stripe, respectively (Fernando et al., 2013). Similarly, Ibrahim et al. (2020) suggested that NPs are expected to be critical components in managing plant diseases and provide an alternative to traditional synthetic fungicides. The authors investigated the inhibition potential of AgNPs synthesized by the endophytic bacterium *Pseudomonas poae* against *Fusarium* head blight. The biosynthesized AgNPs showed strong antifungal activity against *F. graminearum* attributed to inhibitory effects on spore germination, germ tube growth, mycotoxin production, and damage to the cell membrane (Ibrahim et al., 2020). The phytopathogenic fungus *F. graminearum* is a significant obstacle to wheat production worldwide (Dubin et al., 1996). Besides, it can infect other cereal crops such as rice, barley, and maize (Goswami and Kistler, 2004; Yu et al., 2017). This pathogen has caused more than 3.41 million tons of wheat loss in China during 2000–18 (Hongxiang et al., 2008; Chen et al., 2019).

El-Din-Hassan et al., 2018) reported that novel, nontoxic, and eco-friendly copper nanoparticles (CuNPs) were biosynthesized by endophytic actinomycete *S. capillispiralis* Ca-1 strain. The antimicrobial activity of CuNPs was tested against the phytopathogenic fungi *Alternaria* spp., *Aspergillus niger*, *Pythium* spp., and *Fusarium* spp. The results showed suitable antifungal activities at different concentrations of CuNPs. The possible action mechanisms of CuNPs are based on the changing structure and function of fungal cells (Cioffi et al., 2005). Similarly, Grillo et al. (2014) reported that the synthesized CuNPs demonstrated antifungal activity against the crop pathogenic fungi *Curvularia lunata* and *Phoma destructive*.

Endophytic fungi have often been considered as an important source of novel bioactive secondary metabolites with biological activities (Kamalraj and Muthumary, 2012; Subban et al., 2013). As a green nanotechnology approach, fungal endophytes have been reported to biosynthesize AgNPs (Devi and Joshi, 2014; Devi et al., 2014; Qian et al., 2013; Singh et al., 2014). Balakumaran et al. (2015) reported antifungal activity of AgNPs synthesized by the endophytic fungus *Guignardia mangiferae* (Bios PTK 4 strain). The mycosynthesized AgNPs were tested using an in vitro assay against various fungal pathogens. The AgNPs exhibited good antifungal activity against *Colletotrichum* sp., followed by *Rhophitulus solani* and *C. lunata*. The least activity was observed against *Fusarium* sp.

The toxicity of nanoparticles to microorganisms depends on several factors such as shape, nature of the metal, concentration of the nanomaterials (Haris and Ahmad, 2017; Pal et al., 2007; Sinha et al., 2011; Hayden et al., 2012), and microbial characteristics (Chavan et al., 2020). Therefore, predicting whether NPs inhibit or stimulate microbial phytopathogens is not simple. Large-scale studies involving several variables influencing plants' beneficial and harmful microorganisms are required to guarantee the safety and effectiveness of nanoparticles in agriculture (de Moraes et al., 2021).

9.5 Conclusions and perspectives

The search for interesting natural biological activities has been the basis for developing various biotechnological and agricultural applications. The microbial world, in particular endophytes and rhizobacteria, exhibit vast genetic and metabolic biodiversity that has not yet been thoroughly explored. In this context, the potential use of these microorganisms as biofertilizers in agriculture constitutes an economical and eco-friendly way to reduce agrochemicals.

Nanotechnology serves as a promising tool to overcome the obstacles associated with microbial biofertilizers. Many nanomaterials have shown positive effects on the plant growth-promoting microbes of various plant species, making such materials suitable components of biofertilizer formulations. The approach of using nanomaterials in association with beneficial microorganisms in plants can increase microbial viability and resistance in the environment, improve field stability, and, consequently, improve plant growth promotion. In this scenario, nanotechnology has emerged as a new source of quality enrichment for the agricultural sector. The use of nanoparticles can be an effective method to meet the challenges regarding the strength of biofertilizers in natural environments.

Further experimentation involving plant growth-promoting microbes—NPs—plants is necessary to evaluate this complex system at the molecular level. Some factors involved are microorganisms' nature and metabolism, nanomaterials (substance used, synthesis method, shape, size, and concentration), plant species, and environmental interactions.

Acknowledgments

The authors acknowledge the funding from the São Paulo Research Foundation (FAPESP), Grant number 2020/11315-6.

References

Abdalla, S.A., Algam, S.A.A., Ibrahim, E.A., Naim, A.M.E., 2014. In vitro screening of *Bacillus* isolates for biological control of early blight disease of tomato in Shambat soil. World J. Agric. Res. 2 (1), 47–50.

Abdallah, Y., Ogunyemi, S.O., Abdelazez, A., Zhang, M., Hong, X., Ibrahim, E., 2019. The green synthesis of MgO nano-flowers using *Rosmarinus officinalis* L. (Rosemary) and the antibacterial activities against *Xanthomonas oryzae* pv. *oryzae*. Biomed. Res. Int., 5620989. https://doi.org/10.1155/2019/5620989.

Abdel Latef, A.A.H., Srivastava, A.K., El-sadek, M.S.A., et al., 2018. Titanium dioxide nanoparticles improve growth and enhance tolerance of broad bean plants under saline soil conditions. Land Degrad. Dev. 29, 1065–1073. https://doi.org/10.1002/ldr.2780.

Abdel-Aziz, H.M., Hasaneen, M.N., Omer, A.M., 2016. Nano chitosan-NPK fertilizer enhances the growth and productivity of wheat plants grown in sandy soil. Span. J. Agric. Res. 14 (1), 0902.

Ahemad, M., Kibret, M., 2014. Mechanisms and applications of plant growth promoting rhizobacteria: current perspective. J. King Saud Univ. Sci. 26, 1–20. https://doi.org/10.1016/j.jksus.2013.05.001.

Arora, S., Sharma, P., Kumar, S., Nayan, R., Khanna, P.K., Zaidi, M.G.H., 2012. Gold nanoparticle induced enhancement in growth and seed yield of *Brassica juncea*. Plant Growth Regul. 66 (3), 303–310. https://doi.org/10.1007/s10725-011-9649-z.

Azevedo, J.L., Maccheroni, W.J., Pereira, J.O., Araújo, W.L., 2000. Endophytic microorganisms: a review on insect control and recent advances on tropical plants. Electron. J. Biotechnol. 3, 40–75. https://doi.org/10.2225/vol3-issue1-fulltext-4.

Baker, S., Volova, T., Prudnikova, S.V., Satish, S., Prasad, N., 2017. Nanoagroparticles emerging trends and future prospect in modern agriculture system. Environ. Toxicol. Pharmacol. 53, 10–17. https://doi.org/10.1016/j.etap.2017.04.012.

Balakumaran, M.D., Ramachandran, R., Kalaichelvan, P.T., 2015. Exploitation of endophytic fungus, *Guignardia mangiferae* for extracellular synthesis of silver nanoparticles and their in vitro biological activities. Microbiol. Res. 178, 9–17. https://doi.org/10.1016/j.micres.2015.05.009.

Barik, T.K., Kamaraju, R., Gowswami, A., 2012. Silica nanoparticle: a potential new insecticide for mosquito vector control. Parasitol. Res. 111 (3), 1075–1083. https://doi.org/10.1007/s00436-012-2934-6.

Bashan, Y., de Bashan, L.E., Prabhu, S.R., Hernandez, J.P., 2014. Advances in plant growth-promoting bacterial inoculant technology: formulations and practical perspectives (1998–2013). Plant Soil 378, 1–33. https://doi.org/10.1007/s11104-013-1956-x.

Batista, B.D., Lacava, P.T., Ferrari, A., Teixeira-Silva, N.S., Bonatelli, M.L., Tsui, S., Mondin, M., Kitajima, E.W., Pereira, J.O., Azevedo, J.L., Quecine, M.C., 2018. Screening of tropically derived, multi-trait plant growth-promoting rhizobacteria and evaluation of corn and soybean colonization ability. Microbiol. Res. 206, 33–42.

Benhamou, N., Gagné, S., Le Quéré, D., Dehbi, L., 2000. Bacterial-mediated induced resistance in cucumber: beneficial effect of the endophytic bacterium *Serratia plymuthica* on the protection against infection by *Pythium ultimum*. Phytopathology 90 (1), 45–56. https://doi.org/10.1094/PHYTO.2000.90.1.45.

Berg, G., Eberl, L., Hartmann, A., 2005. The rhizosphere as a reservoir for opportunistic human pathogenic bacteria. Environ. Microbiol. 7, 1673–1685. https://doi.org/10.1111/j.1462-2920.2005.00891.x.

Boroumand, N., Behbahani, M., Dini, G., 2020. Combined effects of phosphate solubilizing bacteria and nanosilica on the growth of land cress plant. J. Soil Sci. Plant Nutr. https://doi.org/10.1007/s42729-019-00126-8.

Brauer, V.S., Rezende, C.P., Pessoni, A.M., De Paula, R.G., Rangappa, K.S., Nayaka, S.C., Gupta, V.K., Almeida, F., 2019. Antifungal agents in agriculture: friends and foes of public health. Biomol. Ther. 9, 521. https://doi.org/10.3390/biom9100521.

Breznak, J.A., McFeters, G.A., Rutter, P.R., 2012. Microbial Adhesion and Aggregation: Report of the Dahlen Workshop on Microbial Adhesion and Aggregation, Berlin, 1984. Springer Science & Business Media.

Burgess, B.K., Lowe, D.J., 1996. Mechanism of molybdenum nitrogenase. Chem. Rev. 96, 2983–3012. https://doi.org/10.1021/cr950055x.

Cattelan, A.J., Hartel, P.G., Fuhrmann, J.J., 1999. Screening for plant growth–promoting rhizobacteria to promote early soybean growth. Soil Sci. Soc. Am. J. 63 (6), 1670–1680. https://doi.org/10.2136/sssaj1999.6361670x.

Chavan, S., Sarangdhar, V., Nadanathangam, V., 2020. Toxicological effects of TiO_2 nanoparticles on plant growth promoting soil bacteria. Emerg. Contam. 6, 87–92. https://doi.org/10.1016/j.emcon.2020.01.003.

Chen, H., Yada, R., 2011. Nanotechnologies in agriculture: new tools for sustainable development. Trends Food Sci. Technol. 22 (11), 585–594. https://doi.org/10.1016/j.tifs.2011.09.004.

Chen, L., Heng, J., Qin, S., Bian, K.A., 2018. A comprehensive understanding of the biocontrol potential of *Bacillus velezensis* LM2303 against *fusarium* head blight. PLoS One 13, e0198560. https://doi.org/10.1371/journal.pone.0198560.

Chen, Y., Kistler, H.C., Ma, Z., 2019. *Fusarium graminearum* Trichothecene mycotoxins: biosynthesis, regulation, and management. Annu. Rev. Phytopathol. 57, 15–39. https://doi.org/10.1146/annurev-phyto-082718-100318.

Chhipa, H., 2017. Nanofertilizers and nanopesticides for agriculture. Environ. Chem. Lett. 15 (1), 15–22. https://doi.org/10.1007/s10311-016-0600-4.

Chi, F., Shen, S.H., Cheng, H.P., Jing, Y.X., Yanni, Y.G., Dazzo, F.B., 2005. Ascending migration of endophytic rhizobia, from roots to leaves, inside rice plants and assessment of benefits to rice growth physiology. Appl. Environ. Microbiol. 71, 7271–7278. https://doi.org/10.1128/AEM.71.11.7271-7278.2005.

Cioffi, N., Torsi, L., Ditaranto, N., Tantillo, G., Ghibelli, L., Sabbatini, L., et al., 2005. Copper nanoparticle/polymer composites with antifungal and bacteriostatic properties. Chem. Mater. 17, 5255–5262. https://doi.org/10.1021/cm0505244.

Compant, S., Duffy, B., Nowak, J., Clement, C., Barka, E.A., 2005. Use of plant growth-promoting bacteria for biocontrol of plant diseases: principles, mechanisms of action, and future prospects. Appl. Environ. Microbiol. 71, 4951–4959. https://doi.org/10.1128/AEM.71.9.4951-4959.2005.

Coombs, J.T., Michelsen, P.P., Franco, C.M.M., 2004. Evaluation of endophytic actinobacteria as antagonists of *Gaeumannomyces graminis* var. *tritici* in wheat. Biol. Control 29, 359–366. https://doi.org/10.1016/j.biocontrol.2003.08.001.

Coutinho, H.L.C., De Oliveira, V.M., Moreira, F.M.S., 2000. Systematics of legume nodule nitrogen fixing bacteria. In: Priest, F.G., Goodfellow, M. (Eds.), Applied Microbial Systematics. Springer, Dordrecht, https://doi.org/10.1007/978-94-011-4020-1_5.

Cullen, M.G., Thompson, L.J., Carolan, J.C., Stout, J.C., Stanley, D.A., 2019. Fungicides, herbicides and bees: a systematic review of existing research and methods. PLoS One 14, e0225743. https://doi.org/10.1371/journal.pone.0225743.

Damalas, C.A., Koutroubas, S.D., 2018. Current status and recent developments in biopesticide use. Agriculture 8, 13. https://doi.org/10.3390/agriculture8010013.

de Moraes, A.C.P., da Silva Ribeiro, L., de Camargo, E.R., Lacava, P.T., 2021. The potential of nanomaterials associated with plant growth-promoting bacteria in agriculture. 3Biotech 11 (7), 1–17. https://doi.org/10.1007/s13205-021-02870-0.

De, A., Bose, R., Kumar, A., Mozumdar, S., 2014. Worldwide pesticide use. In: Targeted Delivery of Pesticides Using Biodegradable Polymeric Nanoparticles. Springer, New Delhi, pp. 5–6, https://doi.org/10.1007/978-81-322-1689-6_2.

Devi, L., Joshi, S., 2014. Evaluation of the antimicrobial potency of silver nanoparticles biosynthesized by using an endophytic fungus, *Cryptosporiopsis ericae* PS4. J. Microbiol. 52, 667–674. https://doi.org/10.1007/s12275-014-4113-1.

Devi, L., Bareh, D., Joshi, S.R., 2014. Studies on biosynthesis of antimicrobial silver nanoparticles using endophytic fungi isolated from the ethno-medicinal plant *Gloriosa superba* L. Proc. Natl Acad. Sci. India Sect. B. Biol. Sci. 84, 1091–1099. https://doi.org/10.1007/s40011-013-0185-7.

Dias, A.C.F., Costa, F.E.C., Andreote, F.D., et al., 2009. Isolation of micropropagated strawberry endophytic bacteria and assessment of their potential for plant growth promotion. World J. Microbiol. Biotechnol. 25, 189–195. https://doi.org/10.1007/s11274-008-9878-0.

Dimkpa, C.O., Mclean, J.E., Britt, D.W., Anderson, A.J., 2012. CuO and ZnO nanoparticles differently affect the secretion of fluorescent siderophores in the beneficial root colonizer, *Pseudomonas chlororaphis* O6. Nanotoxicology 6, 635–642. https://doi.org/10.3109/17435390.2011.598246.

Djaya, L., Hersanti, Istifadah, N., Hartati, S., Joni, I.M., 2019. *In vitro* study of plant growth promoting rhizobacteria (PGPR) and endophytic bacteria antagonistic to *Ralstonia solanacearum* formulated with graphite and silica nano particles as a biocontrol delivery system (BDS). Biocatal. Agric. Biotechnol. 19, 101153. https://doi.org/10.1016/j.bcab.2019.101153.

Dubin, H.J., Gilchrist, L., Reeves, J., McNab, A., 1996. *Fusarium* head scab: global status and future prospects; conflicts. In: Conference Proceedings 13–17, Inc., El Batan, Mexico.

Duhan, J.S., Kumar, R., Kumar, N., Kaur, P., Nehra, K., Duhan, S., 2017. Nanotechnology: the new perspective in precision agriculture. Biotechnol. Rep. 15, 11–23.

Eady, R.R., Postgate, J.R., 1974. Nitrogenase. Nature 249, 805–810. https://doi.org/10.1038/249805a0.

El-Bendary, H.M., El-Helaly, A.A., 2013. First record nanotechnology in agricultural: silica nano-particles a potential new insecticide for pest control. Appl. Sci. Rep. 4 (3), 241–246.

El-Din-Hassan, S., Salem, S.S., Fouda, A., Awad, M.A., El-Gamal, M.S., Abdo, A.M., 2018. New approach for antimicrobial activity and bio-control of various pathogens by biosynthesized copper nanoparticles using endophytic actinomycetes. J. Radiat. Res. Appl. Sci. 11 (3), 262–270. https://doi.org/10.1016/j.jrras.2018.05.003.

El-Ghamry, A.M., Mosa, A.A., Alshaal, T.A., El-Ramady, H.R., 2018. Nanofertilizers vs. biofertilizers: new insights. Environ. Biodivers. Soil Secur. 2, 51–72. Available from: https://journals.ekb.eg/article_8905.html.

Feng, Z.V., Gunsolus, I.L., Qiu, T.A., Hurley, K.R., Nyberg, L.H., Frew, H., et al., 2015. Impacts of gold nanoparticle charge and ligand type on surface binding and toxicity to Gram-negative and Gram-positive bacteria. Chem. Sci. 6, 5186–5196. https://doi.org/10.1039/c5sc00792e.

Fernando, L.M., Merca, F.E., Paterno, E.S., 2013. Biogenic synthesis of gold nanoparticles by plant-growth-promoting bacteria isolated from Philippine soils. Philip. Agric. Sci. 96 (2), 129–136. https://doi.org/10.1016/j.btre.2017.03.002.

Fetsiukh, A., Conrad, J., Bergquist, J., Timmusk, S., 2021. Silica particles trigger the exopolysaccharide production of harsh environment isolates of growth-promoting rhizobacteria and increase their ability to enhance wheat biomass in drought-stressed soils. Int. J. Mol. Sci. 22, 6201. https://doi.org/10.3390/ijms22126201.

Ghalamboran, M.R., Ramsden, J.J., 2012. Viability of *Bradyrhizobium japanicum* on soybean seeds enhanced by magnetite nanoparticles during desiccation. Int. J. Biol. Life Sci. 8, 228–233. Available from: http://dspace.lib.cranfield.ac.uk/handle/1826/5564.

Glick, B.R., 2012. Plant growth-promoting bacteria: mechanisms and applications. Scientifica. https://doi.org/10.6064/2012/963401.

Glick, B.R., 2020. Introduction to plant growth-promoting bacteria. In: Beneficial Plant-Bacterial Interactions. Springer, Cham, https://doi.org/10.1007/978-3-030-44368-9_1.

Goswami, R.S., Kistler, H.C., 2004. Heading for disaster: *Fusarium graminearum* on cereal crops. Mol. Plant Pathol. 5, 515–525. https://doi.org/10.1111/j.1364-3703.2004.00252.x.

Gray, E.J., Smith, D.L., 2005. Intracellular and extracellular PGPR: commonalities and distinctions in the plant-bacterium signaling processes. Soil Biol. Biochem. 37, 395–412. https://doi.org/10.1016/j.soilbio.2004.08.030.

Griffin, M.R., 2014. Biocontrol and bioremediation: two areas of endophytic research which hold great promise. In: Verma, V., Gange, A. (Eds.), Advances in Endophytic Research. Springer, https://doi.org/10.1007/978-81-322-1575-2_14.

Grillo, R., Pereira, A.E., Nishisaka, C.S., de Lima, R., Oehlke, K., Greiner, R., et al., 2014. Chitosan/tripolyphosphate nanoparticles loaded with paraquat herbicide: an environmentally safer alternative for weed control. J. Hazard. Mater. 278, 163–171. https://doi.org/10.1016/j.jhazmat.2014.05.079.

Guan, X., Gao, X., Avellan, A., et al., 2020. CuO nanoparticles alter the rhizospheric bacterial community and local nitrogen cycling for wheat grown in a calcareous soil. Environ. Sci. Technol. 54, 8699–8709. https://doi.org/10.1021/acs.est.0c00036.

Gudadhe, N.N., Suthar, K.P., Thanki, J.D., 2018. Synthesis and evaluation of ZnO nanoparticles through green and chemical methods and its effect on rice. In: XXI Biennial National Symposium of Indian Society of Agronomy. MPUAT, Udaipur, pp. 24–26.

Gupta, K., Dubey, N.K., Singh, S.P., Kheni, J.K., Gupta, S., Varshney, A., 2021. Plant growth-promoting rhizobacteria (PGPR): current and future prospects for crop improvement. In: Yadav, A.N., Singh, J., Singh, C., Yadav, N. (Eds.), Current Trends in Microbial Biotechnology for Sustainable Agriculture. Environmental Microbial Biotechnology. Springer, https://doi.org/10.1007/978-981-15-6949-4_9.

Gutiérrez, F.J., Mussons, M.L., Gatón, P., Rojo, R., 2011. Nanotechnology and food. In: Valdez, B., Schorr, M., Zlatev, R. (Eds.), Scientific, Health and Social Aspects of the Food Industry. InTech, pp. 95–128.

Hafez, E.E., Hassan, H.S., Elkady, M., Salama, E., 2014. Assessment of antibacterial activity for synthesized zinc oxide nanorods against plant pathogenic strains. Int. J. Sci. Technol. Res. 3 (9), 318–324.

Haris, Z., Ahmad, I., 2017. Impact of metal oxide nanoparticles on beneficial soil microorganisms and their secondary metabolites. Int. J. Life Sci. Sci. Res. 3, 1020–1030. https://doi.org/10.21276/ijlssr.2017.3.3.10.

Hasaneen, M.N.A., Abdel-Aziz, H.M.M., El-Bialy, D.M.A., Omer, A.M., 2014. Preparation of chitosan nanoparticles for loading with NPK fertilizer. Afr. J. Biotechnol. 13 (31). https://doi.org/10.5897/AJB2014.13699.

Hayat, R., Ahmed, I., Sheirdil, R.A., 2012. An overview of plant growth promoting rhizobacteria (PGPR) for sustainable agriculture. In: Crop Production for Agricultural Improvement. Springer, pp. 557–579, https://doi.org/10.1007/978-94-007-4116-4_22.

Hayden, S.C., Zhao, G., Saha, K., et al., 2012. Aggregation and interaction of cationic nanoparticles on bacterial surfaces. J. Am. Chem. Soc. 134, 6920–6923. https://doi.org/10.1021/ja301167y.

Hongxiang, M., Yao, J., Zhou, M., Zhang, X., Ren, L., Yu, G., 2008. Molecular breeding for wheat fusarium head blight resistance in China. Cereal Res. Commun. 36, 203–212. https://doi.org/10.1556/crc.36.2008.suppl.b.16.

Hossain, A., Hong, X., Ibrahim, E., Li, B., Sun, G., Meng, Y., 2019. Green synthesis of silver nanoparticles with culture supernatant of a bacterium *Pseudomonas rhodesiae* and their antibacterial activity against soft rot pathogen *Dickeya dadantii*. Molecules 24, 2303. https://doi.org/10.3390/molecules24122303.

Hosseinpour, A., Ilhan, E., Özkan, G., et al., 2021. Plant growth-promoting bacteria (PGPBs) and copper (II) oxide (CuO) nanoparticle ameliorates DNA damage and DNA methylation in wheat (*Triticum aestivum* L.) exposed to NaCl stress. J. Plant Biochem. Biotechnol. https://doi.org/10.1007/s13562-021-00713-w.

Ibrahim, E., Fouad, H., Zhang, M., Zhang, Y., Qiu, W., Yan, C., Li, B., Mo, J., Chen, J., 2019. Biosynthesis of silver nanoparticles using endophytic bacteria and their role in inhibition of rice pathogenic bacteria and plant growth promotion. RSC Adv. 9, 29293–29299. https://doi.org/10.1039/c9ra04246f.

Ibrahim, E., Zhang, M., Zhang, Y., Hossain, A., Qiu, W., Chen, Y., Wang, Y., Wu, W., Sun, G., Li, B., 2020. Green-synthesization of silver nanoparticles using endophytic bacteria isolated from garlic and its antifungal activity against wheat *Fusarium* head blight pathogen *Fusarium graminearum*. Nanomaterials 10, 219. https://doi.org/10.3390/nano10020219.

Inui, R., Takeshi, L., Picchi, S., Barbosa, J., Lemos, M., Marcondes, J., Macedo, E., Lemos, D., 2012. Phosphorus solubilizing and IAA production activities in plant growth promoting rhizobacteria from Brazilian soils under sugarcane cultivation. J. Eng. Appl. Sci. 7, 1446–1454.

Jha, C.K., Annapurna, K., Saraf, M., 2012. Isolation of rhizobacteria from *Jatropha curcas* and characterization of produced ACC deaminase. J. Basic Microbiol. 52, 285–295. https://doi.org/10.1002/jobm.201100113.

Kadri, A., Zarai, Z., Chobba, I.B., Gharsallah, N., Damak, M., Békir, A., 2013. Chemical composition and in vitro antioxidant activities of Thymelaea hirsuta L: essential oil from Tunisia. Afr. J. Biotechnol. 10, 2930–2935.

Kalia, A., Kaur, H., 2019. Nano-biofertilizers: harnessing dual benefits of nano-nutrient and bio-fertilizers for enhanced nutrient use efficiency and sustainable productivity. In: Pudake, R., Chauhan, N., Kole, C. (Eds.), Nanoscience for Sustainable Agriculture. Cham, Springer, pp. 51–73.

Kamalraj, S., Muthumary, J., 2012. Screening and identification of potential coelomycetous strain (S) for the production of extracellular antimicrobial metabolites. Asian J. Pharm. Clin. Res. 5, 137–145.

Kanhed, P., Birla, S., Gaikwad, S., Gade, A., Seabra, A.B., Rubilar, O., Duran, N., Rai, M., 2014. *In vitro* antifungal efficacy of copper nanoparticles against selected crop pathogenic fungi. Mater. Lett. 115, 13–17. https://doi.org/10.1016/j.matlet.2013.10.011.

Karunakaran, G., Suriyaprabha, R., Manivasakan, P., et al., 2013. Effect of nanosilica and silicon sources on plant growth promoting rhizobacteria, soil nutrients and maize seed germination. IET Nanobiotechnol. 7, 70–77. https://doi.org/10.1049/iet-nbt.2012.0048.

Kasprowicz, M.J., Gorczyca, A., Szymocha, A., 2015. Physiological effects of nanosilver on vegetative mycelium, conidia and the development of the entomopathogenic fungus, *Isaria fumosorosea*. Biocontrol Sci. Tech. 25, 873–887. https://doi.org/10.1080/09583157.2015.1020284.

Khalil, N.M., El-Ghany, M.N.A., Rodríguez-Couto, S., 2019. Antifungal and antimycotoxin efficacy of biogenic silver nanoparticles produced by *Fusarium chlamydosporum* and *Penicillium chrysogenum* at non-cytotoxic doses. Chemosphere 218, 477–486. https://doi.org/10.1016/j.chemosphere.2018.11.129.

Khan, N., Bano, A., 2015. Role of plant growth promoting rhizobacteria and Ag-nano particle in the bioremediation of heavy metals and maize growth under municipal wastewater irrigation. Int. J. Phytoremediation 18, 211–221. https://doi.org/10.1080/15226514.2015.1064352.

Khan, M.S., Zaidi, A., Wani, P.A., 2009. Role of phosphate solubilizing microorganisms in sustainable agriculture—a review. In: Lichtfouse, E., Navarrete, M., Debaeke, P., Véronique, S., Alberola, C. (Eds.), Sustainable Agriculture. Springer, https://doi.org/10.1007/978-90-481-2666-8_34.

Kibbey, T.C., Strevett, K.A., 2019. The effect of nanoparticles on soil and rhizosphere bacteria and plant growth in lettuce seedlings. Chemosphere 221, 703–707. https://doi.org/10.1016/j.chemosphere.2019.01.091.

Kloepper, J.W., Schroth, M.N., 1978. Plant growth-promoting rhizobacteria on radishes. In: Station de Pathologie Végétale et Phyto-Bactériologie (Ed.), Proceedings of the 4th International Conference on Plant Pathogenic Bacteria, 2, pp. 879–882.

Kottegoda, N., Sandaruwan, C., Priyadarshana, G., Siriwardhana, A., Rathnayake, U.A., Arachchige, D.M.B., et al., 2017. Urea-hydroxyapatite nanohybrids for slow release of nitrogen. ACS Nano 11 (2), 1214–1221. https://doi.org/10.1021/acsnano.6b07781.

Kour, D., Rana, K.L., Yadav, A.N., Yadav, N., Kumar, M., Kumar, V., Vyas, P., Dhaliwal, H.S., Saxena, A.K., 2020. Microbial biofertilizers: bioresources and eco-friendly technologies for agricultural and environmental sustainability. Biocatal. Agric. Biotechnol. 23, 101487. https://doi.org/10.1016/j.bcab.2019.101487.

Kumar, S., Kaushik, N., Proksch, P., 2013. Identification of antifungal principle in the solvent extract of an endophytic fungus *Chaetomium globosum* from *Withania somnifera.* Springerplus 2, 37. https://doi.org/10.1186/2193-1801-2-37.

Kumar, H., Venkatesh, N., Bhowmik, H., Kuila, A., 2018. Metallic nanoparticle: a review. Biomed. J. Sci. Tech. Res. 4 (2), 3765–3775. https://doi.org/10.26717/BJSTR.2018.04.001011.

Kumari, R., Singh, D.P., 2020. Nano-biofertilizer: an emerging eco-friendly approach for sustainable agriculture. Proc. Natl. Acad. Sci. India Sect. B Biol. Sci. 90, 733–741. https://doi.org/10.1007/s40011-019-01133-6.

Kurdish, I.K., 2019. Interaction of microorganisms with nanomaterials as a basis for creation of high-efficiency biotechnological preparations. In: Prasad, R., Kumar, V., Kumar, M., Choudhary, D. (Eds.), Nanotechnology in Bioformulations. Nanotechnology in the Life Sciences. Springer, Cham, pp. 259–287.

Lacava, P.T., Azevedo, J.L., 2013. Endophytic bacteria: a biotechnological potential in agrobiology system. In: Maheshwari, D., Saraf, M., Aeron, A. (Eds.), Bacteria in Agrobiology: Crop Productivity. Springer, Berlin, Heidelberg, https://doi.org/10.1007/978-3-642-37241-4_1.

Lacava, P.T., Azevedo, J.L., 2014. Biological control of insect-pest and diseases by endophytes. In: Verma, V.C., Gange, A.C. (Eds.), Advances in Endophytic Research. Springer, pp. 231–256, https://doi.org/10.1007/978-81-322-1575-2_13.

Leinhos, V., Vacek, O., 1994. Biosynthesis of auxins by phosphate-solubilizing rhizobacteria from wheat (*Triticum aestivum*) and rye (*Secale cereale*). Microbiol. Res. 149, 31–35. https://doi.org/10.1016/S0944-5013(11)80132-1.

Lisboa, P.H.G., de Andrade, P.H.M., Machado, P.C., de Sousa, C.P., Lacava, P.T., 2021. Isolation and in vitro screening of plant growth-promoting rhizobacteria from *Solanum lycocarpum* St. Hil., an endemic plant of the Brazilian tropical savannah. Afr. J. Microbiol. Res. 15 (5), 253–261. https://doi.org/10.5897/AJMR2021.9524.

Luyckx, M., Hausman, J.F., Lutts, S., Guerriero, G., 2017. Silicon and plants: current knowledge and technological perspectives. Front. Plant Sci. 8, 411. https://doi.org/10.3389/fpls.2017.00411.

Malandrakis, A.A., Kavroulakis, N., Chrysikopoulos, C.V., 2019. Use of copper, silver and zinc nanoparticles against foliar and soil-borne plant pathogens. Sci. Total Environ. 670, 292–299. https://doi.org/10.1016/j.scitotenv.2019.03.210.

Maluin, F.N., Hussein, M.Z., Yusof, N.A., Fakurazi, S., Idris, A.S., Zainol Hilmi, N.H., 2019. Preparation of chitosan-hexaconazole nanoparticles as fungicide nanodelivery system for combating ganoderma disease in oil palm. Molecules 24, 2498. https://doi.org/10.3390/molecules24132498.

Masum, M.M.I., Siddiqa, M.M., Ali, K.A., Zhang, Y., Abdallah, Y., Ibrahim, E., 2019. Biogenic synthesis of silver nanoparticles using *Phyllanthus emblica* fruit extract and its inhibitory action against the pathogen *Acidovorax oryzae* strain RS-2 of rice bacterial brown stripe. Front. Microbiol. 10, 820. https://doi.org/10.3389/fmicb.2019.00820.

Mendes, R., Pizzirani-Kleiner, A.A., Araujo, W.L., Raaijmakers, J.M., 2007. Diversity of cultivated endophytic bacteria from sugarcane: genetic and biochemical characterization of *Burkholderia cepacia* complex isolates. Appl. Environ. Microbiol. 73 (22), 7259–7267. https://doi.org/10.1128/AEM.01222-07.

Mendes, R., Garbeva, P., Raaijmarkers, J.M., 2013. The rizhosphere microbiome: significance of plant beneficial, plant pathogenic, and humam pathogenic microorganisms. FEMS Microbiol. Rev. 37, 634–663. https://doi.org/10.1111/1574-6976.12028.

Mishra, S., Keswani, C., Abhilash, P.C., Fraceto, L.F., Singh, H.B., 2017. Integrated approach of agri-nanotechnology: challenges and future trends. Front. Plant Sci. 8, 471. https://doi.org/10.3389/fpls.2017.00471.

Mondal, K.K., Mani, C., 2012. Investigation of the antibacterial properties of nanocopper against *Xanthomonas axonopodis* pv. *punicae*, the incitant of pomegranate bacterial blight. Ann. Microbiol. 62 (2), 889–893. https://doi.org/10.1007/s13213-011-0382-7.

Monowar, T., Rahman, M., Bhore, S.J., Raju, G., Sathasivam, K.V., 2018. Silver nanoparticles synthesized by using the endophytic bacterium Pantoea ananatis are promising antimicrobial agents against multidrug resistant bacteria. Molecules 23 (12), 3220. https://doi.org/10.3390/molecules23123220.

Ogunyemi, S.O., Abdallah, Y., Zhang, M., Fouad, H., Hong, X., Ibrahim, E., 2019. Green synthesis of zinc oxide nanoparticles using different plant extracts and their antibacterial activity against *Xanthomonas oryzae* pv. *oryzae*. Artif. Cells Nanomed. Biotechnol. 47, 341–352. https://doi.org/10.1080/21691401.2018.1557671.

Omomowo, O.I., Babalola, O.O., 2019. Bacterial and fungal endophytes: tiny giants with immense beneficial potential for plant growth and sustainable agricultural productivity. Microorganisms 7 (11), 481. https://doi.org/10.3390/microorganisms7110481.

Orr, D., Lahiri, S., 2014. Biological control of insect pests in crops. In: Integrated Pest Management. Academic Press, pp. 531–548, https://doi.org/10.1016/B978-0-12-398529-3.00028-2.

Pal, S., Tak, Y.K., Song, J.M., 2007. Does the antibacterial activity of silver nanoparticles depend on the shape of the nanoparticle? A study of the gram-negative bacterium *Escherichia coli*. Appl. Environ. Microbiol. 73, 1712–1720. https://doi.org/10.1128/AEM.02218-06.

Palmqvist, N.G.M., Bejai, S., Meijer, J., et al., 2015. Nano titania aided clustering and adhesion of beneficial bacteria to plant roots to enhance crop growth and stress management. Sci. Rep. 5, 10146. https://doi.org/10.1038/srep10146.

Paret, M.L., Vallad, G.E., Averett, D.R., Jones, J.B., Olson, S.M., 2013. Photocatalysis: effect of light-activated nanoscale formulations of TiO2 on *Xanthomonas perforans* and control of bacterial spot of tomato. Phytopathology 103 (3), 228–236. https://doi.org/10.1094/PHYTO-08-12-0183-R.

Pariona, N., Mtz-Enriquez, A.I., Sánchez-Rangel, D., Carrión, G., Paraguay-Delgado, F., Rosas-Saito, G., 2019. Green-synthesized copper nanoparticles as a potential antifungal against plant pathogens. RSC Adv. 9, 18835–18843. https://doi.org/10.1039/C9RA03110C.

Park, M.R., Banks, M.K., Applegate, B., Webster, T.J., 2008. Influence of nanophase titania topography on bacterial attachment and metabolism. Int. J. Nanomedicine 3, 497–504. https://doi.org/10.2147/ijn.s4399.

Pereira, A.E., Grillo, R., Mello, N.F., Rosa, A.H., Fraceto, L.F., 2014. Application of poly (epsilon-caprolactone) nanoparticles containing atrazine herbicide as an alternative technique to control weeds and reduce damage to the environment. J. Hazard. Mater. 268, 207–215. https://doi.org/10.1016/j.jhazmat.2014.01.025.

Pour, M.M., Saberi-Riseh, R., Mohammadinejad, R., Hosseini, A., 2019. Nano-encapsulation of plant growth-promoting rhizobacteria and their metabolites using alginate-silica nanoparticles and carbon nanotube improves UCB1 pistachio micropropagation. J. Microbiol. Biotechnol. 29, 1096–1103. https://doi.org/10.4014/jmb.1903.03022.

Pramanik, P., Krishnan, P., Maity, A., Mridha, N., Mukherjee, A., Rai, V., 2020. Application of nanotechnology in agriculture. In: Dasgupta, N., Ranjan, S., Lichtfouse, E. (Eds.), Environmental Nanotechnology. Volume 4. Environmental Chemistry for Sustainable World. vol. 32. Springer, https://doi.org/10.1007/978-3-030-26668-4_9.

Qian, Y., Yu, H., He, D., Yang, H., Wang, W., Wan, X., Wang, L., 2013. Biosynthesis of silver nanoparticles by the endophytic fungus *Epicoccum nigrum* and their activity against pathogenic fungi. Bioprocess Biosyst. Eng. 36, 1613–1619. https://doi.org/10.1007/s00449-013-0937-z.

Quecine, M.C., Araújo, W.L., Marcon, J., Gai, C.S., Azevedo, J.L., Pizzirani-Kleiner, A.A., 2008. Chitinolytic activity of endophytic streptomyces and potential for biocontrol. Lett. Appl. Microbiol. 47, 486–491. https://doi.org/10.1111/j.1472-765X.2008.02428.

Quecine, M.C., Batista, B.D., Lacava, P.T., Ananda Kumar, P., Org, 2014. Diversity and biotechnological potential of plant-associated endophytic bacteria. In: Biotechnology (Multi-Volume Set). Studium Press LLC, Houston, pp. 377–423.

Rana, K.L., Kour, D., Yadav, N., Yadav, A.N., 2020. Endophytic microbes in nanotechnology: current development, and potential biotechnology applications. In: Kumar, A., Singh, V.K. (Eds.), Microbial Endophytes. Woodhead Publishing, Cambridge, MA, pp. 231–262, https://doi.org/10.1016/B978-0-12-818734-0.00010-3.

Rangaraj, S., Gopalu, K., Rathinam, Y., et al., 2014. Effect of silica nanoparticles on microbial biomass and silica availability in maize rhizosphere. Biotechnol. Appl. Biochem. 61, 668–675. https://doi.org/10.1002/bab.1191.

Rosenblueth, M., Martínez-Romero, E., 2006. Bacterial endophytes and their interactions with hosts. Mol. Plant-Microbe Interact. 19 (8), 827–837. https://doi.org/10.1094/MPMI-19-0827.

Santoyo, G., Moreno-Hagelsieb, G., del Carmen Orozco-Mosqueda, M., Glick, B.R., 2016. Plant growth-promoting bacterial endophytes. Microbiol. Res. 183, 92–99. https://doi.org/10.1016/j.micres.2015.11.008Get.

Seefeldt, L.C., Yang, Z.Y., Lukoyanov, D.A., Harris, D.F., Dean, D.R., Raugei, S., Hoffman, B.M., 2020. Reduction of substrates by nitrogenases. Chem. Rev. https://doi.org/10.1021/acs.chemrev.9b00556.

Sessitsch, A., Reiter, B., Berg, G., 2004. Endophytic bacterial communities of field grown potato plants and their plant growth-promoting and antagonistic abilities. Can. J. Microbiol. 50, 239–249. https://doi.org/10.1139/w03-118.

Sharma, P., Bhatt, D., Zaidi, M.G.H., Saradhi, P.P., Khanna, P.K., Arora, S., 2012. Silver nanoparticle-mediated enhancement in growth and antioxidant status of *Brassica juncea*. Appl. Biochem. Biotechnol. 167 (8), 2225–2233. https://doi.org/10.1007/s12010-012-9759-8.

Sharma, S., Kooner, R., Arora, R., 2017. Insect pests and crop losses. In: Arora, R., Sandhu, S. (Eds.), Breeding Insect Resistant Crops for Sustainable Agriculture. Springer, https://doi.org/10.1007/978-981-10-6056-4_2.

Shukla, S.K., Kumar, R., Mishra, R.K., et al., 2015. Prediction and validation of gold nanoparticles (GNPs) on plant growth promoting rhizobacteria (PGPR): a step toward development of nanobiofertilizers. Nanotechnol. Rev. 4, 439–448. https://doi.org/10.1515/ntrev-2015-0036.

Singh, D., Rathod, V., Ninganagouda, S., Hiremath, J., Singh, A.K., Mathew, J., 2014. Optimization and characterization of silver nanoparticle by endophytic fungi *Penicillium* sp. isolated from *Curcuma longa* (Turmeric) and application studies against MDR *E. coli* and *S. aureus*. Bioinorg. Chem. Appl. 2014, 8. https://doi.org/10.1155/2014/408021.

Singh, I., 2018. Plant growth promoting rhizobacteria (PGPR) and their various mechanisms for plant growth enhancement in stressful conditions: a review. Eur. J. Biol. Res. 8 (4), 191–213. https://doi.org/10.5281/zenodo.1455995.

Singh, S., Gupta, V.K., 2016. Biodegradation and bioremediation of pollutants: perspectives strategies and applications. Int J Pharm. Bio. Sci 10, 53.

Sinha, R., Karan, R., Sinha, A., Khare, S.K., 2011. Interaction and nanotoxic effect of ZnO and Ag nanoparticles on mesophilic and halophilic bacterial cells. Bioresour. Technol. 102, 1516–1520.

Stenberg, J.A., Sundh, I., Becher, P.G., et al., 2021. When is it biological control? A framework of definitions, mechanisms, and classifications. J. Pest. Sci. 94, 665–676. https://doi.org/10.1007/s10340-021-01354-7.

Subban, K., Subramani, R., Johnpaul, M., 2013. A novel antibacterial and antifungal phenolic compound from the endophytic fungus *Pestalotiopsis mangiferae*. Nat. Prod. Res. 27, 1445–1449. https://doi.org/10.1080/14786419.2012.722091.

Taiz, L., Zeiger, E., Møller, I.M., Murphy, A., 2015. Plant Physiology and Development, sixth ed. Sinauer Associates Incorporated.

Tao, G.C., Tian, S.J., Cai, M.Y., Xie, G.H., 2008. Phosphate solubilizing and -mineralizing abilities of bacteria isolated from. Pedosphere 18, 515–523. https://doi.org/10.1016/S1002-0160(08)60042-9.

Thirugnanasambandan, T., 2019. Advances and Trends in Nanobiofertilizers. SSRN, p. 59, https://doi.org/10.2139/ssrn.3306998.

Tian, L., Shen, J., Sun, G., Wang, B., Ji, R., Zhao, L., 2020. Foliar application of SiO_2 nanoparticles alters soil metabolite profiles and microbial community composition in the Pakchoi (*Brassica chinensis* L.) rhizosphere grown in contaminated mine soil. Environ. Sci. Technol. 54 (20), 13137–13146. https://doi.org/10.1021/acs.est.0c03767.

Tilman, D., Cassman, K.G., Matson, P.A., Naylor, R., Polasky, S., 2002. Agricultural sustainability and intensive production practices. Nature 418, 671–677. https://doi.org/10.1038/nature01014.

Timmusk, S., Seisenbaeva, G., Behers, L., 2018. Titania (TiO2) nanoparticles enhance the performance of growth-promoting rhizobacteria. Sci. Rep. 8, 617. https://doi.org/10.1038/s41598-017-18939-x.

Tripathi, S., Sonkar, S.K., Sarkar, S., 2011. Growth stimulation of gram (*Cicer arietinum*) plant by water soluble carbon nanotubes. Nanoscale 3 (3), 1176–1181. https://doi.org/10.1039/C0NR00722F.

Verma, S., Kuila, A., 2019. Bioremediation of heavy metals by microbial process. Environ. Technol. Innov. 14, 100369. https://doi.org/10.1016/j.eti.2019.100369.

Walia, A., Guleria, S., Chauhan, A., Mehta, P., 2017. Endophytic bacteria: role in phosphate solubilization. In: Maheshwari, D., Annapurna, K. (Eds.), Endophytes: Crop Productivity and Protection. Sustainable Development and Biodiversity. 16. Springer, https://doi.org/10.1007/978-3-319-66544-3_4.

Yu, D., Zhang, S., Li, X., Xu, J.-R., Schultzhaus, Z., Jin, Q.A., 2017. Gin4-like protein kinase GIL1 involvement in hyphal growth, asexual development, and pathogenesis in *Fusarium graminearum*. Int. J. Mol. Sci. 18, 424. https://doi.org/10.3390/ijms18020424.

Zand, A.D., Mikaeili Tabrizi, A., Vaezi, H.A., 2020a. Application of titanium dioxide nanoparticles to promote phytoremediation of Cd-polluted soil: contribution of PGPR inoculation. Bioremediat. J. 24, 171–189. https://doi.org/10.1080/10889868.2020.1799929.

Zand, A.D., Tabrizi, A.M., Heir, A.V., 2020b. The influence of association of plant growth-promoting rhizobacteria and zero-valent iron nanoparticles on removal of antimony from soil by *Trifolium repens*. Environ. Sci. Pollut. Res. 27, 42815–42829. https://doi.org/10.1007/s11356-020-10252-x.

Zhang, S., Reddy, M.S., Kloepper, J.W., 2004. Tobacco growth enhancement and blue mold disease protection by rhizobacteria: relationship between plant growth promotion and systemic disease protection by PGPR strain 90–166. Plant Soil 262, 277–288. https://doi.org/10.1023/B:PLSO.0000037048.26437.fa.

CHAPTER 10

Potential of microbial inoculants for the management of agricultural soils contaminated by recalcitrant compounds

Maria Gabriela Azevedo Barros, Rosane dos Santos Grignet, Suzan Prado Bernandes Bernal, Caroline Da Costa Silva Gonçalves, Michel Rodrigo Zambrano Passarini, and Júlia Ronzella Ottoni
Laboratory of Environmental Biotechnology, Federal University of Latin American Integration, Latin American Institute of Life Sciences and Nature, Foz do Iguaçu, Brazil

10.1 Introduction

In recent years, population growth has led to an increase in the demands for food. This fact together with growing industrialization has contributed to the increase in environmental pollution. Anthropogenic activities produce large amounts of toxic pollutants, including dyes, hydrocarbons, drugs, pesticides, and heavy metals, which are inappropriately released into the environment, thus causing degradation of the ecosystem (Yu et al., 2020). Such hazardous pollutants, named "xenobiotics," are produced by industrial, agricultural, and domestic activities. They can accumulate in the environment due to their high affinity to organic substances and produce toxic effects on humans, animals, and the environment (Mishra et al., 2021).

To mitigate the adverse consequences of the presence of xenobiotics within an ecosystem, bioremediation represents an eco-friendly alternative. Bioremediation comprises an efficient, economical, and biological process aimed at reducing/removing contaminants from a given environment. The use of bioinoculants in this process has become widespread (Chaudhary and Shukla, 2019), thus elucidating the advantages of adopting a biological approach.

The use of microbial bioinoculants for bioremediation faces some difficulties. For example, in marine environments, microorganisms are subjected to low nutrient availability, competition for carbon sources with autochthonous organisms, dispersion, dilution, variations in pH and temperature, and

Microbial Inoculants
https://doi.org/10.1016/B978-0-323-99043-1.00018-9

the availability of pollutants, whereas in soil, the introduction of microorganisms can impact the indigenous microbial community and geochemical composition in the long term. In situ bioremediation can be more efficient when microorganisms are immobilized, thus protecting both the microbes and the environment during bioaugmentation in areas experiencing adverse environmental conditions (Dellagnezze et al., 2016; Valdivia-Rivera et al., 2021).

Microbial communities inhabiting environments with harsh circumstances, including contaminated environments, adapt to the adverse conditions and can utilize xenobiotic pollutants such as carbon and nitrogen as sources to maintain their metabolic activities. In addition, microorganisms present high genetic variability and adaptability, functional diversity, enormous catabolic potential and degradation pathways, rapid reproduction, and diverse metabolic pathways, thus offering potential for the remediation of contaminated sites. Moreover, the improvement of techniques in molecular biology, biotechnology, and bioinformatics has allowed access to information at the gene level with the purpose of optimizing bioremediation processes (Mishra et al., 2021).

Science has tried to understand how the relationships and dynamics between microorganisms and contaminants occur in order to improve remediation efficiency. As such, microbial bioinoculants composed of unique microbial species, consortia of microorganisms, or isolated enzymes are resources that have been widely studied and applied in the recovery of soil health (Scagliola et al., 2021; Janczak et al., 2020).

In this context, this chapter seeks to explore how some xenobiotic contaminants can affect the quality of cultivable soil and the potential of microorganisms and their metabolites to reduce the pollutants present, with an objective to restore the soil's health. This chapter will also address the presence of pesticides, microplastics, hydrocarbons, and urban wastes and the role of microbial bioinoculants in the removal of these compounds from the environment.

10.2 Bioinoculants in pesticide decontamination

In the last 60 years, the world has witnessed the Green Revolution, a period in which there was a significant increase in agricultural productivity with a consequent scarcity of land resources. A rise in crop yields occurred, among other factors, through the increased use of synthetic pesticides. However, the excessive and inadvertent use of these chemical compounds resulted in deterioration of the biological, physical, and chemical health of the soil, leading to a negative impact on productivity (Basu et al., 2021).

Managing the quality of agricultural soil is essential to ensure sustainable agricultural production and conservation of biodiversity. The use of chemical fertilizers and pesticides has proven to be efficient in guaranteeing agricultural production, but, in return, they cause undesirable effects on the environment and health, in addition to the effects on soil communities (Prashar and Shah, 2016). An efficient strategy of bioremediation targeting pesticides must consider the metabolic and physiological characteristics of microorganisms, and, therefore, the use of different omics tools such as genomics, transcriptomics, proteomics, and metabolomics is essential to obtain information about the genes and proteins involved in pesticide degradation, generated metabolites, and cell tactics to overcome the stresses caused by the presence of pesticides (Rodríguez et al., 2020). Based on this background and its consequences, this section aims to provide information on how bioinoculants can help maintain soil health and agricultural productivity in a sustainable manner as well as the bioremediation of pesticides in soil.

10.2.1 Pesticides: Definition, characteristics, degradation, and consequences to living organisms and the environment

Pesticides are organic mixtures of broad-spectrum chemicals designed to prevent plant diseases, pests, and weeds and are classified as insecticides, fungicides, herbicides, nematicides, rodenticides, and plant growth regulators (Rajmohan et al., 2020; Kumar et al., 2021). They are formulated to control the target organism without interfering with other macro/microorganisms and the environment, but their intensive use in recent decades has led to the presence of these compounds in soil (Rajmohan et al., 2020; Vryzas, 2018). The contaminations caused by pesticides are considered long-term and persistent, given the recalcitrant and potentially toxic nature of xenobiotic substances (Chowdhury et al., 2008). The degradation of xenobiotics occurs by biotic and photolytic factors (Vryzas, 2018) (Fig. 1).

Despite their susceptibility to decomposition, extensive inputs of xenobiotics can result in accumulation of toxic compounds in soil. The presence of xenobiotic residues in the environment has raised concerns among the public and the scientific community (Vryzas, 2018). Humans may be exposed to pesticides through ingestion of contaminated water and food, inhalation of air containing pesticide residues, or direct contact in occupational, agricultural, and domestic use (Raffa and Chiampo, 2021).

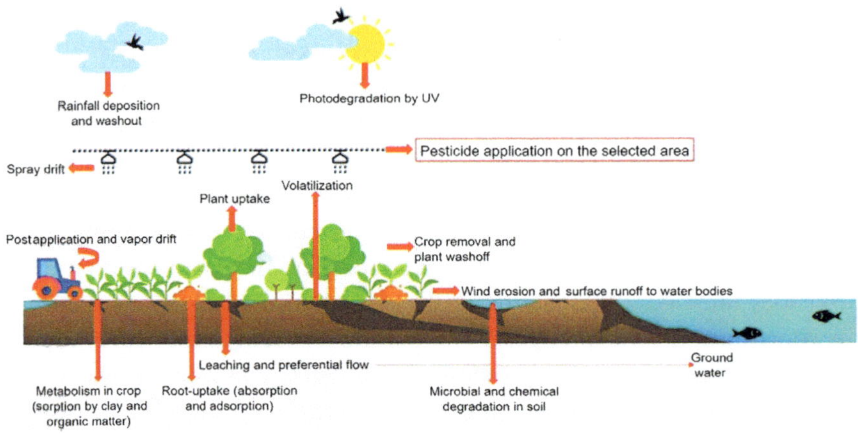

Fig. 1 Fate and transport of pesticides in the environment (Sun et al., 2020).

The presence of pesticides in soils also directly or indirectly affects the activities of soil enzymes and the physiological activities of nontarget microorganisms inhabiting the ground, eventually leading to losses in soil fertility (Anjum et al., 2012). Moreover, inadequate soil fertility demands the use of agrochemicals and pesticides, which, in turn, leave more residues in the soil and affect its balance, subsequently resulting in increased demands (Scagliola et al., 2021).

To manage contaminated agricultural soils affected by intensive exploration and to achieve sustainable production, the use of chemical fertilizers and pesticides must be replaced by environmentally friendly agricultural practices. In addition, the concepts associated with sustainable agriculture, mitigation of environmental impacts, and recovery of degraded and contaminated areas have gained more attention in the agricultural market. In order to provide food security to all, agriculture is facing the challenge of increasing food production without causing damage to the environment. In this context, the use of microbial inoculants has emerged as an efficient alternative (Santos et al., 2019).

Biodegradation of pesticides may occur through microorganisms or the microbial consortia (Kumar et al., 2021). Partial breakdown forming less complex and toxic organic compounds is called partial biodegradation, whereas the complete degradation of pesticides into inorganic compounds is called biomineralization ($CO_2 + H_2O$) (Odukkathil and Vasudevan, 2013).

Biodegradability of pesticides depends on their molecular structure. After transformation or degradation by microorganisms, the resulting products serve as a source of carbon, nitrogen, minerals, and as the final electron acceptor in the respiratory chain (Odukkathil and Vasudevan, 2013; Mishra et al., 2021). Biodegradation of pesticides is influenced by factors such as concentration (Anjum et al., 2012), content of organic matter in the soil (Odukkathil and Vasudevan, 2013), soil moisture and pH (Anjum et al., 2012), temperature, and other climatic factors (Chowdhury et al., 2008).

10.2.2 Bioremediation potential of microorganisms

Microbial metabolism is the most efficient and accepted mechanism for persistent pesticide degradation. Microbiological processes are a versatile, efficient, safe, cost-effective, and eco-friendly alternative to bioremediation (Kumar et al., 2021; Rodríguez et al., 2020). By modifying the structure of pollutants, microorganisms and their enzymes can reduce the toxicity of these compounds (Raffa and Chiampo, 2021). The isolation of microbial strains or microbial consortia capable of biodegrading pesticides has been the focus of many studies. Although noncultivable microorganisms and their potential are not accessed by cultivation-dependent methods, omics tools have evolved as the most efficient approach to solving the associated limitations (Rodríguez et al., 2020).

The microbial communities present in soil constitute an essential element for the agroecosystem and for the maintenance of soil health. Microbes are key counterparts in the bioremediation of contaminated soils (Prashar and Shah, 2016). Additionally, microorganisms can metabolize most of the recalcitrant pesticides either as an energy source or by cometabolism with other substrates (Kumar et al., 2021). In this context, a better understanding of the factors involved in the growth, dynamics, functions, and metabolism of indigenous microbial communities inhabiting contaminated sites is necessary to increase the efficiency of the bioremediation process (Desai et al., 2010). Moreover, genomic, transcriptomic, proteomic, and metabolomic techniques together with bioinformatics tools have allowed unraveling the molecular mechanisms involved in xenobiotic transformations, including pesticides (Desai et al., 2010; Rodríguez et al., 2020). The metagenomic approach accesses all the information of the microbial community from any of the contaminated environment and guides the formulation of a successful bioremediation approach through bioaugmentation or biostimulation based on ex situ or in situ treatment (Nandy et al., 2021). In this sense, the formulation of efficient bioinoculants may promote the biodegradation of complex substances and participate in bioremediation processes in soils contaminated with xenobiotic substances, including pesticides.

10.2.3 Microbial bioinoculants in pesticide degradation: Microbial bioremediation and molecular tools

Microorganisms that can be used for the purpose of degrading toxic contaminants and pesticides in soil are called bioinoculants (Chaudhary and Shukla, 2019). Soil-inhabiting microorganisms, when in contact with pesticides, can develop the ability to degrade contaminants by using them as a carbon source and as electron donors. However, not all soils have microbial communities with the ability to degrade contaminants. Under such conditions, another alternative is the external addition (bioinoculation) of pesticide-degrading microorganisms (Parte et al., 2017). In spite of being promising, microbial bioremediation strategies are usually limited by the lack of information about the factors associated with the regulation of the growth and metabolism of indigenous microorganisms (Rodríguez et al., 2020). Therefore, a multidisciplinary approach can be more efficient in bioremediation processes (Nandy et al., 2021).

The use of microbial consortia from the indigenous soil microbiota is more effective in pesticide degradation as compared to nonindigenous microorganisms due to their adaptation to native soil conditions. In their habitats, microorganisms coexist with other organisms, and, in many cases, their viability depends on each other (Raffa and Chiampo, 2021). The isolation and purification of indigenous microorganisms provide information about the metabolites produced and biodegradation processes; however, in vitro cultivation conditions are not able to reproduce the environmental conditions (Chandran et al., 2020).

These microorganisms possess greater potential for degrading pesticides and act synergistically in a catabolic cooperation, as the interactions between microorganisms in the system are complex; however, information about a single microorganism may not be enough to understand the processes occurring during bioremediation (Rodríguez et al., 2020). A proper bioremediation process must be structured based on information about microorganisms and their relationship with the contaminated environment that they thrive in, and the assessment of this information can be obtained through genomics, transcriptomics, proteomics, and metabolomics. Genomic tools provide genetic data of the entire material, including mRNA, which indicates the microbially expressed genes during exposure to pesticides. This approach generally includes genome/metagenome sequencing and bioinformatics analysis (Dangi et al., 2019).

The response of microorganisms to stress such as that caused by the presence of pesticides can be accelerated by the manipulation of genes associated with contamination degradation pathways (Anode and Onguso, 2021). Through molecular biology and genetic engineering, it has become possible to produce microbial strains with a high efficiency in pollutant degradation. Under optimum microbiological activities and environmental conditions, effective strains convert complex macromolecules and smaller molecules are assimilated by other species (Kumar and Sharma, 2019). The first use of genetically engineered microorganisms to degrade pesticides started with the detoxification of organophosphate pesticides by cloning and expressing genes encoding hydrolases in *P. pseudoalcaligenes*, *Escherichia coli*, *Streptomyces lividans*, *Yarrowia lipolytica*, and *Pichia pastoris* (Ortiz-Hernández et al., 2013).

Bioaugmentation is an approach that can be used in pesticide biodegradation systems and contaminated areas and depends on the number of microorganisms and the total biomass added to the soil. For this reason, the way in which microorganisms are delivered is also important for successful degradation of pesticides. The use of immobilized cells has shown advantages as they protect microorganisms from possible adverse conditions present in the environment and do not compete with indigenous microorganisms, in addition to increased stability of cells, DNA, and plasmids (Kumar and Sharma, 2019).

The use of the microbial consortia has been proven to be efficient in the degradation of several pesticides. Furthermore, isolates of bacteria and fungi belonging to diverse species show pesticide-degrading abilities. Members of Actinobacteria, Bacteroidetes, Ascomycota, Basidiomycota, Chlorophyta, Cyanobacteria, Firmicutes, and Proteobacteria are identified as degraders of carbamate, organochlorine, organophosphate, and pyrethroid pesticides. Within these groups, the important ones are *Achromobacter*, *Arthrobacter*, *Aspergillus*, *Bacillus*, *Burkholderia*, *Chlamydomonas*, *Chlorella*, *Methylobacterium*, *Micrococcus*, *Paenibacillus*, *Penicillium*, *Pseudomonas*, *Rhizobium*, *Trichoderma*, and *Xanthomonas* sp., capable of degrading 118 pesticides (Kumar et al., 2021).

Soil-inhabiting fungi are quite promising in pesticide degradation by the production of enzymes involved in contaminant degradation and modulation of soil properties. In addition, fungi are more resistant to pollutants even at high concentrations, and a combination of fungi and bacteria can improve pesticide degradation. Most notably, fungi are able to transform these pollutants into molecules that can be easily accessed by bacteria (Raffa and Chiampo, 2021).

With the emergence of new systemic fields (including systems and synthetic biology and metabolic engineering), new approaches to the bioremediation of pollutants, including pesticides, have become available.

Computational tools allow access to important information that makes it possible to predict natural or synthetic pathways or a set of pathways suitable for the degradation of a specific xenobiotic in selected microbial hosts. Systems biology provides crucial information for metabolic engineering of microorganisms, thus enhancing pesticide degradation capacity (Dangi et al., 2019). Through omics approaches, it has now become possible to understand the molecular mechanisms, genes, proteins, and metabolic pathways involved in pesticide degradation as well as the physiological changes and cell strategies to survive in the presence of pesticides. In addition, the use of the proteomics approach toward pesticide biodegradation is gaining attention as it allows the identification of enzymes in the process, main metabolic pathways, energy sources during biodegradation, and proteins expressed after pesticide exposure (Rodríguez et al., 2020).

Microorganisms and their broad functional, genetic, and metabolic capacities represent an important tool in pesticide bioremediation. Microbial bioinoculants have been developed for this purpose using isolated strains, microbial consortia, and, more recently, omics tools and systemic biology. The latter approaches allow broad access to genetic, functional, and metabolic pathways. The obtained data can be used in a targeted and personalized manner by the development of engineered microorganisms with an increased capacity for pesticide biodegradation. With the help of innovations in the area together with other measures focused on the recovery and preservation of environmental resources, it is possible to envision a future with productive and sustainable agriculture.

10.3 Bioinoculants in microplastic decontamination

Since the 1950s, the consumption of plastics has dramatically increased. The characteristics that stimulate the use of this material, such as low production cost, physical and mechanical resistance, and insusceptibility to air and water, are responsible for making plastics an environmental problem. Considered a recalcitrant contaminant, plastics take more than 200 years to degrade and are touted as a ubiquitous material as it is already present even in the deepest parts of the ocean (Huang et al., 2020). However, before their decomposition, plastics are often broken down into small pieces over time and give rise to microplastics (<5 mm) (Rillig et al., 2021).

In the early 2000s, the amount of plastic volume in our oceans became a major problem, and, more recently, the identification of microplastics in fish has shown the magnitude of this contamination. However, soil is considered to be more contaminated. The estimated volume of microplastics varies and is generally high in industrialized areas, but this contamination is expected to be 4–23 times higher than the volume found in the oceans (Horton et al., 2017; Ya et al., 2021).

The entry route of microplastics into soil is diverse (Fig. 2). Currently, plastics are associated with reused water, utilized in crop irrigation, and the more commonly present are acrylic and polyamide fibers derived from the textile industry. Among the classes of fragments, polypropylene (PP) is frequently found, although the amount of this material varies between regions (Zhou et al., 2020).

Initial studies have indicated that microplastics can alter soil dynamics, thereby impacting agroecosystems and terrestrial biodiversity. The presence of these materials is a concern for the nutritional quality of the soil, since studies have associated the presence of microplastics with the reduction of essential compounds such as nitrogen (N), phosphorus (P), carbon (C), and potassium (K) (Machado et al., 2019).

Yu et al. (2020) verified that the introduction of microplastics into the soil induces its acidification and reduces the long-term activity of the enzymes present in the environment. Low pH interferes in the soil interaction with heavy metals (Zhang et al., 2020). The authors showed that the presence of microplastics and consequent pH reduction led to less cadmium adsorption, potential risk of water contamination, and accumulation in seeds. These findings highlight the impact of microplastics present in the soil.

Additionally, microplastics can influence soil water retention. Polyester microfibers can clog pores up to 30 μm and impact the transfer of water and gases in the soil. Water retention is linked to pore size, and smaller ones are more easily clogged by microplastics (Zhang et al., 2019). Machado et al. (2018) analyzed impacts on the soil in the presence of different microplastics, i.e., PE (polyethylene), polyester, and polyacrylic. The presence of microplastics in soil increases water retention and reduces soil density and microbial activity. Even though there are few studies associating modifications in soil characteristics with microbial features, they showed that these compounds can impact the soil and demonstrated that this interaction is of great importance to the soil's health.

Microbial decomposition of polymeric materials can be divided into four steps: biodeterioration, biofragmentation, assimilation, and mineralization (Amobonye et al., 2021), and microplastics are formed in the two first steps. In addition, when evaluating the plastic degradation potential of microorganisms in soil, some environmental parameters need

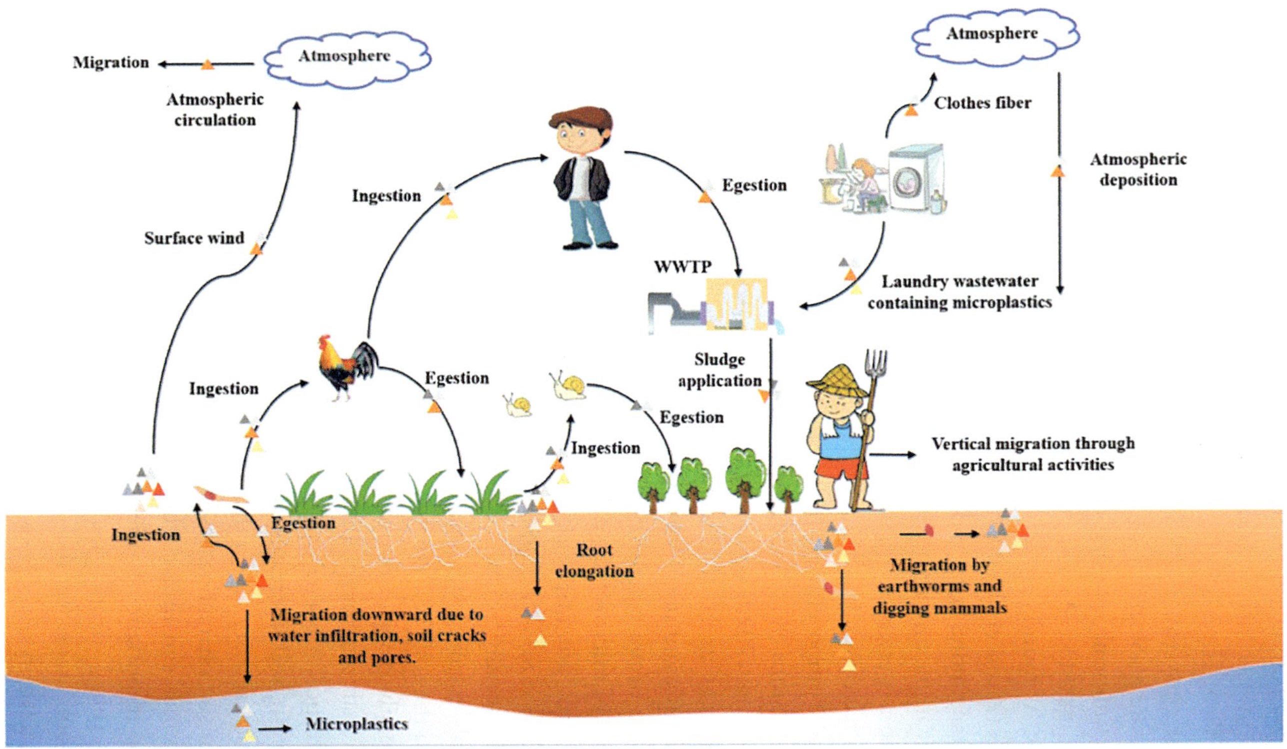

Fig. 2 The migration pathways of microplastics in the environment (Ya et al., 2021).

to be considered, such as pH, humidity, oxygen content, and temperature, which can significantly influence the activity of microbial inoculants (Janczak et al., 2020).

In this manner, the use of soil-inhabiting microorganisms can be an advantageous option. The mycorrhizal bacterium *Rhodococcus rhodochrous* has been considered as an interesting alternative for the assimilation of polyethylene intermediates and represents a promising option as a microbial bioinoculant for the degradation of this polymer (Gravouil et al., 2017). The microbial consortia were tested to degrade PET (polyethylene terephthalate) and PLA (polylactic acid) by Janczak et al. (2020). The consortia consisted of rhizospheric microbes, including the bacterium *Arthrobacter sulfonivorans* and the fungus *Laccaria laccata*. The presence of the microbial consortia enhanced plastic degradation and contributed to plant growth. Furthermore, the authors, through the analysis of the genomes of these microorganisms, obtained fundamental data needed for the future development of a biofertilizer capable of contributing to the bioremediation of plastic contamination.

The microbial consortium of *Pseudomonas* sp. (ADL15) and *Rhodococcus* sp. (ADL36) isolated from the Antarctic continent was able to consume about 20% of polypropylene (PP) after 40 days of evaluation even at low temperatures and showed great potential as microbial inoculants (Habib et al., 2020).

Moreover, the use of genetic engineering techniques to construct catabolic pathways for recalcitrant compounds has led to important advances in the application and development of microbial inoculants. The marine bacterium *Ideonella sakaiensis* 201-F6 was able to degrade PET almost completely after 6 weeks at 30°C (Yoshida et al., 2016). After a draft genome sequence, a putative lipase sharing 51% amino acid sequence identity and a catalytic site with PET hydrolase from *Thermobifida fusca* (TfH) was identified. The incubation of a PET film with purified recombinant *I. sakaiensis* proteins (30°C, 18 h) released mono(2-hydroxyethyl) terephthalic acid (MHET) as the major product and small amounts of TPA and bis(2-hydroxyethyl) TPA (BHET), suggesting the presence of proteins hydrolyzing PET (Moog et al., 2019).

So, it is notable that soil contamination by plastics, including micro- or nano-variations, requires further studies to assess the presence of these materials in the soil and to develop solutions that reduce or completely mitigate these soil pollutants. Hence, the use of microorganisms as inoculants may offer a way to the management of soils impacted by plastics.

10.4 Bioinoculants in industrial and urban waste decontamination

Industrial activities, municipalities, and agricultural areas are major generators of organic wastes, which pose risks to the environment, human health, and beneficial microorganisms present in the soil. These residues have the potential to pollute the environment, and their adequate management is of great concern. However, if properly treated, these residues can be converted into a nutrient-rich product, which can be applied to the land for nutrients and organic matter supply to plants and the soil (Ahmad et al., 2007).

The best way to manage organic solid wastes for agricultural application is composting, a biological process in which these wastes are bioconverted into a stable, sanitized, humus-rich product that can be used as an organic amendment and bring about advantages from both economic and productivity viewpoints (Ahmad et al., 2007; Sayara et al., 2020). This system offers a sustainable solution for the disposal of urban wastes, as it generates a fertilizer while conserving natural resources, thus protecting the soil and reducing the environmental impact. The generated fertilizer (biofertilizer) can replace inorganic fertilizers, which depreciate soil through continuous and intense use. In addition, the biofertilizer is rich in organic matter, macro- and micronutrients, and provides advantages such as the reestablishment of soil organic carbon and nitrogen and improvement in soil aggregation, which in the long term improves soil fertility and productivity (Sayara et al., 2020).

During the composting process, the microbial consortium plays different roles. Bacteria normally initiate the composting process due to their high metabolic versatility and other inherent characteristics such as high tolerance to variations in the environment. Afterward, the fungi begin their action by decomposing woody substances in addition to waxes, proteins, hemicelluloses, lignin, and pectin. Finally, *Actinomycetes* act by degrading organic acids, sugars, starches, hemicelluloses, celluloses, proteins, polypeptides, amino acids, and even lignins that are prevalent in the later stages of composting. At this point, moisture levels decrease and the pH becomes less acidic (Jaiswal and Shukla, 2020).

The inoculation of microorganisms in compost favors the biodegradation of organic matter, speeds up the process, and provides a better-quality final product (Ahmad et al., 2007). In a compost pile, a variety of microbial inoculants grow in response to different levels of temperature, humidity, oxygen, and pH. Compounds such as wheat straw, plant root residues, and bran are commonly degraded by the fungus *Phanerochaete chrysosporium*. Municipal solid waste and dry grass are decomposed by ammonia-oxidizing bacteria, nitrifying bacteria, and fungi and bacteria that decompose cellulose and lignin (Jaiswal and Shukla, 2020; Palaniveloo et al., 2020). There are many efficient hydrolytic bacteria and other physiologically important microorganisms present in the compost, and several isolates of the genes *Bacillus* and *Pseudomonas* have already been described based on their hydrolytic potential (Pan et al., 2012). Usually inoculated in the consortia, other microorganisms that stand out are thermophilic bacteria, *Pichia kudriavzevii*, *Trametes versicolor*, *Trichuris spiralis*, *Paecilomyces fusisporus*, *Trichoderma harzianum*, *Aspergillus* spp., *Penicillium* spp., *Pleurotus sajor*, and nitrifying bacteria (Ahmad et al., 2007; Xi et al., 2015). The quality and composition of the final compost product depend on the composition of the waste used and the microorganisms that participate in the process.

The use of inoculants containing cellulose degraders such as *Trichoderma viride*, *Trichuris spiralis*, *Aspergillus niger*, *Paecilomyces fusisporus*, *Aspergillus heteromorphus*, *Aspergillus terreus*, *Aspergillus flavus*, and *Rhizomucor pusillus* is common and aims to improve the processes that occur in composting. Fungi that degrade lignin such as *Clavaris* sp., *Cephalosporium* sp., and *Humicola* sp. are also added to improve and complement the processes occurring in this system (Bhattacharjya et al., 2021).

The availability of nutrients and the stability of the final product mainly depend on the conditions of the composting process and the type of organic raw material used. Nitrogen and carbon are the most important elements of the biocompost, and the C/N ratio should be in the range of 25–30. In addition, the final product should contain about 33.3% of organic matter and 35% of water, which are the recommended levels for marketed composts. Other parameters to be considered are pH, which should be close to 7.5, salt content, which should be 3.89 g/L of fresh matter, phosphate (P_2O_5) in about 0.62% of DM (dry matter), and potassium, with a content of approximately 1.01% of DM (Sayara et al., 2020).

The importance of the participation of bioinoculants in the transformations that occur in the composting of urban wastes is incontestable, with microorganisms being the main actors in the degradation of organic compounds. This process generates a product rich in nutrients and organic matter that can be used as a fertilizer in crops, which further emphasizes the important role of microorganisms in transforming waste into a highly useful product without harming the environment.

10.5 Bioinoculants in hydrocarbon decontamination

The contamination and degradation of the environment by hydrocarbons originated from anthropogenic activities, and the inappropriate disposal of residues and effluents is a problem that emerged in the 21st century (Timmusk et al., 2017; Ahmad et al., 2018).

Among different classes listed as environmental contaminants are polycyclic aromatic hydrocarbons (PAHs) and monoaromatics (benzene, toluene, ethylbenzene, and xylenes), collectively known as BTEX. PAHs are highly recalcitrant due to their structural complexity, low water solubility, and tendency to adsorb into the solid phase of soil. BTEX, despite being more recalcitrant, is more soluble in water and therefore has greater mobility in environmental compartments, leading to higher migration potential in groundwater and the rapid spread of contamination plumes (Oghoje et al., 2020).

The pollution caused by hydrocarbons is of great concern due to a substantial impact on the environment and on the health of living beings (Liu et al., 2018). The pollution potential of these substances has led to the development of different soil and groundwater remediation methods. Physicochemical treatments are not sustainable, present a high implementation cost, and generate secondary pollutants, whereas biological methods, known as "bioremediation," present advantages such as a lower production cost and easy implementation, sustainability, and use renewable resources (Bisht et al., 2015; Mushtaq et al., 2021).

Microorganisms have been utilized to remove pollutants from the environment in order to reduce toxic effects. Both living microorganisms and their metabolites are applied in either natural or artificial form to remove, reduce, or immobilize pollutants present in the environment (Ahmad et al., 2018). Focusing on mineralization or conversion into compounds that are less toxic, bioremediation is being considerably utilized for the biodegradation of xenobiotics, including hydrocarbons, thereby reducing their impact on the ecosystem (Dzionek et al., 2016).

Many microorganisms have already been identified as hydrocarbon degraders; however, the search for more efficient microorganisms has inspired research using the metagenomic approach, which consists of a direct analysis of all genomes contained in one environment without the need for cultivation. This technology allows the identification of genes, which leads to the production of enzymes responsible for the most varied functions, including those related to the degradation of hydrocarbons (Ufarté et al., 2015).

10.5.1 Monoaromatic hydrocarbons: BTEX

BTEX hydrocarbons are considered the most persistent organic pollutants in the environment. BTEX compounds are capable of diluting and adhering to soil particles and move easily in water, thus causing harmful effects on living beings due to their toxicity, mutagenicity, and carcinogenicity (Bolden et al., 2015).

Contamination by BTEX compounds in an environment is frequently associated with oil spills, atmospheric deposition, disposal of effluents from chemical and pharmaceutical industries, agrochemicals, and combustion engines. These compounds represent the most mobile and water-soluble fractions of crude oil and its by-products (Wongbunmak et al., 2020).

Due to the environmental and health problems caused by these hydrocarbons, different technologies are being applied, focusing on their mitigation based on bioremediation, including biostimulation and bioaugmentation (Brzeszcz et al., 2020). In the bioremediation process, microorganisms use hydrocarbon pollutants as carbon sources and as energy for their growth (Wang et al., 2019).

Since indigenous microorganisms can naturally degrade hydrocarbons, the isolation and the characterization of microorganisms with the ability to degrade BTEX in their natural habitat is a key tool for successful biodegradation (Wongbunmak et al., 2020). However, the bioremediation of BTEX present in the soil may be affected by factors such as the concentration and variation of BTEX molecules, diversity, and interaction of the microbial community present in the soil, oxygen availability, pH, temperature, and substrates associated with the soil (Dzionek et al., 2016).

Among bacteria able to degrade BTEX compounds in the environment, *Acinetobacter*, *Archaeoglobus*, *Aromatoleum*, *Bacillus*, *Halomonas*, *Pseudomonas*, *Rhodococcus*, and *Sphingobacterium* are well-acknowledged (Varjani, 2017).

The ability to degrade BTEX compounds has been recently identified in the bacterium *Bacillus amyloliquefaciens* subspecies *plantarum* W1 strain. Additionally, the metabolites formed during the biodegradation of each compound and their respective paths of degradation were identified, hinting at the potential of bacteria to be applied in in situ bioremediation (Wongbunmak et al., 2020).

Leal et al. (2018), aiming to develop a bioinoculant to degrade hydrocarbons present in the soil, accomplished the enrichment of urban solid residues with a hydrocarbonclastic microbial population in the presence of gasoline as an additive. The authors verified that more than 99% of hydrocarbons present in clay soil contaminated with diesel and gasoline were removed by the inoculants, thus emphasizing the potential of microorganisms in the bioremediation of contaminated soil.

Bacosa et al. (2021) developed a bacterial consortium obtained from the Yabase oil field in Akita, Japan, and found their ability to degrade aliphatic (octane) and aromatic hydrocarbon mixtures. Among the tested isolates, *Burkholderia* sp. B5 was able to degrade all the BTEX components more rapidly than octane as it contained enzymes capable of degrading both groups of hydrocarbons. The preference for aromatics may assist in the development of bioinoculants using the enzymes recovered from bacteria with a focus on bioremediation of areas contaminated with BTEX.

10.5.2 Polycyclic aromatic hydrocarbons (PAHs)

PAHs are formed by two or more benzene rings known as polyaromatics and include naphthalene, phenanthrene, anthracene, pyrene, chrysene, fluoranthene, and benzo(*a*)pyrene (Varjani, 2017). These compounds, when present in the environment, are slowly degraded and may cause harmful effects on human health and other organisms. Besides, they are semivolatile and remain in the atmosphere in the form of gases or particles. They can also accumulate in soil and water or can be transported over long distances (Yan et al., 2012).

The physicochemical properties of PAHs (lipophilicity, low water solubility, and high environmental persistence) are associated with their wide distribution in the environment and increased risk of contamination by these hydrocarbons. The harmful effects of these pollutants include increased incidences of lung, breast, skin, esophageal, and prostate cancers (Zhao et al., 2012), and this has encouraged the development or improvement of efficient methods for total or partial removal of these molecules (Dzionek et al., 2016). Several microorganisms have already demonstrated the ability to degrade PAHs, elucidating the potential of using microorganisms in biodegradation processes (Ghosal et al., 2016).

Different bacterial groups are reported to be competent in PAH degradation processes. Species of the genera *Pseudomonas*, *Mycobacterium*, *Ochrobactrum*, *Ralstonia*, *Comamonas*, *Rhodococcus*, *Sphingomonas*, *Staphylococcus*, *Acinetobacter*, *Aeromonas*, *Beijerinckia*, *Flavobacterium*, *Nocardia*, *Corynebacterium*, *Stenotrophomonas*, *Paracoccus*, *Burkholderia*, *Microbacterium*, and *Gordonia* are the most efficient in PAH degradation (Jacques et al., 2007; Ghosal et al., 2016).

When compared to pure cultures, the microbial consortia are more efficient in the degradation of these contaminants, as they accept a wider variety of PHAs as a carbon source and their mineralization occurs more easily as the metabolic pathways are complementary. Additionally, the presence of intermediate compounds may favor PAH decomposition in the environment. The isolates from the genera *Pseudomonas*, *Sphingomonas*, *Nocardia*, *Beijerinckia*, *Rhodococcus*, and *Mycobacterium* were capable of completely mineralize anthracene with dihydriol as an initial oxygenated intermediate (Ramesha et al., 2017).

Bioremediation processes performed by fungi mainly occur through the action of ligninolytic enzymes, which are responsible for the total or partial degradation of different organic pollutants, including PAHs. The activity of these enzymes over PAHs is due to the similarities between the two classes of molecules, which are highly nonpolar compounds formed by several aromatic rings (Drevinskas et al., 2016).

Among the main groups of fungi-degrading PAHs are the *Phanerochaete*, *Fusarium*, *Candida*, *Penicillium*, *Pleurotus*, and *Bjerkandera* genera (Jacques et al., 2007). Additionally, products derived from microbial metabolism may play an important role in the biodegradation of these compounds, such as the lignocellulosic extract from *Phanerochaete chrysosporium*, which was used in the commercial preparation of the laccase enzyme for the removal of naphthalene, phenanthrene, and pyrene. This extract was responsible for accelerating the degradation of these PAHs, thus demonstrating that the products derived from microorganisms also have potential application in agricultural soils (Taha et al., 2018).

Different biotechnological tools have contributed to the remediation of PAHs, including functional metagenomics, used to prospect enzymes that degrade contaminants. As around 99% of microbial species have not yet been characterized, this approach allows access to the genetic information of the entire microbial community present in the contaminated site without previous cultivation, thereby expanding the perspective of finding genes that encode enzymes with the potential to degrade PAHs (Ufarté et al., 2015). Microorganisms thriving in the presence of pollutants may contain genes that allow them to metabolize these compounds, and, through metagenomics, it is possible to identify the key genes involved in the metabolic pathways of hydrocarbons, thus bringing about new insights into the production of bioinoculants for bioremediation.

10.6 Future perspectives

Soil contamination by substances harmful to both the environment and living organisms is a reality that needs to be managed. Soil health depends on its good physicochemical and biological conditions, which are constantly threatened by the use of polluting compounds. Microorganisms constitute a large group of organisms capable of acting on these pollutants and are essential for soil remediation. Studies using microbial bioinoculants with the ability to degrade pesticides, hydrocarbons, and microplastics and to transform urban wastes have shown efficiency and agility in the processes. Microbial bioinoculants, whether isolated, in consortia, or transformed microorganisms or their metabolites, are powerful tools in the fight against soil pollutants and represent a resource with great potential and several advantages in soil remediation. Omics tools are a great approach to search for genes that encode enzyme-degrading xenobiotics, including those discussed in this chapter. These technologies have greatly allowed accessing information on noncultivable organisms, thus highlighting the potential of combining omics and synthetic biology in the formulation of bioinoculants for the purpose of soil bioremediation.

References

Ahmad, R., Jilani, G., Arshad, M., Zahir, Z.A., Khalid, A., 2007. Bio-conversion of organic wastes for their recycling in agriculture: an overview of perspectives and prospects. Ann. Microbiol. 57 (4), 471–479. https://doi.org/10.1007/BF03175343.

Ahmad, M., Pataczek, L., Hilger, T.H., Zahir, Z.A., Hussain, A., Rasche, F., Schafleitner, R., Solberg, S.Ø., 2018. Perspectives of microbial inoculation for sustainable development and environmental management. Front. Microbiol. 9, 2992. https://doi.org/10.3389/fmicb.2018.02992.

Amobonye, A., Bhagwat, P., Singh, S., Pillai, S., 2021. Plastic biodegradation: frontline microbes and their enzymes. Sci. Total Environ. 759, 143536. https://doi.org/10.1016/j.scitotenv.2020.143536.

Anjum, R., Rahman, M., Masood, F., Malik, A., 2012. Bioremediation of pesticides from soil and wastewater. In: Environmental Protection Strategies for Sustainable Development. Springer, Dordrecht, pp. 295–328.

Anode, S., Onguso, J., 2021. Current methods of enhancing bacterial bioremediation of pesticide residues in agricultural farmlands. In: Microbial Rejuvenation of Polluted Environment. Springer, Singapore, pp. 167–187.

Bacosa, H.P., Mabuhay-Omar, J.A., Balisco, R.A.T., Omar, D.M., Inoue, C., 2021. Biodegradation of binary mixtures of octane with benzene, toluene, ethylbenzene or xylene (BTEX): insights on the potential of *Burkholderia, Pseudomonas* and *Cupriavidus* isolates. World J. Microbiol. Biotechnol. 37, 122. https://doi.org/10.1007/s11274-021-03093-4.

Basu, A., Prasad, P., Das, S.N., Kalam, S., Sayyed, R.Z., Reddy, M.S., El Enshasy, H., 2021. Plant growth promoting rhizobacteria (PGPR) as green bioinoculants: recent developments, constraints, and prospects. Sustainability 13 (3), 1140. https://doi.org/10.3390/su13031140.

Bhattacharjya, S., Das, S., Amat, D., 2021. Potential of microbial inoculants for organic waste decomposition and decontamination. In: Biofertilizers. Woodhead Publishing, pp. 103–132, https://doi.org/10.1016/B978-0-12-821667-5.00027-0.

Bisht, S., Pandey, P., Bhargava, B., Sharma, S., Kumar, V., Sharma, K.D., 2015. Bioremediation of polyaromatic hydrocarbons (PAHs) using rhizosphere technology. Braz. J. Microbiol. 46 (1), 7–21. https://doi.org/10.1590/S1517-838246120131354.

Bolden, A.L., Kwiatkowski, C.F., Colborn, T., 2015. New look at BTEX: are ambient levels a problem? Environ. Sci. Technol. 49 (9), 5261–5276. https://doi.org/10.1021/es505316f.

Brzeszcz, J., Kapusta, P., Steliga, T., Turkiewicz, A., 2020. Hydrocarbon removal by two differently developed microbial inoculants and comparing their actions with biostimulation treatment. Molecules 25 (3), 661. https://doi.org/10.3390/molecules25030661.

Chandran, H., Meena, M., Sharma, K., 2020. Microbial biodiversity and bioremediation assessment through omics approaches. Front. Environ. Chem. 1, 9. https://doi.org/10.3389/fenvc.2020.570326.

Chaudhary, T., Shukla, P., 2019. Bioinoculants for bioremediation applications and disease resistance: innovative perspectives. Indian J. Microbiol. 59 (2), 129–136. https://doi.org/10.1007/s12088-019-00783-4.

Chowdhury, A., Pradhan, S., Saha, M., Sanyal, N., 2008. Impact of pesticides on soil microbiological parameters and possible bioremediation strategies. Indian J. Microbiol. 48 (1), 114–127.

Dangi, A.K., Sharma, B., Hill, R.T., Shukla, P., 2019. Bioremediation through microbes: systems biology and metabolic engineering approach. Crit. Rev. Biotechnol. 39 (1), 79–98. https://doi.org/10.1080/07388551.2018.1500997.

Dellagnezze, B.M., Vasconcellos, S.P., Angelim, A.L., Melo, V.M.M., Santisi, S., Cappello, S., Oliveira, V.M., 2016. Bioaugmentation strategy employing a microbial consortium immobilized in chitosan beads for oil degradation in mesocosm scale. Mar. Pollut. Bull. 107 (1), 107–117. https://doi.org/10.1016/j.marpolbul.2016.04.011.

Desai, C., Pathak, H., Madamwar, D., 2010. Advances in molecular and "-omics" technologies to gauge microbial communities and bioremediation at xenobiotic/anthropogen contaminated sites. Bioresour. Technol. 101, 1558–1569. https://doi.org/10.1016/j.biortech.2009.10.080.

Drevinskas, T., Mickienė, R., Maruška, A., Stankevičius, M., Tiso, N., Mikašauskaitė, J., et al., 2016. Downscaling the in vitro test of fungal bioremediation of polycyclic aromatic hydrocarbons: methodological approach. Anal. Bioanal. Chem. 408 (4), 1043–1053. https://doi.org/10.1007/s00216-015-9191-3.

Dzionek, A., Wojcieszyńska, D., Guzik, U., 2016. Natural carriers in bioremediation: a review. Electron. J. Biotechnol. 23, 28–36. https://doi.org/10.1016/j.ejbt.2016.07.003.

Ghosal, D., Ghosh, S., Dutta, T.K., Ahn, Y., 2016. Current state of knowledge in microbial degradation of polycyclic aromatic hydrocarbons (PAHs): a review. Front. Microbiol. 7, 1369. https://doi.org/10.3389/fmicb.2016.01369.

Gravouil, K., Ferru-Clément, R., Colas, S., Helye, R., Kadri, L., Bourdeau, L., Moumen, B., Mercier, A., Ferreira, T., 2017. Transcriptomics and lipidomics of the environmental strain *Rhodococcus ruber* point out consumption pathways and potential metabolic bottlenecks for polyethylene degradation. Environ. Sci. Technol. 51 (9), 5172–5181. https://doi.org/10.1021/acs.est.7b00846.

Habib, S., Iruthayam, A., Abd Shukor, M.Y., Alias, S.A., Smykla, J., Yasid, N.A., 2020. Biodeterioration of untreated polypropylene microplastic particles by Antarctic bacteria. Polymers 12 (11), 2616. https://doi.org/10.3390/polym12112616.

Horton, A.A., Walton, A., Spurgeon, D.J., Lahive, E., Svendsen, C., 2017. Microplastics in freshwater and terrestrial environments: evaluating the current understanding to identify the knowledge gaps and future research priorities. Sci. Total Environ. 586, 127–141. https://doi.org/10.1016/j.scitotenv.2017.01.190.

Huang, Y., Liu, Q., Jia, W., Yan, C., Wang, J., 2020. Agricultural plastic mulching as a source of microplastics in the terrestrial environment. Environ. Pollut. 260, 114096. https://doi.org/10.1016/j.envpol.2020.114096.

Jacques, R.J.S., Bento, F.M., Antoniolli, Z.I., Camargo, F.A.D.O., 2007. Biorremediação de solos contaminados com hidrocarbonetos aromáticos policíclicos. Ciência Rural 37 (4), 1192–1201. https://doi.org/10.1590/S0103-84782007000400049.

Jaiswal, S., Shukla, P., 2020. Alternative strategies for microbial remediation of pollutants via synthetic biology. Front. Microbiol. 11, 808. https://doi.org/10.3389/fmicb.2020.00808.

Janczak, K., Dąbrowska, G.B., Raszkowska-Kaczor, A., Kaczor, D., Hrynkiewicz, K., Richert, A., 2020. Biodegradation of the plastics PLA and PET in cultivated soil with the participation of microorganisms and plants. Int. Biodeterior. Biodegrad. 155, 105087. https://doi.org/10.1016/j.ibiod.2020.105087.

Kumar, A., Sharma, S., 2019. Microbes and Enzymes in Soil Health and Bioremediation. Springer, pp. 353–366, https://doi.org/10.1007/978-981-13-9117-0.

Kumar, M., Yadav, A.N., Saxena, R., Paul, D., Tomar, R.S., 2021. Biodiversity of pesticides degrading microbial communities and their environmental impact. Biocatal. Agric. Biotechnol. 31, 101883. https://doi.org/10.1016/j.bcab.2020.101883.

Leal, A.J., Rodrigues, E.M., Fernandes, R.D.C.R., Borges, A.C., Júlio, A.D.L., de Souza Freitas, F., Tótola, M.R., 2018. Microbial inoculants development for bioremediation of gasoline and diesel contaminated soil. Open Access Lib. J. 5 (4), 1–17. https://doi.org/10.4236/oalib.1104449.

Liu, Z., Liu, Y., Zeng, G., Shao, B., Chen, M., Li, Z., et al., 2018. Application of molecular docking for the degradation of organic pollutants in the environmental remediation: a review. Chemosphere 203, 139–150. https://doi.org/10.1016/j.chemosphere.2018.03.179.

Machado, A.A.S., Lau, C.W., Till, J., Kloas, W., Lehmann, A., Becker, R., Rillig, M.C., 2018. Impacts of microplastics on the soil biophysical environment. Environ. Sci. Technol. 52 (17), 9656–9665. https://doi.org/10.1021/acs.est.8b02212.

Machado, A.A.S., Lau, C.W., Kloas, W., Bergmann, J., Bachelier, J.B., Faltin, E., Becker, R., Gorlich, A.S., Rillig, M.C., 2019. Microplastics can change soil properties and affect plant performance. Environ. Sci. Technol. 53 (10), 6044. https://doi.org/10.1021/acs.est.9b01339.

Mishra, S., Lin, Z., Pang, S., Zhang, W., Bhatt, P., Chen, S., 2021. Recent advanced technologies for the characterization of xenobiotic-degrading microorganisms and microbial communities. Front. Bioeng. Biotechnol. 9, 31.

Moog, D., Schmitt, J., Senger, J., Zarzycki, J., Rexer, K.-H., Linne, U., Erb, T., Maier, U. G., 2019. Using a marine microalga as a chassis for polyethylene terephthalate (PET) degradation. Microb. Cell Factories 18, 171. https://doi.org/10.1186/s12934-019-1220-z.

Mushtaq, Z., Mushtaq, H., Faizan, S., Parray, M.A., 2021. Microbial degradation of organic constituents for sustainable development. In: Microbiota and Biofertilizers. Vol. 2. Springer, Cham, pp. 103–117, https://doi.org/10.1007/978-3-030-61010-4_5.

Nandy, S., Andraskar, J., Lanjewar, K., Kapley, A., 2021. Challenges in bioremediation: from lab to land. In: Bioremediation for Environmental Sustainability. Elsevier, pp. 561–583, https://doi.org/10.1016/B978-0-12-820524-2.00023-7.

Odukkathil, G., Vasudevan, N., 2013. Toxicity and bioremediation of pesticides in agricultural soil. Rev. Environ. Sci. Biotechnol. 12 (4), 421–444. https://doi.org/10.1007/s11157-013-9320-4.

Oghoje, S.U., Ukpebor, J.E., Ukpebor, E.E., Ejeomo, C., 2020. Comparison of the effects of two forms of organic stimulation on the bioremediation of monocyclic-aromatic hydrocarbon in soils. J. Chem. Soc. Nigeria 45 (3), 555–566.

Ortiz-Hernández, M.L., Sánchez-Salinas, E., Dantán-González, E., Castrejón-Godínez, M. L., 2013. Pesticide biodegradation: mechanisms, genetics and strategies to enhance the process. In: Biodegradation-Life of Science, pp. 251–287, https://doi.org/10.5772/56098.

Palaniveloo, K., Amran, M.A., Norhashim, N.A., Mohamad-Fauzi, N., Peng-Hui, F., Hui-Wen, L., et al., 2020. Food waste composting and microbial community structure profiling. PRO 8 (6), 723. https://doi.org/10.3390/pr8060723.

Pan, I., Dam, B., Sen, S.K., 2012. Composting of common organic wastes using microbial inoculants. 3 Biotech 2 (2), 127–134. https://doi.org/10.1007/s13205-011-0033-5.

Parte, S.G., Mohekar, A.D., Kharat, A.S., 2017. Microbial degradation of pesticide: a review. Afr. J. Microbiol. Res. 11 (24), 992–1012. https://doi.org/10.5897/AJMR2016.8402.

Prashar, P., Shah, S., 2016. Impact of fertilizers and pesticides on soil microflora in agriculture. In: Sustainable Agriculture Reviews. Springer, Cham, pp. 331–361.

Raffa, C.M., Chiampo, F., 2021. Bioremediation of agricultural soils polluted with pesticides: a review. Bioengineering 8 (7), 92. https://doi.org/10.3390/bioengineering8070092.

Rajmohan, K.S., Chandrasekaran, R., Varjani, S., 2020. A review on occurrence of pesticides in environment and current technologies for their remediation and management. Indian J. Microbiol. 60, 125–138. https://doi.org/10.1007/s12088-019-00841-x.

Ramesha, V., Praveen, H.G., Madhushree, A., 2017. Polycyclic hydrocarbon degradation by bacteria: a review. Trends Biosci. 10 (14), 2465–2473.

Rillig, M.C., Leifheit, E., Lehmann, J., 2021. Microplastic effects on carbon cycling processes in soils. PLoS Biol. 19 (3), e3001130. https://doi.org/10.1371/journal.pbio.3001130.

Rodríguez, A., Castrejón-Godínez, M.L., Salazar-Bustamante, E., Gama-Martínez, Y., Sánchez-Salinas, E., Mussali-Galante, P., et al., 2020. Omics approaches to pesticide biodegradation. Curr. Microbiol. 77 (4), 545–563. https://doi.org/10.1007/s00284-020-01916-5.

Santos, M.S., Nogueira, M.A., Hungria, M., 2019. Microbial inoculants: reviewing the past, discussing the present and previewing an outstanding future for the use of beneficial bacteria in agriculture. AMB Express 9 (1), 1–22. https://doi.org/10.1186/s13568-019-0932-0.

Sayara, T., Basheer-Salimia, R., Hawamde, F., Sánchez, A., 2020. Recycling of organic wastes through composting: process performance and compost application in agriculture. Agronomy 10, 1838. https://doi.org/10.3390/agronomy10111838.

Scagliola, M., Valentinuzzi, F., Mimmo, T., Cesco, S., Crecchio, C., Pii, Y., 2021. Bioinoculants as promising complement of chemical fertilizers for a more sustainable agricultural practice. Front. Sustain. Food Syst. 4, 305. https://doi.org/10.3389/fsufs.2020.622169.

Sun, Y., Kumar, M., Wang, L., Gupta, J., Tsang, D.C., 2020. Biotechnology for soil decontamination: opportunity, challenges, and prospects for pesticide biodegradation. In: Bio-Based Materials and Biotechnologies for Eco-Efficient Construction. Woodhead Publishing, pp. 261–283, https://doi.org/10.1016/B978-0-12-819481-2.00013-1.

Taha, M., Shahsavari, E., Aburto-Medina, A., Foda, M.F., Clarke, B., Roddick, F., Ball, A. S., 2018. Bioremediation of biosolids with *Phanerochaete chrysosporium* culture filtrates enhances the degradation of polycyclic aromatic hydrocarbons (PAHs). Appl. Soil Ecol. 124, 163–170. https://doi.org/10.1016/j.apsoil.2017.11.002.

Timmusk, S., Behers, L., Muthoni, J., Muraya, A., Aronsson, A.C., 2017. Perspectives and challenges of microbial application for crop improvement. Front. Plant Sci. 8, 49. https://doi.org/10.3389/fpls.2017.00049.

Ufarté, L., Laville, É., Duquesne, S., Potocki-Veronese, G., 2015. Metagenomics for the discovery of pollutant degrading enzymes. Biotechnol. Adv. 33 (8), 1845–1854. https://doi.org/10.1016/j.biotechadv.2015.10.009.

Valdivia-Rivera, S., Ayora-Talavera, T., Lizardi-Jiménez, M.A., García-Cruz, U., Cuevas-Bernardino, J.C., Pacheco, N., 2021. Encapsulation of microorganisms for bioremediation: techniques and carriers. Rev. Environ. Sci. Biotechnol. 1-24. https://doi.org/10.1007/s11157-021-09577-x.

Varjani, S.J., 2017. Microbial degradation of petroleum hydrocarbons. Bioresour. Technol. 223, 277–286. https://doi.org/10.1016/j.biortech.2016.10.037.

Vryzas, Z., 2018. Pesticide fate in soil-sediment-water environment in relation to contamination preventing actions. Curr. Opin. Environ. Sci. Health 4, 4–9. https://doi.org/10.1016/j.coesh.2018.03.001.

Wang, X., Xue, L., Chang, S., He, X., Fan, T., Wu, J., et al., 2019. Bioremediation and metabolism of clothianidin by mixed bacterial consortia enriched from contaminated soils in Chinese greenhouse. Process Biochem. 78, 114–122. https://doi.org/10.1016/j.procbio.2018.12.031.

Wongbunmak, A., Khiawjan, S., Suphantharika, M., Pongtharangkul, T., 2020. BTEX biodegradation by Bacillus amyloliquefaciens subsp. plantarum W1 and its proposed BTEX biodegradation pathways. Sci. Rep. 10 (1), 1–13. https://doi.org/10.1038/s41598-020-74570-3.

Xi, B., He, X., Dang, Q., Yang, T., Li, M., Wang, X., Li, D., Tang, J., 2015. Effect of multi-stage inoculation on the bacterial and fungal community structure during organic municipal solid wastes composting. Bioresour. Technol. 196, 299–405. https://doi.org/10.1016/j.biortech.2015.07.069.

Ya, H., Jiang, B., Xing, Y., Zhang, T., Lv, M., Wang, X., 2021. Recent advances on ecological effects of microplastics on soil environment. Sci. Total Environ. 798, 149338. https://doi.org/10.1016/j.scitotenv.2021.149338.

Yan, L., Li, X., Chen, J., Wang, X., Du, J., Ma, L., 2012. Source and deposition of polycyclic aromatic hydrocarbons to Shanghai, China. J. Environ. Sci. 24 (1), 116–123. https://doi.org/10.1016/S1001-0742(10)60638-9.

Yoshida, S., Hiraga, K., Takehana, T., Taniguchi, I., Yamaji, H., Maeda, Y., Toyohara, K., Miyamoto, K., Kimura, Y., Oda, K., 2016. A bacterium that degrades and assimilates poly(ethylene terephthalate). Science 351 (6278), 1196–1199. https://doi.org/10.1126/science.aad6359.

Yu, H., Fan, P., Hou, J., Dang, Q., Cui, D., Xi, B., Tan, W., 2020. Inhibitory effect of microplastics on soil extracellular enzymatic activities by changing soil properties and direct adsorption: an investigation at the aggregate-fraction level. Environ. Pollut. 267, 115544. https://doi.org/10.1016/j.envpol.2020.115544.

Zhang, G.S., Zhang, F.X., Li, X.T., 2019. Effects of polyester microfibers on soil physical properties: perception from a field and a pot experiment. Sci. Total Environ. 670, 1–7. https://doi.org/10.1016/j.scitotenv.2019.03.149.

Zhang, S., Han, B., Sun, Y., Wang, F., 2020. Microplastics influence the adsorption and desorption characteristics of Cd in an agricultural soil. J. Hazard. Mater. 388, 121775. https://doi.org/10.1016/j.jhazmat.2019.121775.

Zhao, P., Fu, J., Yao, B., Song, Y., Mi, L., Li, Z., et al., 2012. In vitro malignant transformation of human bronchial epithelial cells induced by benzo (a) pyrene. Toxicol. in Vitro 26 (2), 362–368. https://doi.org/10.1016/j.tiv.2011.12.013.

Zhou, B., Wang, J., Zhang, H., Shi, H., Fei, Y., Huang, S., Tong, Y., Wen, D., Luo, Y., Barceló, D., 2020. Microplastics in agricultural soils on the coastal plain of Hangzhou Bay, East China: multiple sources other than plastic mulching film. J. Hazard. Mater. 388, 121814. https://doi.org/10.1016/j.jhazmat.2019.121814.

CHAPTER 11

Tropical biomes as microbial sources for efficient biocatalysts to environmental purposes

Suzan Pantaroto de Vasconcellos[a], Lidiane Maria dos Santos Lima[a], Mariana Rocha Roswell[a], and Paulo Teixeira Lacava[b]
[a]Laboratory of Health and Environment—LABMSMA, Department of Pharmaceutical Sciences, Federal University of São Paulo (UNIFESP), Diadema, SP, Brazil
[b]Laboratory of Microbiology and Biomolecules—LaMiB, Department of Morphology and Pathology, Center for Biological and Health Sciences, Federal University of São Carlos, São Carlos, SP, Brazil

11.1 Biomes and biotechnological potential of environmental microorganisms

The search for biotechnological innovations is increasing, as more investigations and prospecting for new biological sources in nature, including microbial enzymes, are becoming necessary. Brazil is one of the most biodiverse countries in the world (Mittermeier et al., 2005), highlighting the international scenario of biotechnological applications.

It is already known that microorganisms can inhabit different environments in the biosphere. These habitats can have a set of abiotic factors such as climate, humidity, soil, and light that allow and increase the adaptation and survival of microbial life (Gupta et al., 2017). Many of these ecosystems are reservoirs of microorganisms with high metabolic flexibility and physiological tolerance to environmental changes, especially fungi and actinomycetes (Procópio and Barreto, 2021).

In addition, fungi and actinomycetes have an important ecological function in the Brazilian biomes in the decomposition of organic compounds, soil maintenance, and the production of secondary metabolites with pharmacological and agricultural importance, including antibiotics and antitumors, fertilizers, pesticides, and biological controllers (Vargas Hoyos et al., 2021; Noriler et al., 2018). Intrinsic characteristics of the Brazilian biomes and some examples of microorganisms with a biotechnological application are covered in this chapter.

Microbial Inoculants
https://doi.org/10.1016/B978-0-323-99043-1.00003-7

Tropical biomes, such as mangroves, are characterized as coastal ecosystems present in tropical and subtropical regions. They are sparsely distributed around the world and can be found in many countries, including India, China, Malaysia, Thailand, Nigeria, Australia, New Zealand, and Brazil (Cadamuro et al., 2021; Thatoi et al., 2013). Mangroves are considered a transition ecosystem comprising marine and terrestrial habitats, characterized by adverse conditions, including extensive salinity, tidal effects, anaerobic clay soil, high temperature, and humidity (Nicolleti et al., 2018). This ecosystem can also provide organic matter in sediments and vegetal material, always rich in lignocellulose. Furthermore, mangroves also have unique profiles allowing diverse groups of microorganisms, associated not only with sediment material but also with their vegetation (Capdeville et al., 2019; Thatoi et al., 2013). Consequently, it is possible to find several reports on fungi, actinomycetes, and bacteria from mangrove habitats, connected to lignocellulosic biomass degradation and production of bioactive molecules for agricultural, environmental, medical, and industrial uses (Behera et al., 2017).

It is important to mention the contributions of mangrove microorganisms for sustainable agriculture development. *Azotobacter vinelandii* and *Bacillus megaterium* are commonly associated with mangrove vegetation, specifically at the plant roots, promoting nitrogen fixation, phosphate solubilization, and production of ammonia, as well as supplementing indole acetic acid, as a growth increment (Kathiresan and Selvam, 2006). Therefore the use of these bacteria as inoculants in agriculture is currently described as a sustainable technique to increase productivity (Thatoi et al., 2013). Brazilian mangroves are primarily made up of three tree species, *Rhizophora mangle*, *Laguncularia racemosa*, and *Avicennia* sp. (Dias et al., 2009), from which several diverse endophytic bacteria have been isolated (Castro et al., 2014). In this scenario, Castro et al. (2014) performed a diversity analysis on endophytic bacteria obtained from the branches of Brazilian mangrove forests. *Bacillus* was the most frequently isolated genus, comprising 42% of the species isolated. However, other common endophytic genera, such as *Curtobacterium*, *Enterobacter*, and *Pantoea*, were also identified. Additionally, in Brazil, Castro et al. (2018) evaluated a large number of endophytic bacterial strains from three different plant species, namely, *R. mangle*, *L. racemosa*, and *Avicennia* sp., to examine phosphate solubilization. All 115 strains examined produced a halo during the phosphate solubilization test in vitro. The endophytic strain MCR1.48 (*Enterobacter* sp.), which has a high P solubilization index, was selected for in vivo assays in *Acacia polyphylla*. These authors selected the commonly used reforestation tree *A. polyphylla*, which has a few published studies involving inoculation by bacteria of agronomic interest and used for the reforestation of degraded areas in Brazil (Rao et al., 2007). Inoculation with *Enterobacter* sp. strain MCR1.48 increased the dry mass of *A. polyphylla* shoots and roots, suggesting that the presence of the endophyte generates important benefits that promote the growth and fitness of this plant.

Regarding fungi, there are many strains of endophytic fungi isolated from the mangrove rhizosphere, including producers of bioinsecticide with high cytotoxicity against pathogens, and also inhibition of bacterial, protozoan, and viral development. In this way, bioactive molecules from endophytic fungi from mangroves can be considered as efficient tools to solve many problems, including also bacterial resistance and mitigation of environmental impacts caused by large amounts of chemicals, including pesticides and fertilizers (Cadamuro et al., 2021). There are also reports about fungi associated with common vegetation of mangrove trees, which have an important physiological function contributing to the tolerance to biotic and abiotic stress. Among endophytic fungi, the main descriptions concern *Aspergillus*, *Penicillium*, *Fusarium*, and *Trichoderma* that occur in mangrove forests (Thatoi et al., 2013).

Endophytes, as biological control agents, are considered indirect mechanisms through which competition, antibiosis, and systemic resistance against phytopathogens are induced (Azevedo et al., 2000; Lacava and Azevedo, 2014; Quecine et al., 2014). Endophytic microorganisms colonize ecological niches similar to those colonized by phytopathogens, and interactions between these two groups are likely. Research studies have shown that some endophytes isolated from plant tissues exhibit potential as biocontrol agents against phytopathogens (Azevedo et al., 2000; Ryan et al., 2008), increase plant growth, and hasten plant development (Lacava and Azevedo, 2013; Ryan et al., 2008). The capability of biological control by endophytes via indirect growth promotion has generated considerable interest with regard to understanding the mechanisms that these bacteria use for biocontrol and the use of these microorganisms for their potential to produce biopesticides rather than using agrochemicals (Azevedo et al., 2000; Lacava and Azevedo, 2013, 2014).

The biological control method using endophytic fungi capable of inhibiting phytopathogens seems to be an eco-friendlier and more sustainable tool in agricultural management than using agrochemicals that negatively impact the environment, humans, and other organisms (Tewari et al., 2019). In this context, Moreira et al. (2020) evaluated the antagonistic potential of the *Diaphorte* sp. FS-94(4) strain isolated from *Avicennia nitida*, in Brazilian mangrove forests (Sebastianes et al., 2013), against the phytopathogens *Colletotrichum* sp., *Fusarium oxysporum*, *Phytophthora sojae*, and *Rhizopus microspores*. The antagonism index (AI) ranged from 20% to 62%. The endophyte produced in vitro cellulase, suggesting that this enzyme is involved in 430 suppressing phytopathogens by breaking down the cell wall.

Regarding actinomycetes, the genus *Streptomyces* has already been isolated from mangroves in Asia, including soil and plant samples (Law et al., 2019). Recently, rare actinobacteria including *Micromonospora* sp., *Pseudonocardia* sp., *Verrucosispora* sp., *Nocardiopsis* sp., *Actinophytocola* sp., *Dietzia* sp., *Gordonia* sp., *Micrococcus* sp., *Mycobacterium* sp., *Nocardia* sp., *Saccharopolyspora* sp., and *Rhodococcus* sp. derived from mangrove have been highlighted, because they are considered a new source of active metabolites due to their uniquely structured biomolecules (Abidin et al., 2020; Azman et al., 2015).

Brazilian mangroves are found from the north to the south along the extensive coastline (Ottoni et al., 2021). According to Castro et al. (2018), this ecosystem is a rich source of endophytic bacteria and fungi, constituting an underexplored microbial biodiversity. Hence, they are an opportunity to discover new sources of microbial bioactives with biotechnological and agricultural applications. However, the diversity of microorganisms from mangroves varies according to vegetation and site. In the region Cananeia and Bertioga, in Brazil, endophytic bacteria of the genus *Bacillus* sp. were predominant, while *Ochrobactrume* and *Microbacterium* were found only in branches of *R. mangle* (Castro et al., 2018). In this same site also were isolated endophytic fungi such as *Colletotrichum*, *Diaporthe*, *Fusarium*, *Trichoderma*, and *Xylaria* (Taub, 2019). Martinho et al. (2019) also investigated some endophytic fungi isolated from the Cananeia mangrove and described their production of enzymes and bioactive secondary metabolites. According to the authors, isolates of *Fusarium sambucinum*, *Diaporthe* sp., *Fusarium* sp., and *Hypocrea lixii* showed efficient ligninolytic activities. The species *Aspergillus awamori*, *Fusarium chlamydosporum*, *Xylaria enteroleuca*, and *Trichoderma camerunense* could also exhibit some emulsifying and surfactant abilities. Regarding actinomycetes, the genera *Brevibacterium*, *Dermabacter*, *Kocuria*, *Kytococcus*, *Microbacterium*, *Nesterenkonia*, and *Rothia* were isolated from mangrove forests in Brazil (Thatoi et al., 2013).

Another tropical biome is the savanna, which is found on the African and American continents. This ecosystem is characterized by a peculiar vegetation type and forests, displaying a weathered and acidic soil, with high rates of clay. In general, the weather is tropical, with a temperature around 30°C, displaying two main seasons, a wet summer and a mild winter (Castro et al., 2016). It is also considered a biodiversity hotspot, as a real source of fauna, flora, and microorganisms. Regarding microbial communities from the African savanna, some bacterial strains are already described in the literature (*Bacillus* sp., *Pantoea agglomerans*, *Caballeronia concitans*, *Pseudomonas* sp., and *Stenotrophomonas maltophilia*), all of them with abilities to fix nitrogen and having an important role in the protection, nutrition, and growth of vegetation (Maquia et al., 2020).

Cerrado, also known as the Brazilian savanna, represents 22% of the country's territory, being considered as the second-most extensive biome in Brazil, representing 30% of its total biodiversity (Françoso et al., 2015; Klink and Machado, 2005; Silva et al., 2012). This biome can show some intrinsic species and it is also considered a special hotspot for microbial biodiversity recovery. For many years, Cerrado has been used for agricultural proposals, as it presents adequate soil and weather conditions for many cultures. The Cerrado has approximately 6000 vascular plant species, featuring the highest floral biodiversity among the world's savannas, and has a high diversity of microorganisms (de Sousa et al., 2017; Lacava and de Sousa, 2016). In this sense, it is important to mention their inherent microbial communities, responsible for Cerrado importance in agriculture, and the necessity to improve the application of sustainable techniques, like the use of the fungus *Trichoderma harzianum* for biological control of the gray mold *Sclerotium rolfsii*, which causes rotting of the collar region in seedlings (Auler et al., 2013). *Bacillus thuringiensis* is another classical biological controller, applied in the caterpillar *Spodoptera frugiperda*, which attacks the military rice crop, for example (Valicente and Barreto, 2003).

Among the diverse trees in the Cerrado, research groups have considered two species: *Stryphnodendron adstringens* (Mart.) Coville (Fabaceae) and *Solanum lycocarpum* Saint-Hill (Solanaceae). *S. adstringens* trees have the potential for medicinal applications. Their bark and fruit contain antiinflammatory (Lima et al., 1998), antiviral (Felipe et al., 2006), and antimicrobial properties (Almeida et al., 2017). *S. lycocarpum* produces secondary metabolites that exhibit antioxidant, antibacterial, cytotoxic, and antiinflammatory properties (Costa et al., 2015). Lisboa et al. (2021) reported that bacterial strains were isolated from rhizosphere samples of *S. lycocarpum* and were tested in vitro for direct mechanisms of plant growth promotion, such as phosphate solubilization. Among the 131 strains evaluated, the calcium phosphate solubilizing capacity trials showed that 45.8% of the strains exhibited a positive reaction. Some bacterial strains identified as *Bacillus* sp., *Burkholderia* sp., and *Microbacterium* sp. showed high solubilization potential. According to the authors, this research is the first report on the characterization, identification, and isolation of rhizospheric bacteria from *S. lycocarpum* in Brazil.

The reduced availability of phosphorus in potentially productive acidic soils such as in the Brazilian Cerrado requires plants to associate symbiotically with microorganisms that are adapted to those soil conditions (Zapata and Axmann, 1995) and that alter complexes that adsorb phosphate under acidic conditions, including $FePO_4$ and $AlPO_4$ (Andrade et al., 2003). The plant growth-promoting effects of microbes include increased plant height and biomass of shoots, stems, and roots; lignification of the xylem vessels; and increased crop yield (Azevedo and Quecine, 2019). In this context, phosphorus deficiency limits agricultural production, especially in acidic soils such as those of the Brazilian savanna (Nakayama et al., 1998). In this biome, the water-soluble P is transformed into iron phosphate ($FePO_4$) and aluminum phosphate ($AlPO_4$) (Silva et al., 2011), which are moderately soluble complexes (Chai et al., 2011; Yadav et al., 2010). In this context, Vitorino et al. (2012) reported the capacity of endophytic fungi isolated from roots of *Hyptis marrubioides* Epling, a Brazilian Cerrado species, to solubilize calcium phosphate in GELP medium (Sylvester-Bradley et al., 1982) and iron phosphate in modified Reyes basal medium. Of six fungal strains analyzed by solubilization assays, none of the fungi tested solubilized $CaHPO_4$. In contrast, all of these fungi demonstrated the capacity to solubilize high levels of $FePO_4$. These authors suggested that none of the fungi strains testing solubilized $CaHPO_4$ may be related to the fact that these fungi were isolated from the root system of *H. marrubioides* Epling, a plant adapted to acidic soil, in which P is bound to Fe^{3+} and to Al^{3+} (Khan and Joergensen, 2009).

Torres (2018) reported the isolation and identification of 66 endophytic fungi from leaves of *S. adstringens* and *S. lycocarpum* (Fig. 1), isolated from Brazilian savanna species (Bogas et al., 2020), collected in the savanna area of São Paulo State, Brazil. Additionally, this author reported the ability of these endophytic fungal strains to solubilize inorganic phosphate (Fig. 2). The author observed the potential to solubilize inorganic phosphate with a mean phosphate solubilization index ranging from 1.46 to 1.93 and 1.19 to 2.61 by endophytic strains. In a similar study, Sharma and Roy (2015) reported that fungal endophytes were isolated from the root, stem, and leaves of the plant *Amaranthus spinosus*, another species occurring in the savanna biome, and stated that endophytic fungal isolates of the plant *A. spinosus* showed a positive test for phosphate solubilization. The

Fig. 1 Fungal morphological groups retrieved in the culture media isolation procedure. Endophytic fungi (morphological groups) isolated from leaves of *Stryphnodendron adstringens* and *Solanum lycocarpum*, examples of endemic Brazilian tropical savanna trees (22°01′03″S, 47°53′27″W). *(Adapted from Torres, F.L., Isolamento, caracterização e potencial biotecnológico de fungos endofíticos associados às plantas de Cerrado. (Dissertation). São Carlos (SP), Universidade Federal de São Carlos, Brazil, 2018.)*

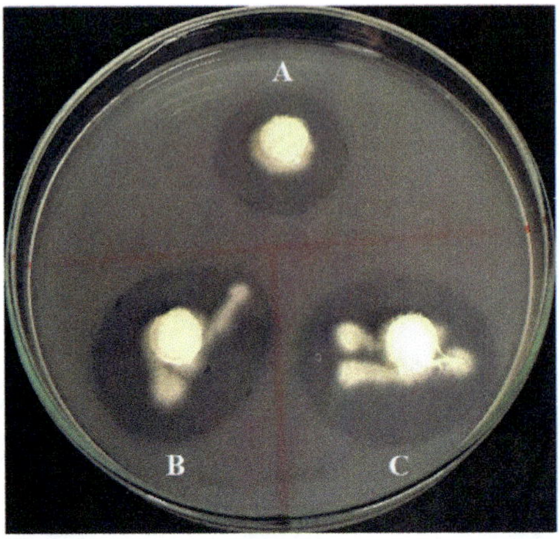

Fig. 2 In vitro phosphate solubilization by endophytic fungi strains. A clear halo around endophytic colonies A, B, and C indicated a positive result according to Berraquero et al. (1976). *(Adapted from Torres, F.L., Isolamento, caracterização e potencial biotecnológico de fungos endofíticos associados às plantas de Cerrado. (Dissertation). São Carlos (SP), Universidade Federal de São Carlos, Brazil, 2018.)*

phosphate solubilization efficiency was found to be highest for the fungal genus *Aspergillus* isolated from the stem of the plant.

Tropical forests are predominantly distributed in South America, Africa, and South Asia, near to the equatorial region. Besides having a hot climate and abundant rain, tropical forests favor the development of a thick vegetation bed in the canopy. The microbial community in the soil is responsible for the decomposition of organic matter, which contributes to sustaining the forest (Pereira et al., 2017). In addition, there is enormous biodiversity of plants, animals, and microorganisms in this environment. However, the biodiversity pattern of tropical forests differs from one location to another, including the diversity of microorganisms, especially endophytic species (Tripathi et al., 2016).

The biomes of the Amazon and Atlantic Forest located in Brazil are considered to be true reservoirs of biological diversity, including fauna and flora species, due to their geography between the tropics. These findings also extend microbial life because there are reports of environmental fungi isolated from soil samples from the Atlantic Forest (Lima et al., 2021). This study presented some enzymatic diversity (cellulases, xylanases, and proteases) from fungi strains, including *Aspergillus niger*, *Aspergillus fumigatus*, *Trichoderma koningii*, and *Geotrichum candidum*. Preeminent activities of lignin peroxidases were described from *Mucor* sp., *Byssochlamys nivea*, and

Paecilomyces saturatus strains, constituting the first report about this enzymatic ability in *Mucor* sp.

There is an improvement in the use of fungi for human food, as in the case of the mushroom from *Pleurotus* genera, which is increasing and could generate the commercialization of around 42 billion dollars each year, with an estimated growth of up to US$ 62.2 billions in 2023 (Sheldrake, 2020). Although some species of the same genera, such as *Pleurotus ostreatoroseus*, can also be correlated to other functions, like enzymatic lignin degradation, being described as white rot species from tropical and subtropical regions (Sheldrake, 2020). The ability of fungi to degrade substrates, like lignin polymer, is due to its special ability to produce extracellular enzymes (Fonseca et al., 2015; Omrane et al., 2019). Lignin is like a cement that, in association with hemicellulose and cellulose, can form lignocellulosic material, which is responsible for the cellular structure of vegetables (Manan and Webb, 2017). Because of the special ability of fungi to produce lignocellulolytic enzymes, they are frequently used in bioprocesses related to bioenergy and biofuels, as well as specific molecules from lignocellulosic substrates (Magalhães et al., 2019; Ravindran and Jaiswal, 2016).

Dry or semiarid environments worldwide are normally characterized by high temperature and salinity, low water availability, and, consequently, arid soil. Such conditions cause abiotic stress for the soil microbiota and its rhizosphere. However, microorganisms and vegetation can also exhibit adaptive mechanisms to survive in such adverse environments. An example can be the action of bacteria in plant assistance, promoting the production of phytohormones (like indole acetic acid), as well as realizing nitrogen fixation and biological control of pathogens (Bonatelli et al., 2021). Several reports about bacteria from soil and rhizosphere, stems, leaves, and nodules in semiarid regions point out the benefits of the microbiome in vegetation. Niu et al. (2018) described a dry tolerant crop and the isolation of bacterial strains from its roots and rhizosphere, which could potentially decrease water stress in the evaluated agroecosystem (Niu et al., 2018). In the same way, rhizobacteria from roots of the wheat crops were also isolated from saline conditions and were described as halotolerant microorganisms, allowing their application in agriculture systems with some adverse conditions (Sorty et al., 2016).

In this sense, Brazil stands out for having several tropical ecosystems like rainforests, Cerrado, and the Caatinga, a real semiarid ecosystem, uniquely found in the northeastern region of the country. Fig. 3 presents the main Brazilian tropical biomes previously mentioned.

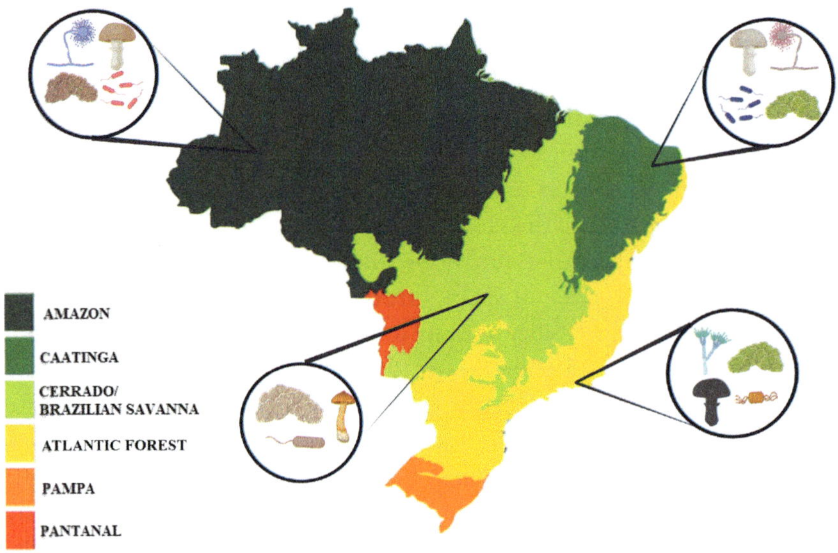

Fig. 3 Representation of the demographic division of biomes, which are represented by the Amazon, Caatinga, Cerrado, Atlantic Forest, considered tropical or subtropical forests. These biomes have great biodiversity of microorganisms with biotechnological potential. *(Copyright figure, using BioRenger.)*

Caatinga is a dry tropical forest, exclusively in Brazil, characterized by arid soil, elevated salinity, unstable pH, high temperatures, and dry spells (Marinho et al., 2019). These characteristics offer unique conditions for the development of a singular and notable microbial community, mainly associated with the rhizosphere of plants (Silva et al., 2015). Despite the existing interest in biodiversity conservation of Caatinga, this ecosystem has been explored for anthropic activities like agriculture and forestry; consequently, there is a decrease in biodiversity, green cover, and soil fertility, bringing a desertification risk (Marinho et al., 2019; Sousa et al., 2014).

Despite the adverse environmental conditions of Caatinga, this ecosystem has native plant species well adapted to survive both water and nutrient stress (Bonatelli et al., 2021). In addition, there are reports of arbuscular mycorrhizal fungi, associated with contributing to the assimilation and absorption of nutrients and also water in plants (Frosi et al., 2016).

Marinho et al. (2019) reported a diversity of arbuscular mycorrhizal fungi species from Caatinga. According to the authors, the most abundant genera were *Acaulospora* (20 species) and *Glomus* (16), followed by *Racocetra* (6), *Clarideoglomus* (5), *Gigaspora* and *Rhizoglomus* (4), *Scutellospora*, *Dentiscutata* and *Fuscutata* (3), *Ambispora*, *Entrophospora*, *Cetraspora*, *Dominikia*, *Paradentiscutata*, *Sclerocystis* (2), and *Bulbospora*, *Funneliformis*, *Intraornatospora*, and *Orbispora* (1). The variety of species of arbuscular mycorrhizal fungi is due to the variation in vegetation, soil, and climatic conditions, which can influence the maintenance of different habitats in Caatinga (Pontes et al., 2017).

Associations between bacteria and plants from arid environments such as Caatinga have also been reported in the literature. Bonatelli et al. (2021) evidenced the potential of bacteria associated with native caatinga plants, as inoculants to promote plant growth under drought conditions. Santos et al. (2020) recovered bacterial strains from the rhizosphere of cactus in the Brazilian Caatinga, revealing the potential of bacteria such as *Bacillus* spp. to promote maize growth under drought conditions.

11.2 Bioinoculants from enzymes of environmental origin

Bioinoculants are microorganisms that live in association with root plants or soil matrices (Ali et al., 2019). This symbiotic interaction favors crops' productivity in agrosystems and also can help to regulate water and salinity stress. Bioinoculants are known for their action under crops as biofertilizers, for example, solubilizing phosphate and other important nutrients. In addition, they are a biostimulator tool promoting the production of phytohormones, besides a biocontrol agent against phytopathogens (Tallapragada and Seshagiri, 2017).

The use of bioinoculants is a sustainable technique in modern agriculture. It is known that intensive land use and rudimentary agricultural practices can lead to organic matter depletion that affects the soil microbiota, in addition to the low productivity of food crops (Odoh, 2017). Moreover, there are phytopathogens such as viruses, bacteria, fungi, nematodes, or protozoa that cause diseases in plants, especially those intended for food crops. In this sense, the efficient management of soil can improve fertility and the elimination of diseases and pathogens in agriculture (Tallapragada and Seshagiri, 2017). For this, agrochemicals (synthetic fertilizers and pesticides) have been adopted to replace missing nutrients and eliminate pathogens. However, some implications arising from these practices have contributed to environmental degradation, since these compounds when improperly used cause the contamination of soil and water and consequently of food (Vyas, 2018).

Alternatively, the use of bioinoculants can be an attractive tool to alleviate some agricultural limitations, such as salinity and soil dryness

(Tallapragada and Seshagiri, 2017). The presence of a soil microbiome in the rhizosphere of plants brings physiological and metabolic benefits, promoting crop growth and increasing productivity. These microorganisms also produce enzymes that help the development of plants (Ali et al., 2019). Therefore, it is an eco-friendly and sustainable approach for agricultural systems (Sharma et al., 2016). Furthermore, bioinoculants can allow establishment of a model of sustainable agriculture, showing an important performance against pathogens as well as supporting vegetal growth, promoting nutrient accessibility, production of phytohormones, increasing productivity simultaneously with the improvement of soil fertility (Itelima et al., 2018; Kour et al., 2020; Ramasamy et al., 2020).

In this sense, bioinoculants are classified according to their role, as biofertilizers, biostimulants, and bioprotectants. In the case of biofertilizers, the application, whether directly to the soil or seeds, using microorganisms contributes to the colonization of rhizosphere plants. In addition, the biological activity promoted by the enzymatic arsenal from the soil microbiome will also increase the availability and assimilation of nutrients (Ali et al., 2019; Mazid et al., 2011). This microbial arsenal provides mutualistic interaction arising from the use of these bioinoculants, bringing benefits for soil fertility and not having adverse effects on the global ecosystem (Sahoo et al., 2014).

Biofertilizers can include nitrogen fixators, potassium and phosphorus solubilizers, growth-promoting rhizobacteria, endo and ectomycorrhizal fungi, cyanobacteria, and other helpful microorganisms (Itelima et al., 2018; Kour et al., 2020; Ramasamy et al., 2020). Phosphate solubilization promoted by phosphatase enzymes from *Trichoderma*, *Pseudomonas*, *Rhizobium*, *Micrococcus*, and *Erwinia* genera can favor plant growth (Ali et al., 2019). Phosphorus, an essential macronutrient for the biogeochemical vegetal cycle, is normally obtained as a synthetic phosphate fertilizer. However, phosphate solubilizing microorganisms can synthesize acid phosphatases that act under the mineralization of the phosphorus. These enzymes are produced when there is a low inorganic phosphorus availability (Ventura et al., 2018). Among phosphate solubilizing microorganisms, the following genera are the main ones having this ability: *Pseudomonas*, *Bacillus*, *Rhizobium*, *Burkholderia*, *Enterobacter*, and *Streptomyces* (Ali et al., 2019).

It is important to mention that another essential nutrient for plant metabolism is nitrogen, which is also one of the main components of proteins and in chlorophyll biosynthesis. Once again, microorganisms play an essential role in the availability of nitrogen as a nutrient for plants (Sharma et al., 2016). Free-living bacteria in soil or associated with plants have special abilities for nitrogen-fixing, due to their nitrogenase enzymatic system. *Azotobacter*, *Azospirillum*, and *Rhizobium* are bacteria already described in their nitrogenase system specialties (Ali et al., 2019). Most nitrogen present in the soil is found in organic compounds, and then urease is the enzyme that catalyzes the reaction to the formation of carbon dioxide and ammonia (Ali et al., 2019). Hence, its activity modulates the nitrogen absorption by plants after the application of urea as fertilizer. Urease can be produced by fungi, bacteria, yeast, actinomycetes, algae, and vegetation residues (Follmer, 2008; Lanna et al., 2010). Analysis of urease activity in the soil will provide data about its fertility (Lanna et al., 2010; Saviozzi et al., 2001).

Similarly, microbial proteases in the soil can also help the mineralization process of organic matter and nitrogen (Liu et al., 2021). Proteases are known as enzymes that hydrolyze proteins into peptides and amino acids (Vranova et al., 2013). When in soil, extracellular microbial proteases act under nitrogen organic form, like nitrogenated compounds, promoting their hydrolyzation into oligopeptides, to be readily available for plants (Greenfield et al., 2020). Examples of microbial proteases have been found not only in fungi, but also in bacterial strains: *Aspergillus*, *Penicillium*, *Bacillus*, and *Pseudomonas* (Ali et al., 2019; Rudenskaya and Pupov, 2008; Wenig et al., 2004).

Hence, other microbial enzymes, when in the soil, like peroxidases, amylases, and cellulases, also can act as biofertilizers, improving fertility and, consequently, the development and growth of plants. Peroxidases are enzymes involved in the lignocellulosic biomass degradation and/or biotransformation. Thereby lignin peroxidases and manganese peroxidases have a role key in the decomposition of organic matter in the soil, being assisted by auxiliary enzymes such as aryl alcohol oxidase and glyoxal oxidase (Ali et al., 2019). In the same way, amylases and cellulases are also recognized for their hydrolytic action under amylaceous and cellulosic substrates in the soil. For this, some techniques used and available from rural production have already been mentioned. Others of these techniques have also been used to help and improve forage conservation and the silage process, as well as for maintaining the nutritional quality of the soil (Allen et al., 2003; Santos et al., 2017). A wide variety of additives have been recommended to improve the nutritional quality of soil by silage processes, and the most commonly used inoculants are bacteria, mainly *Lactobacillus plantarum*, *Lactobacillus casei*, *Enterococcus faecium*, and *Pediococcus* spp., *Streptococcus faecium*, *Streptococcus faecalis*, *Lactobacillus* sp., *Lactobacillus buchneri*, and some Propionibacteria, combined with hemicellulase, cellulase, amylase, α-galactosidase, and β-mannanase enzyme complexes (Bernardes et al., 2005; Kung Jr et al., 2007; Rodrigues et al., 2004).

11.3 Nanotechnology and increased activity of microbial enzymes in soil

The productivity of crops in agriculture has been significantly improved by the use of agrochemicals. Among these, pesticides and synthetic fertilizers have been intensively adopted in agrosystems (Meena et al., 2020). However, most of these agrochemicals when applied are dispersed to non-point sources, including many environmental compartments, through leaching and volatilization (Usman et al., 2020). Consequently, toxic effects and degradation of the environment occur. An alternative is the application of nanotechnology, which can contribute to the development and improvement of sustainable agriculture, allowing the productivity of crops and, in the same way, mitigating impacts on the environment (Singh et al., 2021).

From this perspective, nanofertilizers, nanopesticides, and nanomaterials are essential problem-solving tools for traditional agriculture, such as low-yield crop farming, infertile soils, and also phytopathogens (Chhipa, 2019). It is worth highlighting the role of nanotechnology in several areas, including medicine, engineering, environment, energy, food, and agriculture (Achari and Kowshik, 2018). In addition, nanomaterials are nanoscale structures (<100 nm) with special physicochemical properties, for instance, the high surface area-to-volume ratio in nanoparticles is an attractive feature that becomes very important, giving these materials some specific characteristics for certain applications (Ferreira and Rangel, 2009; Mejias et al., 2021).

In this sense, nanofertilizers are a real and efficient way to apply nanotechnology in agriculture. They constitute nanoscale materials produced from nanoparticles, in addition to macro- and micronutrients released in a controlled process (slow and steady) for crops (Adisa et al., 2019; Shang et al., 2019). The advantages of this approach are the use of smaller amounts of nanofertilizers compared to synthetic ones, including the decreasing of leaching levels, runoff, and gas emissions to the atmosphere (Manjunatha et al., 2016). Consequently, nanofertilizers reduce environmental risks since they have a prolonged and effective action (Jia et al., 2020; Mejias et al., 2021).

Examples of nanofertilizers prepared with nanoparticles (NPs), wherein essential macronutrients such as calcium, iron, and magnesium have been constantly reported in the literature. Magnesium plays an important role in plants, especially in metabolic processes and chlorophyll. The application of Mg NPs at a concentration ranging from 0.2 to $0.5\,\mathrm{g\,L^{-1}}$ increased the enzymatic activities of dehydrogenase, esterase, acid phosphatase, alkaline phosphatase, and nitrate reductase in cultures of *Triticum aestivum* L. (wheat) and *Vigna unguiculata* subsp. unguiculata (black-eyed peas) (Achari and Kowshik, 2018; Delfani et al., 2014; Rathore and Tarafdar, 2015). In the same way, micronutrients are also necessary for the healthy development of crop farming. One of these essential micronutrients is zinc, a cofactor for enzymes and a component of chlorophyll synthesis. The application of zinc nanoparticles (ZnO NPs) at a concentration of $10\,\mathrm{mg\,L^{-1}}$ resulted in an increased microbial community in roots and the activities of enzymes such as acid phosphatase, alkaline phosphatase, phytase, and dehydrogenase in crops of *Vigna radiate* (mung bean) (Raliya et al., 2016).

Silver nanoparticles (Ag-NPs) exhibit potent bactericidal properties against a wide range of microorganisms, including *Acinetobacter baumannii*, *B. subtilis*, *E. coli*, *P. aeruginosa*, and *S. aureus*. However, these biological effects are potentialized or influenced by nanoparticle morphology. NPs that exhibit larger surfaces offer more contact area with microorganisms (Siddiqi et al., 2018; Vanlalveni et al., 2021). Roca et al. (2015) demonstrated that the photocatalytic and antibacterial properties of α-Ag_2WO_4 crystals are facet dependent. Ibrahim et al. (2019) reported the biosynthesis of sphere-shaped Ag-NPs by *Bacillus siamensis* endophytically isolated from the medicinal health plant *Coriandrum sativum*. Ag-NPs strongly inhibited bacterial growth, biofilm formation, and swimming motility of the rice phytopathogen *Xanthomonas oryzae* pv. *oryzae* and *Acidovorax oryzae* and promoted a significant increase in root length, shoot length, fresh weight, and dry weight of rice seedlings. Furthermore, green synthesized Ag-NPs by the garlic (*Allium sativum*) endophyte *Pseudomonas poae* showed strong inhibition of the wheat pathogen *Fusarium graminearum* in terms of mycelium growth, spore germination, length of the germ tubes, and mycotoxin production (Ibrahim et al., 2020).

11.4 Carbon sequestration and soil quality improvement

The increase of atmospheric CO_2 contributes to an increase in environmental temperature of more than 2–4°C, causing climate changes that are responsible for catastrophic disorders (Qafoku, 2015). In the natural biogeochemical carbon cycle, the atmospheric CO_2 came from soil organic carbon, where it is kept like a reservoir, to be converted into organic compounds, like sugars and cellulose, which are responsible for the structure of live biomass (Ahmed et al., 2020).

Biochar is the result of thermochemical conversion through pyrolysis of biomass (Oliveira et al., 2017) subjected to high temperatures under limited O_2 conditions (Zhang et al., 2017). The product has minerals, volatile organic compounds, and the presence of free radicals. These materials can change the soil microbial activity and enzymatic activity that catalyzes several processes related to biogeochemical cycles (Martins Filho et al., 2020; Zhu et al., 2017).

According to many reports described in the literature, activated carbon, like biochars, attracts special interest as efficient support for microbial enzymes, including lignin peroxidases and multicopper oxidases (laccases), to improve soil quality, in the same way that allows its function as a natural carbon reservoir (Martins Filho et al., 2020; Zhang et al., 2017; Zhu et al., 2017).

Many authors have described efficient strategies to increase the capacity of soil to promote atmospheric carbon sequestration, using biotechnological systems (Scholes et al., 2014). It is also known that part of this carbon is converted to humic substances, which is the main form to maintain the soil carbon sequestration mechanism (Miralles et al., 2014). The main structural molecules that are components of humic substances are phenols, carbohydrates, quinones, lignins, and cutins, and it is important to mention that microbial laccase, lignin peroxidase, and manganese peroxidase are the key enzymes to promote humic substance's synthesis, and also their biotransformation (Chen et al., 2011; Moreno and Benitez, 2016). In this sense, it could be possible to infer that those enzymatic microbial systems could be efficient biotechnological methods to provide benefits to agricultural ecosystem services to improve soil capacity as an atmospheric carbon reservoir, moving to increment humification processes by microbial catabolism and anabolism stimulation. However, according to Zhang et al. (2018), it is also important to select an appropriate transport or carrier to optimize results, including here the use of nanomaterials, or natural immobilizers, like biochar or zeolites, considering pore size, economic viability, stability, and surface area, as well as environmental applicability, including here its biodegradability.

References

Abidin, Z.A.Z., Malek, N.A., Zin, N.H.M., Chowdhury, A.J.K., 2020. Rare actinomycetes from Kuantan Mangrove Forest Sediment. Journal Clean WAS 4 (2), 79–83.

Achari, G.A., Kowshik, M., 2018. Recents developments on nanotechnology in agriculture: plant mineral nutrition, health, and interactions with soil microflora. J. Agric. Food Chem. 66, 8647–8661.

Adisa, I.O., Pullagurala, V.L.R., Peralta-Videa, J.R., Dimkpa, C.O., Elmer, W.H., Gardea-Torresdey, J.L., White, J.C., 2019. Recent advances in nano-enabled fertilizers and pesticides: a critical review of mechanisms of action. Environ. Sci. Nano 6, 2002–2030.

Ahmed, A.A.Q., Odelade, K.A., Babalola, O.O., 2020. Microbial inoculants for improving carbon sequestration in agroecosystems to mitigate climate change. In: Leal Filho, W. (Ed.), Handbook of Climate Change Resilience. Springer, Cham, pp. 381–401.

Ali, J., Sharma, D.C., Bano, A., Gupta, A., Sharma, S., Bajpai, B., Pathak, N., 2019. Exploiting microbial enzymes for augmenting crop production. Enzymes in Food Biotechnology, 513–519.

Allen, M.S., Coors, J.G., Roth, G.W., 2003. Corn silage. In: Buxton, D.R., Muck, R.E., Harisson, J.H. (Eds.), Silage Science and Technology. American Society of Agronomy, Madison, WI, pp. 547–608.

Almeida, A.C., Andrade, V.A., Fonseca, F.S.A., Macêdo, A.A., Santos, R.L., Colen, K.G.F., Martins, E.R., Marcelo, N.A., 2017. Acute and chronic toxicity and antimicrobial activity of the extract of *Stryphnodendron adstringens* (Mart.) Coville. Pesq. Vet. Bras. 37, 840–846.

Andrade, F.V., Mendonça, E.S., Alvarez, V.H., Novais, R.F., 2003. Adição de ácidos orgânicos e húmicos em latossolos e adsorção de fosfato. R Bras Ci Solo 27, 1003–1011.

Auler, A.C.V., Carvalho, D.D.C., de Mello, S.C.M., 2013. Antagonismo de *Trichoderma harzianum* a *Sclerotium rolfsii* nas culturas do feijoeiro e soja. Revista Agro@mbiente On-line 7 (3), 359–365.

Azevedo, J.L., Quecine, M.C., 2019. Biodiversity and biotechnological applications of microorganisms associated with tropical plants. In: Kumar, V., Prasad, R., Kumar, M., Choudhary, D. (Eds.), Microbiome in Plant Health and Disease. Springer, Singapore, pp. 293–313.

Azevedo, J.L., Maccheroni, W., Pereira, J.O., Araújo, W.L., 2000. Endophytic microorganisms: a review on insect control and recent advances on tropical plants. Electron J Biotechnol 3, 40–65.

Azman, A.S., Othman, I., Velu, S.S., Chan, K.G., Lee, L.H., 2015. Mangrove rare actinobacteria: taxonomy, natural compound, and discovery of bioactivity. Front. Microbiol. 6, 856.

Behera, B.C., Sethi, B.K., Misha, R.R., Dutta, S.K., Thatoi, H.N., 2017. Microbial cellulases—diversity & biotechnology with reference to mangrove environment: a review. Journal of Genetic Engineering and Biotechnology 15 (1), 197–210.

Bernardes, T.F., Reis, R.A., Moreira, A.L., 2005. Perfil fermentativo e microbiológico das silagens do capim marandu aditivadas com polpa cítrica peletizada. Scientia Agricola 62 (3), 214–220.

Berraquero, F.R., Baya, A.M., Cormenzana, A.R., 1976. Establishment of indices for the study of phosphate solubilization by soil bacteria. Ars Pharmacéutica. 17, 399–406.

Bogas, A.C., Torres, F.L., Sousa, C.P., Lacava, P.T., 2020. Bioactivity of endophytes from the Brazilian tropical savannah. Acta Sci. Microbiol. 3, 15–22.

Bonatelli, M.L., Lacerda-Júnior, G.V., dos Reis Junior, F.B., Fernandes-Júnior, P.I., Melo, I. S., Quecine, M.C., 2021. Beneficial plant-associated microorganisms from semiarid regions and seasonally dry environments: a review. Front. Microbiol. 11, 553223.

Cadamuro, R.D., Silveira Bastos, I.M.A., Silva, I.T., da Cruz, A.C.C., Robl, D., Sandjo, L. P., Alves Jr., S., Lorenzo, J.M., Rodríguez-Lázaro, D., Treichel, H., Steindel, M., Fongaro, G., 2021. Bioactive Compounds from mangrove endophytic fungus and their uses for microorganism control. J. Fungi 7 (6), 1–19.

Capdeville, C., Pommier, T., Gervaix, J., Fromard, F., Rols, J.L., Leflaive, J., 2019. Mangrove facies drives resistance and resilience of sediment microbes exposed to anthropic disturbance. Front. Microbiol. 9, 3337.

Castro, R.A., Quecine, M.C., Lacava, P.T., Batista, B.D., Luvizotto, D.M., Marcon, J., Ferreira, A., Melo, I.S., Azevedo, J.L., 2014. Isolation and enzyme bioprospection of endophytic bacteria associated with plants of Brazilian mangrove ecosystem. SpringerPlus 3, 382.

Castro, A.P., Sartori da Silva, M.R.S., Quirino, B.F., Da Cunha Bustamante, M.M., Krüger, R.H., 2016. Microbial diversity in Cerrado biome (neotropical savanna) soils. PLoS One 11 (2), e0148785.

Castro, R.A., Dourado, M.N., de Almeida, J.R., Lacava, P.T., Nave, A., de Melo, I.S., Azevedo, J.L., Quecine, M.L., 2018. Mangrove endophyte promotes reforestation tree (*Acacia polyphylla*) growth. Braz. J. Microbiol. 49, 59–66.

Chai, B., Wu, Y., Liu, P., Liu, B., Gao, M., 2011. Isolation and phosphate-solubilizing ability of a fungus, *Penicillium* sp. from soil of an alum mine. J Basic Microbiol 51, 5–14.

Chen, M., Zeng, G., Tan, Z., Jiang, M., Li, H., Liu, L., Zhu, Y., Yu, Z., Wei, Z., Liu, Y., Xie, G., 2011. Understanding lignin-degrading reactions of ligninolytic enzymes: binding affinity and interactional profile. PLoS One 6 (9).

Chhipa, H., 2019. Applications of nanotechnology in agriculture. In: Gurtler, V., Ball, A.S., Soni, S. (Eds.), Methods in Silkworm Microbiology. Elsevier, pp. 115–142.

Costa, G.A., Morais, M.G., Saldanha, A.A., Assis Silva, I.C., Aleixo, Á.A., Ferreira, J.M., Soares, A.C., Duarte-Almeida, J.M., Lima, L.A., 2015. Antioxidant, antibacterial, cytotoxic, and anti-inflammatory potential of the leaves of *Solanum lycocarpum* A. St. Hil. (Solanaceae). Evidence-based complementary and alternative medicine: eCAM 2015, 315987.

de Sousa, C.P., Serrano, N.F.G., Lacava, P.T., 2017. Endophytic microorganisms of the Tropical Savannah: a promising source of bioactive molecules. In: de Azevedo, J., Quecine, M. (Eds.), Diversity and Benefits of Microorganisms From the Tropics. Springer, Cham, pp. 37–56.

Delfani, M., Baradarn-Firouzabadi, M., Farrokhi, N., Makarian, H., 2014. Some physiological responses of black-eyed pea to iron and magnesium nanofertilizers. Commun. Soil Sci. Plant Anal. 45, 530–540.

Dias, A.C.F., Andreote, F.D., Dini-Andreote, F., Lacava, P.T., Sá, A.L.B., Melo, I.S., Azevedo, J.L., Araújo, W.L., 2009. Diversity and biotechnological potential of culturable bacteria from Brazilian mangrove sediment. World J. Microbiol. Biotechnol. 25, 1305–1311.

Felipe, A.M., Rincão, V.P., Benati, F.J., Linhares, R.E., Galina, K.J., de Toledo, C.E., Lopes, G.C., de Mello, J.C., Nozawa, C., 2006. Antiviral effect of *Guazuma ulmifolia* and *Stryphnodendron adstringens* on poliovirus and bovine herpesvirus. Biological & Pharmaceutical Bulletin 29 (6), 1092–1095.

Ferreira, H.S., Rangel, M.C., 2009. Nanotecnologia: aspectos gerais e potencial de aplicação em catálise. Quím. Nova 32 (7).

Follmer, C., 2008. Insights into the role and structure of plant ureases. Phytochemistry 69 (1), 18–28.

Fonseca, T.R.B., Silva, T.A., Alecrim, M.M., Cruz Filho, R.F., Teixeira, M.F.S., 2015. Cultivation and nutritional studies of an edible mushroom from North Brazil. African Journal of Microbiology Research 9 (30), 1814–1822.

Françoso, R.D., Brandão, R., Nogueira, C.C., Salmona, Y.B., Machado, R.B., Colli, G.R., 2015. Habitat loss and the effectiveness of protected areas in the Cerrado Biodiversity Hotspot. Natureza & Conservação 1 (1), 35–40.

Frosi, G., Barros, V.A., Oliveira, M.T., Santos, M., Ramos, D.G., Maia, L.C., Santos, M.G., 2016. Symbiosis with AMF and leaf Pi supply increases water deficit tolerance of woody species from seasonal dry tropical forest. Journal of Plant Physiology 207, 84–93.

Greenfield, L.M., Hill, P.W., Paterson, E., Baggs, E.M., Jones, D.L., 2020. Do plants use root-derived proteases to promote the uptake of soil organic nitrogen. Plant Soil 456, 355–367.

Gupta, A., Gupta, R., Singh, R.L., 2017. Microbes and environment. In: Singh, R. (Ed.), Principles and Applications of Environmental Biotechnology for a Sustainable Future. Applied Environmental Science and Engineering for a Sustainable Future. Springer, Singapore, pp. 43–84.

Ibrahim, E., Fouad, H., Zhang, M., Zhang, Y., Qiu, W., Yan, C., Li, B., Mo, J., Chen, J., 2019. Biosynthesis of silver nanoparticles using endophytic bacteria and their role in inhibition of rice pathogenic bacteria and plant growth promotion. RSC Adv. 9 (50), 29293–29299.

Ibrahim, E., Zhang, M., Zhang, Y., Hossain, A., Qiu, W., Chen, Y., Wang, Y., Wu, W., Sun, G., Li, B., 2020. Green-synthesization of silver nanoparticles using endophytic bacteria 804 isolated from garlic and its antifungal activity against wheat fusarium head blight pathogen *Fusarium graminearum*. Nanomaterials (Basel) 10 (2), 219.

Itelima, J.U., Bang, W.J., Onyimba, I.A., Sila, M.D., Egbere, O.J., 2018. Bio-fertilizers as key player in enhancing soil fertility and crop productivity: a review. Direct Research Journal of Agriculture and Food Science 6, 73–83.

Jia, C., Lu, P., Zhang, M., 2020. Preparation and characterization of environmentally friendly controlled release fertilizers coated by leftovers-based polymer. Processes 8 (4), 417.

Kathiresan, K., Selvam, M., 2006. Evaluation of beneficial bacteria from mangrove soil. Botanica Marina 49 (1), 86–88.

Khan, K.S., Joergensen, R.G., 2009. Changes in microbial biomass and P fractions in biogenic household waste compost amended with inorganic P fertilizers. Bioresour. Technol. 100, 303–309.

Klink, C.A., Machado, R.B., 2005. Conservation of the Brazilian cerrado. Conserv. Biol. 19 (3), 707–713.

Kour, D., Rana, K.L., Yadav, A.N., Yadav, N., Kumar, M., Kumar, V., Vyas, P., Dhaliwal, H.S., Saxena, A.K., 2020. Microbial biofertilizers: bioresources and eco-friendly technologies for agricultural and environmental sustainability. Biocatal. Agric. Biotechnol. 23, 101487.

Kung Jr., L., Schmidt, R.L., Ebling, T.E., Hu, W., 2007. The effect of *Lactobacillus buchneri* 40788 on the fermentation and aerobic stability of ground and whole high-moisture corn. J. Dairy Sci. 90, 2309–2314.

Lacava, P.T., Azevedo, J.L., 2013. Endophytic bacteria: a biotechnological potential in agrobiology system. In: Maheshwari, D.K., Sarah, M., Aeron, A. (Eds.), Bacteria in Agrobiology: Crop Productivity. Springer, Berlin, pp. 1–4.

Lacava, P.T., Azevedo, J.L., 2014. Biological control of insect-pest and diseases by endophytes. In: Vijay, V.C., Alan, C.G. (Eds.), Advances in Endophytic Research Advances. Springer, New Delhi, pp. 231–243.

Lacava, P.T., de Sousa, C.P., 2016. Role of endophytic actinomycetes in crop protection: plant growth promotion and biological control. In: Subramaniam, G., Arumugam, S., Rajendran, V. (Eds.), Plant Growth Promoting Actinobacteria. Springer, Singapore.

Lanna, A.C., Silveira, P.M., Silva, M.B., Ferraresi, T.M., Kliemann, H.J., 2010. Atividade de urease no solo com feijoeiro influenciada pela cobertura vegetal e sistemas de plantio. R. Bras. Ci. Solo. 34, 1933–1939.

Law, J.W.F., Pusparajah, P., Ab Mutalib, N.S., 2019. A review on mangrove actinobacterial diversity: the roles of *Streptomyces* and novel species discovery. Prog. Microbes Mol. Biol. 2 (1), a0000024.

Lima, J.C.S., Lima, J.C.D.S., Martins, D.T.O., de Souza Jr, P.T., 1998. Experimental evaluation of stem bark of *Stryphnodendron adstringens* (Mart.) Coville for anti-inflammatory activity. Phytother. Res. 12, 218–220.

Lima, L.M.S., Okamoto, D.N., Passarini, M.R.Z., Gonçalves, S.S., Goldman, G.H., Silveira, M.A.V., Ramos, P.L., Cruz, J.B., Juliano, M., Marcondes, M.F.M., Vasconcellos,

S.P., 2021. Enzymatic diversity of filamentous fungi isolated from forest soil incremented by sugar cane solid waste. Environ. Technol. 43 (20), 3037–3046.

Lisboa, P.H.G., de Andrade, P.H.M., Machado, P.C., de Sousa, C.P., Lacava, P.T., 2021. Isolation and in vitro screening of plant growth-promoting rhizobacteria from *Solanum lycocarpum* St. Hil., an endemic plant of the Brazilian tropical savannah. Afr. J. Microbiol. Res. 15 (5), 253–261.

Liu, J., Liu, W., Xing, S., Zhang, X., He, H., Chen, J., Bielicki, K., Zhou, M., 2021. Diversity of protease-producing bacteria in the soils of the South Shetland Islands,Antarctica. Antonie van Leeuwenhoek 114 (4), 457–464.

Magalhães, A.A.S., Silva, T.A., Teixeira, M.F.C., Cruz Filho, R.F., Silva, S.D., Gomes, D. M.D., Pereira, J.O., 2019. Produção e caracterização de enzimas proteolíticas de *Lentinus crinitus* (L.) Fr. 1825 DPUA 1693 do bioma amazônico (Polyporaceae). Bol. Mus. Para. Emílio Goeldi. Cienc. Nat. Belém. 14 (3), 453–461.

Manan, M.A., Webb, C., 2017. Design aspects of solid state fermentation as applied to microbial bioprocessing. Appl. Biotechnol. Bioeng. 4 (1), 511–532.

Manjunatha, S.B., Biradar, D., Aladakatti, Y., 2016. Nanotechnology and its applications in agriculture: a review. J. Farm Sci. 29, 1–13.

Maquia, I.S., Fareleira, P., Videira e Castro, I., Brito, D.R.A., Soares, R., Chaúque, A., Ferreira-Pinto, M.M., Lumini, E., Berruti, A., Ribeiro, N.S., Marques, I., Ribeiro-Barros, A.I., 2020. Mining the microbiome of key species from African savanna woodlands: potential for soil health improvement and plant growth promotion. Microorganisms 8 (9), 1291.

Marinho, F., Oehl, F., Silva, I.R., Coyne, D., Veras, J.S.N., Maia, L.C., 2019. High diversity of arbuscular mycorrhizal fungi in natural and anthropized sites of a Brazilian tropical dry forest (Caatinga). Fungal Ecol. 40, 82–91.

Martinho, V., Dos Santos Lima, L.M., Barros, C.A., Ferrari, V.B., Passarini, M.R.Z., Santos, L.A., De Souza Sebastianes, F.L., Lacava, P.T., Vasconcellos, S.P., 2019. Enzymatic potential and biosurfactant production by endophytic fungi from mangrove forest in Southeastern Brazil. AMB Express 9, 130.

Martins Filho, A.P., de Medeiros, E.V., Lima, J.R.S., Duda, G.P., Silva, W.M., Antonino, A. C.D., Silva, J.S.A., Oliveira, J.B., Hammecker, C., 2020. Impact of coffee biochar on soil carbon, microbial biomass and enzymatic activities in semiarid sandy soil cultivated with maize. Revista Brasileira de Geografia Física 13 (2), 903–914.

Mazid, M., Khan, T.A., Mohammad, F., 2011. Potencial de NO e H_2O_2 como moléculas de sinalização na tolerância ao estresse abiótico em plantas. J. Industr. Res. Technol. 1, 56–68.

Meena, R.S., Kumar, S., Datta, R., Lal, R., Vijayakumar, V., Brtnicky, M., Sharma, M.P., Yadav, G.S., Jhariya, M.K., Jangir, C.K., Pathan, S.I., Dokulilova, T., Pecina, V., Marfo, T.D., 2020. Impact of agrochemicals on soil microbiota and management: a review. Land 9, 34.

Mejias, J.H., Salazar, F., Pérez-Amaro, L., Hube, S., Rodriguez, M., Alfaro, M., 2021. Nanofertilizers: a cutting-edge approach to increase nitrogen use efficiency in grasslands. Front. Environ. Sci. 9, 635114.

Miralles, I., Piedra-Buena, A., Almendros, G., González-Vila, F.J., González-Pérez, J.A., 2014. Pyrolytic appraisal of the lignin signature in soil humic acids: assessment of its usefulness as carbon sequestration marker. J. Anal. Appl. Pyrol. 113, 107–115.

Mittermeier, R.A., de Fonseca, G.A.B., Rylands, A.B., Brandon, K., 2005. Uma breve história da conservação da biodiversidade no Brasil. Megadiversidade 1 (1), 14–21.

Moreira, C.C., Luna, G.L.F., Soriano, B., Cavicchioli, R., Bogas, A.C., de Sousa, C.P., Anibal, F.F., Lacava, P.T., 2020. Leishmanicidal, cytotoxic, antimicrobial and enzymatic activities of *Diaporthe* species, a mangrove-isolated endophytic fungus. Afr. J. Microbiol. Res. 14 (9), 516–524.

Moreno, B., Benitez, E., 2016. Impact of agricultural management on bacterial laccase-encoding genes with possible implications for soil carbon storage in semi-arid Mediterranean olive farming. PeerJ 4, e2257.
Nakayama, L.H.I., Caceres, N.T., Alcarde, J.C., Malavolta, E., 1998. Eficiência relativa de fontes de fósforo de diferentes solubilidades na cultura do arroz. Sci. Agr. 55, 183–190.
Nicolleti, R., Salvatore, M.M., Andolfi, A., 2018. Secondary metabolites of mangrove-associated strains of talaromyces. Mar. Drugs 16 (1), 12.
Niu, X., Song, L., Xiao, Y., Ge, W., 2018. Drought-tolerant plant growth-promoting rhizobacteria associated with foxtail millet in a semi-arid agroecosystem and their potential in alleviating drought stress. Front. Microbiol. 8, 2580.
Noriler, S.A., Savi, D.C., Aluizio, R., Palácio-Cortes, A.M., Possiede, Y.M., Glienke, C., 2018. Bioprospecting and structure of fungal endophyte communities found in the Brazilian biomes, Pantanal, and Cerrado. Front. Microbiol. 9 (1526), 1–14.
Odoh, C.K., 2017. Plant growth promoting rhizobacteria (PGPR): a bioprotectant bioinoculant for sustainable agrobiology. A review. Int. J. Adv. Res. Biol. Sci. 4 (5), 123–142.
Oliveira, F.R., Patel, A.K., Jaisi, D.P., Adhikari, S., Lu, H., Khanal, S.K., 2017. Environmental application of biochar: current status and perspectives. Bioresour. Technol. 246, 110–122.
Omrane, B.M., Mechri, S., Jaouadi, N.Z., Elhoul, M.B., Rekik, H., Sayadi, S., Bejar, S., Kechaou, N., Jaouadi, B., 2019. Purification and biochemical characterization of a novel thermostable protease from the oyster mushroom *Pleurotus sajor-caju* strain CTM10057 with industrial interest. BMC Biotechnol. 19 (1), 1–18.
Ottoni, F.P., Huhges, R.M., Katz, A.M., Rangel-Pereira, F.S., Bragança, P.H.N., Fernandes, R., Palmeira-Nunes, A.R.O., Nunes, J.L.S., Santos, R.R., Pioeski, N.M., Rodrigues-Filho, J.L., 2021. Brazilian mangroves at risk. Biota Neotrop. 21 (2).
Pereira, P.H.F., Macrae, A., Reinert, F., de Souza, R.F., Coelho, R.R.R., Pötter, G., Klenk, H.P., Labeda, D.P., 2017. *Streptomyces odonnellii* sp. nov., a proteolytic streptomycete isolated from soil under cerrado (savanna) vegetation cover. Int. J. Syst. Evol. Microbiol. 67, 5211–5215.
Pontes, J.S., Oehl, F., Marinho, F., Coyne, D., Silva, D.K.A., Yano-Melo, A.M., Maia, L. C., 2017. Diversity of arbuscular mycorrhizal fungi in Brazil's Caatinga and experimental agroecosystems. Biotropica 49 (3), 413–427.
Procópio, L., Barreto, C., 2021. The soil microbiomes of the Brazilian Cerrado. J. Soils Sediments 21, 2327–2342.
Qafoku, N.P., 2015. Climate change effects on soils: accelerated weathering, soil carbon, and elemental cycling. In: Sparks, D.L. (Ed.), Advances in Agronomy. vol. 131. Academic Press, Elsevier, USA, pp. 111–172.
Quecine, M.C., Batista, B.D., Lacava, P.T., 2014. Diversity and biotechnological potential of plant associated endophytic bacteria. In: Kumar, A.P., Govil, J.N. (Eds.), Biotechnology: Plant Biotechnology. 2. Studium Press LLC, Houston, pp. 377–423.
Raliya, R., Tarafdar, J.C., Biswas, P., 2016. Enhancing the mobilization of native phosphorus in the mung bean rhizosphere using ZnO nanoparticles synthesized by soil fungi. J. Agric. Food Chem. 64, 3111–3118.
Ramasamy, M., Geetha, T., Yuvaraj, M., 2020. Role of biofertilizers in plant growth and soil health. In: Rgobelo, E.C., Serra, A.P.S. (Eds.), Nitrogen Fixation. IntechOpen, pp. 1–11.
Rao, K.P.C., Verchot, L.V., Joshi, L.M., 2007. Adaptation to climate change through sustainable management and development of agroforestry systems. J SAT Agric Res. 4 (1), 1–30.
Rathore, I., Tarafdar, J.C., 2015. Perspectives of biosynthesized magnesium nanoparticles in foliar application of wheat plant. J. Bionanosci. 9, 209–214.
Ravindran, R., Jaiswal, A.K., 2016. Microbial enzyme production using lignocellulosic food industry wastes as feedstock: a review. Bioengineering 3 (30), 1–22.

Roca, R.A., Sczancoski, J.C., Nogueira, I.C., Fabbro, M.T., Alves, H.C., Gracia, L., Santos, L.P.S., Sousa, C.P., Andrés, J., Luz Jr., G.E., Longo, E., Cavalcante, L.S., 2015. Facet-dependent photocatalytic and antibacterial properties of α-Ag2WO4 crystals: combining experimental data and theoretical insights. Catal. Sci. Technol. 5 (8), 4091–4107.
Rodrigues, P.H.M., Almeida, L.F.S., Lucci, C.S., Melotti, L., Lima, F.R., 2004. Efeitos da adição de inoculantes microbianos sobre o perfil fermentativo da silagem de alfafa adicionada de polpa cítrica. Revista Brasileira de Zootecnia 33 (6), 1646–1653.
Rudenskaya, G.N., Pupov, D.V., 2008. Cysteine proteinases of microorganisms and viruses. Biochemistry 73, 1–13.
Ryan, R.P., Germaine, K., Franks, A., Ryan, D.J., Dowling, D.N., 2008. Bacterial endophytes: recent developments and applications. FEMS Microbiol. Lett. 278, 1–9.
Sahoo, R., Ansari, M., Pradhan, M., Dangar, T.K., Mohanty, S., Tuteja, N., 2014. Phenotypic and molecular characterization of efficient native *Azospirillum* strains from rice fields for crop improvement. Protoplasma 251, 943–953.
Santos, G., Moraes, J.M.M., Nussio, L.G., 2017. Custo e Análise de Sensibilidade na Produção de Silagem. Revista iPecege 3, 39–48.
Santos, A.F.J., Morais, J.S.D., Miranda, J.S., Moreira, Z.P.M., Feitoza, A.F.A., Leite, J., Fernandes-Júnior, P.I., 2020. Cacti-associated rhizobacteria from Brazilian Caatinga biome induce maize growth promotion and alleviate abiotic stress. Rev. Bras. Ciências Agrárias. 15, e8221.
Saviozzi, A., Levi-Minzi, R., Cardelli, R., Riffaldi, R., 2001. A comparison of soil quality in adjacent cultivated, forest and native grassland soils. Plant Soil. 233 (2), 251–259.
Scholes, R.J., Palm, C.A., Hickman, J.E., 2014. Agriculture and Climate Change Mitigation in the Developing World. CCAFS Working Paper No. 61, CGIAR Research Program on Climate Change, Agriculture and Food Security (CCAFS), Copenhagen.
Sebastianes, F.L.S., Romão-Dumaresq, A.S., Lacava, P.T., Harakava, R., Azevedo, J.L., de Melo, I.S., Pizzinari-Kleiner, A.A., 2013. Species diversity of culturable endophytic fungi from Brazilian mangrove forests. Current Genetic 59, 153–166.
Shang, Y., Hasan, M.K., Ahammed, G.J., Li, M., Yin, H., Zhou, J., 2019. Applications of nanotechnology in plant growth and crop protection: a review. Molecules 24, 2558–2580.
Sharma, S., Roy, S., 2015. Isolation and identification of a novel endophyte from a plant *Amaranthus spinosus*. Int. J. Curr. Microbiol. Appl. Sci. 4, 785–798.
Sharma, A., Vaishnav, A., Jamali, H., Srivastava, A.K., Saxema, A.K., Srisvastava, A.K., 2016. Halophilic bacteria: potential bioinoculants for sustainable agriculture and environment management under salt stress. In: Choudhary, D., Varma, A., Tuteja, N. (Eds.), Plant-Microbe Interaction: An Approach to Sustainable Agriculture. Springer, Singapore, pp. 297–325.
Sheldrake, M., 2020. Entangled Life: How Fungi Make Our Worlds, Change Our Minds and Shape Our Futures (Hardback). 1, p. 5.
Siddiqi, K.S., Husen, A., Rao, R.A.K., 2018. A review on biosynthesis of silver nanoparticles and their biocidal properties. J. Nanobiotechnol. 16 (1), 14.
Silva, A.C.S., Chagas Junior, A.F., Oliveira, L.A., Chagas, L.F.B., 2011. Ocorrência de bactérias solubilizadoras de fosfato nas raízes de plantas de importância econômica em Manaus e Rio Preto da Eva, Amazonas. J. Biotec. Biodivers. 2, 37–42.
Silva, S.M.F.Q., Pinheiro, S.M.B., Queiroz, M.V.F., Pranchevicius, M.C., Castro, J.G.D., Perim, M.C., Carreiro, S.C., 2012. Atividade in vitro de extratos brutos de duas espécies vegetais do cerrado sobre leveduras do gênero Candida. Ciência & Saúde Coletiva 17, 1649–1656.
Silva, I.L., Coelho, L.C.B.B., Silva, L.A.O., 2015. Biotechnological potential of the Brazilian caatinga biome. Adv. Res. 5 (1), 1–17.
Singh, H., Sharma, A., Bhardwaj, S.K., Arya, S.K., Bhardwaj, N., Katri, M., 2021. Recent advances in the applications of nano-agrochemicals for sustainable agricultural development. Environ. Sci. Process. Impacts 23, 213.

Sorty, A.M., Meena, K.K., Choudhary, K., Bitla, U.M., Minhas, P.S., Krishnani, K.K., 2016. Effect of plant growth promoting bacteria associated with halophytic weed (*Psoralea corylifolia* L) on germination and seedling growth of wheat under saline conditions. Appl. Biochem. Biotechnol. 180, 872–882.

Sousa, C.S., Menezes, R.S.C., Samapio, E.V.S.B., Lima, F.S., Maia, L.C., Oehl, F., 2014. Arbuscular mycorrhizal fungi in successional stages of Caatinga in the semiarid region of Brazil. Ciênc. Florest. 24 (1), 137–148.

Sylvester-Bradley, R., Asakawa, N., La Torraca, S., Magalhães, F.M.M., Oliveira, L.A., Pereira, R.M., 1982. Quantitative survey of phosphate solubilizing microorganisms in the rhizosphere of grasses and legumes in the Amazon. Acta Amazon. 12, 15–22.

Tallapragada, P., Seshagiri, S., 2017. Application of bioinoculants for sustainable agriculture. In: Kumar, V., Kumar, M., Sharma, S., Prasad, R. (Eds.), Probiotics and Plant Health. Springer, Singapore, pp. 473–495.

Taub, L., 2019. Bioprospecção e Caracterização de Fungos Endofíticos Produtores de Compostos Bioativos Isolados de Dalbergia (Dissertation). Alfenas (MG), Universidade Federal de Alfenas, Brazil.

Tewari, S., Shrivas, V.L., Hariprasad, P., Sharma, S., 2019. Harnessing endophytes as biocontrol agents. In: Ansari, R., Mahmood, I. (Eds.), Plant Health Under Biotic Stress. Springer, Singapore, pp. 189–218.

Thatoi, H., Behera, B.C., Mishra, R.R., Dutta, S.K., 2013. Biodiversity and biotechnological potential of microorganisms from mangrove ecosystems: a review. Ann. Microbiol. 63, 1–19.

Torres, F.L., 2018. Isolamento, caracterização e potencial biotecnológico de fungos endofíticos associados às plantas de Cerrado (Dissertation). São Carlos (SP), Universidade Federal de São Carlos, Brazil.

Tripathi, B.M., Song, W., Slik, J.W.F., Sukri, R.S., Jaafar, S., Dong, K., Adams, J.M., 2016. Distinctive tropical forest variants have unique soil microbial communities, but not always low microbial diversity. Front. Microbiol. 7, 376.

Usman, M., Farooq, M., Wakeel, A., Nawaz, A., Cheema, S.A., Rehman, H., Ashraf, I., Sanaullah, M., 2020. Nanotechnology in agriculture: current status, challenges and future opportunities. Sci. Total Environ. 721, 137778.

Valicente, F.H., Barreto, M.R., 2003. *Bacillus thuringiensis* survey in Brazil: geographical distribution and insecticidal activity against Spodoptera frugiperda (JE Smith) (Lepidoptera: Noctuidae). Neotrop. Entomol. 32 (4), 639–644.

Vanlalveni, C., Lallianrawna, S., Biswas, A., Selvaraj, M., Changmai, B., Rokhum, S.L., 2021. Green synthesis of silver nanoparticles using plant extracts and their antimicrobial activities: a review of recent literature. RSC Adv. 11 (5), 2804–2837.

Vargas Hoyos, H.A., Chiaramonte, J.B., Barbosa-Casteliani, A.G., Fernandez Morais, J., Perez-Jaramillo, J.E., Nobre Santos, S., Nascimento Queiroz, S.C., Soares Melo, I., 2021. An actinobacterium strain from soil of cerrado promotes phosphorus solubilization and plant growth in soybean plants. Front. Bioeng. Biotechnol. 9, 579906.

Ventura, B.S., Almeida, A.F.N., Bortolini, J.G., Stoffel, S.C.G., Lovato, P.E., Soares, C.R.F. S., 2018. Colonização micorrízica, fosfatase ácida e fósforo do solo sob a influência do manejo e da cobertura vegetal. Reunião Sul Brasileira de Ciência do Solo, Xanxerê.

Vitorino, L.C., Silva, F.G., Soares, M.A., Souchie, E.L., Costa, A.C., Lima, W.C., 2012. Solubilization of calcium and iron phosphate and *in vitro* production of indoleacetic acid by endophytic isolates of *Hyptis marrubioides* Epling (Lamiaceae). Int. Res. J. Biotechnol. 3, 47–54.

Vranova, V., Rejsek, K., Formanek, P., 2013. Proteolytic activity in soil: a review. Appl. Soil Ecol. 70, 23–32.

Vyas, P., 2018. Endophytic microorganisms as bio-inoculants for sustainable agriculture. In: Singh, J., Sharma, D., Kumar, G., Sharma, N. (Eds.), Microbial Bioprospecting for Sustainable Development. Springer, Singapore, pp. 41–60.

Wenig, K., Chatwell, L., von Pawel-Rammminger, U., Björck, L., Huber, R., Sondermann, P., 2004. Structure of the streptococcal endopeptidase IdeS, a cysteine proteinase with strict specificity for IgG. PNAS 101 (50), 17371–17376.

Yadav, V., Kumar, M., Deep, D.K., Kumar, H., Sharma, R., Tripathi, T., Tuteja, N., Saxena, A.K., Johri, A.K., 2010. A phosphate transporter from the root endophytic fungus *Piriformospora indica* plays a role in phosphate transport to the host plant. J. Biol. Chem. 285, 26532–26544.

Zapata, F., Axmann, H., 1995. 32P isotopic techniques for evaluating the agronomic effectiveness of rock phosphate material. Nutr. Cycl. Agroecosys. 41, 189–195.

Zhang, M., Cheng, G., Feng, H., Sun, B., Zhao, Y., Chen, H., Zhang, A., 2017. Effects of straw and biochar amendments on aggregate stability, soil organic carbon, and enzyme activities in the Loess Plateau, China. Environ. Sci. Pol. Res. 24.

Zhang, T., Song, Y., Zhao, Y., Zhang, B., 2018. Effect of hybrid sizing with nano-SiO_2 on the interfacial adhesion of carbon fibers/nylon 6 composites. Colloids Surfaces A 553, 125–133.

Zhu, L.X., Xiao, Q., Cheng, H.Y., Shi, B.J., Shen, Y.F., Li, S.Q., 2017. Seasonal dynamics of soil microbial activity after biochar addition in a dryland maize field in North-Western China. Ecol. Eng. 104, 141–149.

CHAPTER 12

Role of microbial biofilms in bioremediation: Current perspectives

Manju[a], Rajnish Kumar Verma[b], Yogita Bohra[b], Ajay Kumar Gautam[c], Sushma[d], Subhi Avasthi[e], and Deepali Ashok[f]

[a]Department of Life-Sciences, RIMT University, Mandi Gobindgarh, Punjab, India
[b]Department of Plant Pathology, Punjab Agricultural University, Ludhiana, Punjab, India
[c]Patanjali Herbal Research Department, Patanjali Research Institute, Haridwar, Uttarakhand, India
[d]Department of Biosciences, Chandigarh University, Gharuan, Punjab, India
[e]School of Studies in Botany, Jiwaji University, Gwalior, Madhya Pradesh, India
[f]Department of Botany, Government Degree College Drang at Narla, Mandi, Himachal Pradesh, India

12.1 Introduction

Each year, humans are releasing a large number of organic and inorganic substances into the environment. These releases are deliberate and well-regulated in certain circumstances, such as industrial emissions, but they are unintentional in others, such as chemical or oil spills. Most of these chemicals are both poisonous and persistent in land and marine environments. These chemicals lead to contamination of the groundwater and soil surface if their accumulation is above the permissible limit. Such problems of contamination are worldwide (Cairney, 1993). Contaminated lands are now generally recognized as a matter of possible concern to human health, and recent discoveries have prompted international initiatives to remediate many of these sites (Kensa, 2011). Bioremediation is gaining traction as a cost-effective strategy in integrated environmental restoration initiatives, due to advances in microbial variety and biotechnology. The main reasons for limiting water and soil pollution and considering bioremediation are, first and foremost, public health concerns, second, environmental protection, and third, decontamination costs (Alori and Babalola, 2018). The utilization of biological systems for the removal or reduction of contamination from air, soil, and water is known as bioremediation. This procedure entails exposing an organism taken from the study environment or imported from another system to a specific contaminant to either decrease or eliminate the hazardous component (Akpor and Muchie, 2010; Ezeonu et al., 2012;

Microbial Inoculants
https://doi.org/10.1016/B978-0-323-99043-1.00001-3

Ibrahim et al., 2014; Srinivasan et al., 2014; Tarangini, 2009). As a result, bioremediation is the use of biosystems like bacteria, plants, and animals to minimize the potential toxicity of pollutants in the environment by degrading, converting, and immobilizing these dangerous compounds into less harmful forms (Ajao et al., 2011; Ezeonu et al., 2012).

The temptation to find appropriate and more competent methods to efficiently detoxify the environment is even stronger now, given that persistent contaminants that have accumulated are continually being disposed of. Indigenous bacterial populations in the environment can metabolize and oxidize persistent organic pollutants and toxins. However, their less abundance and low activity in the ecosystem, as well as issues accessing pollutant or nutrient restrictions in the environment, prohibit the processes from occurring as quickly as expected, preventing the proposed cleanup goals from being met (de Lipthay et al., 2003; Petrie et al., 2003). Biofilm communities provide the participating microorganisms with a useful structure, the ability to share nutrients and genetic information, and the capacity to defend themselves from elements of the surrounding environment, such as predation and chemical and shear pressures (von Canstein et al., 2002; Boles et al., 2004; Vu et al., 2009). Biofilms can also be used as biomarkers of stream water quality, such as from mine drainage. Biofilms are important players in biofilm-mediated remediation and ecosystem monitoring because of their persistence and varied structural and metabolic features. Not only are bacterial communities used to combat organic pollutants but they are also used to combat metal toxicity. Passive oxidation of arsenic and iron by employing biofilms has been successfully utilized at gold quartz mining sites (Guezennec et al., 2012). Selenium levels were lowered and then concentrated on biofilms in nutrient-containing tubes (Williams et al., 2013), and biofilm enzymes were used as biomarkers of stream water quality in coal mine drainage regions (Pool et al., 2013). These examples demonstrate the wide range of applications of biofilm-mediated pollution cleanup and monitoring services. Therefore, this chapter highlights the basic concept of biofilms and their achievements as bioremediation participants.

12.2 Role of microbial populations in bioremediation

Microbes can be isolated from nearly any type of environment. They adapt and grow at subzero temperatures, intense heat, desert conditions, water, with an abundance of oxygen, and anaerobic circumstances, with the presence of dangerous substances, or in any waste stream. An energy source and a carbon supply are the two most important necessities. Microbes and other biological systems can be employed to degrade or remediate environmental problems due to their flexibility. *Pseudomonas*, *Alcaligenes*, *Sphingomonas*, *Rhodococcus*, and *Mycobacterium* are some of the examples of aerobic bacteria known for their ability to degrade contaminants. Pesticides and hydrocarbons, including alkanes and polyaromatic chemicals, have been reported to be degraded by these bacteria on numerous occasions. Most of these bacteria rely only on pollutants for carbon and energy. Anaerobic bacteria are less commonly used than are aerobic bacteria. The use of anaerobic bacteria for bioremediation of polychlorinated biphenyls (PCBs) in river sediments, as well as for dechlorination of the solvents trichloroethylene (TCE) and chloroform, is gaining popularity. The fungus *Phanerochaete chrysosporium* has the potential to degrade a variety of persistent pollutants in the environment (Kensa, 2011).

12.3 Biofilms

The production of biofilms is a well-known process. For more than 40 years, the ubiquity and importance of biofilm formation in numerous processes have prompted substantial research on this topic. Biofilms are microbial cell clusters that adhere to a surface. They are usually found in every moist environment where there is adequate nutrient flow and surface attachment is possible. A single bacterial species can produce a biofilm, but it can also be made up of a variety of bacteria, fungi, algae, and protozoa. Different microbial species participating in biofilm consortia, each with their metabolic degradation pathway, may be capable of degrading multiple contaminants both individually and collectively (Horemans et al., 2013; Gieg et al., 2014). Many sorts of positive/synergistic and negative/antagonistic interactions occur in a multispecies biofilm (Burmølle et al., 2014). Water, which is linked to the capsules of microbial cells, makes up around 97% of the biofilm matrix, whereas solvents, whose physical properties (such as viscosity) are influenced by the solutes dissolved in them, make up the remaining 3% (Sutherland, 2001). In addition to water, the biofilm matrix is made up of secreted polymers, ingested nutrients and metabolites, cell lysis products, and even particulate debris and detritus from the nearby surroundings (Flemming, 1995).

Most bacteria in natural environments exist in the form of biofilms, enclosed in an extracellular polymeric substance (EPS) matrix, which provide a favorable structure to biofilm-forming microbes in bioremediation

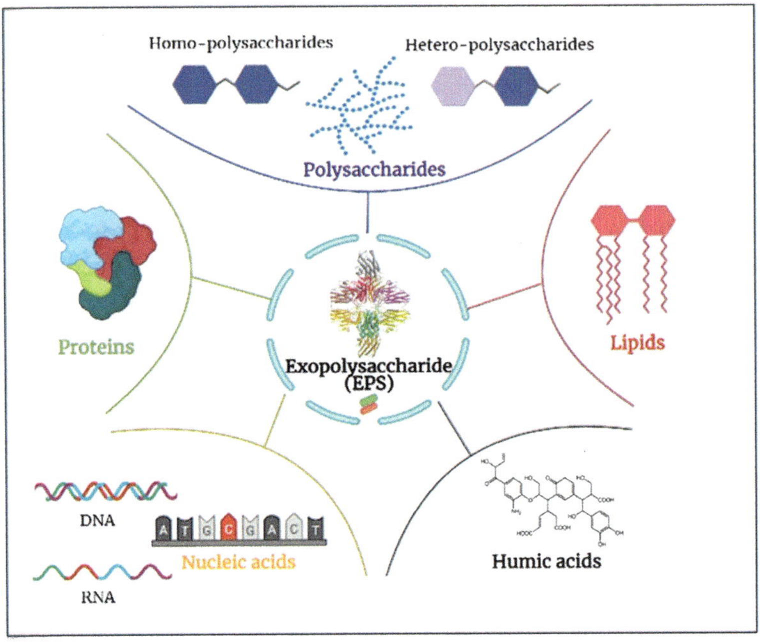

Fig. 1 General constituents of EPS in biofilm.

(Flemming and Wingender, 2001, 2010; More et al., 2014). Humic compounds, lipids, nucleic acids, polysaccharides, proteins, and water are all found in EPSs in both bound and secreted forms (Fig. 1) (Branda et al., 2005). The components of EPSs vary from species to species and are highly influenced by factors such as environmental stress, growth circumstances, and the surfaces on which biofilms are produced. Depending on the environmental conditions in which microorganisms grow, bacterial biofilms and EPS production can fluctuate in the overall structure and contents (Kreft and Wimpenny, 2001; Jung et al., 2013; Miqueleto et al., 2010). As a result, all the major types of macromolecules (i.e., DNA, polysaccharides, RNA, and proteins) can be recognized inside a biofilm environment, in addition to peptidoglycans, lipids, phospholipids, and other cell components (Sutherland, 2001). The inclusion of uronic acids (D-glucuronic, D-galacturonic, and mannuronic acids) or ketal-linked pyruvates confers anionic features on biofilms (Flemming, 1995). Mass movement in biofilms is influenced by biofilm structure, which is dependent on substrate availability locally. Transport within pores and water channels, as well as diffusion in denser aggregates, promotes solute transport in biofilms (Horn and

Morgenroth, 2006). As a result of the multiple microenvironments that coexist inside it, the matrix has a high degree of microheterogeneity.

12.4 Biofilm-forming microbes

12.4.1 Bacteria

Almost 99.9% of bacteria may produce biofilms on a variety of surfaces, both living and nonliving. *Bacillus denitrificans*, *B. cereus* SJV, *Comamonas denitrificans* 123, *Acinetobacter calcoaceticus* ATCC 23055, *Zoogloea ramigera*, *Brachymonas denitrificans* (Andersson et al., 2008), *B. subtilis* (Li et al., 2013), and *Listeria monocytogenes* (do Valle Gomes and Nitschke, 2012) have all been identified as bacteria that can produce biofilms. Bioremediation capability has been documented for a variety of biofilm-forming bacterial species, including *Cronobacter sakazakii*, *Enterobacter agglomerans*, *Pantoea agglomerans* (Turki et al., 2017), *Pseudomonas aeruginosa* (Crouzet et al., 2014), *L. monocytogenes* (Colagiorgi et al., 2017), and *Pseudomonas* spp. (Douterelo et al., 2018).

12.4.2 Fungi

Fungi also adhere to the basic requirements for biofilm development, which include adhesion to a surface or to one another and encapsulation in EPSs. Recently, it has been discovered that filamentous fungi have amphipathic-type hydrophobins that aid in fungal biofilm attachment to the surface (Pérez-Mendoza et al., 2011). These hydrophobins act as morphogenetic-type cues, enabling fungal spores to adhere to hydrophobic surfaces. Hydrophobins of eight cysteine residue patterns are secreted by *Aspergillus fumigates*, creating significant adhesion forces during spore adherence to the surface (Dupres et al., 2010). *Candida albicans* (Rajendran et al., 2016), *Penicillium rubrum*, *Fusarium* spp. (Mailafia and Agbede, 2016), *Acremonium*, and *Neocosmospora* are some of the fungal species that can produce biofilms (Douterelo et al., 2018).

12.4.3 Algae

Microalgae are the dominant population in algal biofilms, and they may colonize any surface with enough moisture nutrients. Algal biofilm formation is almost similar to bacterial biofilm creation. Algal biofilms can demonstrate flexibility to environmental changes, maintain population stability on a surface, and detach from it in the form of a single colony or clumps (Boelee et al., 2014). The self-purification process of water is a symbiotic relationship between algae and bacteria, which functions as a common cleanser in natural water. In a study, the algal bacterial symbiosis system outperformed the standard activated sludge method in removing total nitrogen, ammonium nitrogen, and phosphorus from home wastewater, even though algae require a certain aeration rate for growth (Tang et al., 2015).

12.5 Biofilm development

Biofilm development is a significant microbial proliferation mechanism, which is different from growing planktonic organisms in the environment (Butler and Boltz, 2014). Multiple factors through different pathways under different conditions play a role in the creation of specific biofilm architectures. Optimizing various steps is a prerequisite for biofilm development. Chemotaxis and flagellar-dependent motility are essential factors in biofilm development in microorganisms (Pratt and Kolter, 1999). In the beginning, microbes reach the surface and their motility is slowed down. Then, they develop a brief association with the surface and settle down. A microcolony develops, and, then, a three-dimensional biofilm structure is formed from which colonies detach occasionally (Jamal et al., 2018). The pili and flagella; cell surface appendages help microbes in attachment to the surface (Qi and Christopher, 2019). Recent research in a variety of bacterial species, including *Bacillus subtilis*, *Burkholderia cenocepacia*, *Clostridium difficile*, *Escherichia coli*, *Pseudomonas fluorescens*, *P. aeruginosa*, *Pseudomonas putida*, *Salmonella enterica*, *Vibrio cholerae*, and *Yersinia pestis*, has suggested that an increase in the intracellular secondary messenger c-di-GMP causes biofilm development (Ross et al., 1987; Garcia et al., 2004; Tischler and Camilli, 2004; Kirillina et al., 2004; Simm et al., 2004; Hickman et al., 2005; Gjermansen et al., 2005, 2006, 2010; Kulasakara et al., 2006; Goymer et al., 2006; Lim et al., 2006; Monds et al., 2007; Ryan et al., 2009; Hengge, 2009; Fazli et al., 2011; Chen et al., 2012; Purcell et al., 2012; Massie et al., 2012; O'Connor et al., 2012; Valentini and Filloux, 2016).

A microcolony is the basic unit of a biofilm, but its structure varies widely depending on the bacterial species that form the biofilm. During biofilm formation, bacterial subpopulations, such as motile and nonmotile subpopulations, can develop and interact, and these interactions can impact the biofilm's structure. A biofilm's structure can be flat, fluffy, smooth or rough, filamentous, or mushroom-shaped (Ghanbari et al., 2016). *P. putida*, for example, has been shown to create loose projecting microcolonies in a flow chamber under identical conditions, but *Pseudomonas knackmussii* (previously known as *Pseudomonas* sp. B13) develops spherical microcolonies. Furthermore, when the two species were grown together in dual-species biofilms, they established their typical microcolony structures without interfering with one another (Tolker-Nielsen et al., 2000). Therefore, the structure, design, and organization of different biofilms depend on the biofilm-forming bacterial species. Biofilm development in flow chambers in *P. putida* is mostly driven by the major adhesive protein LapA (Gjermansen et al., 2005, 2010; Nilsson et al., 2011), whereas biofilm production in flow chambers by other *Pseudomonads*, such as *P. aeruginosa*, is primarily governed by the exopolysaccharides Psl and Pel (Wozniak et al., 2003; Jackson et al., 2004; Matsukawa and Greenberg, 2004). Different architectures of microcolonies may result from differences in the extracellular matrix components that connect bacteria in biofilms. When *P. aeruginosa* is grown in flow chambers irrigated with a glucose medium, it develops mushroom-shaped microcolonies, whereas when it is grown in flow chambers watered with citrate medium, it forms flat biofilms (Klausen et al., 2003). The shift in the nutritional source results in a change in the shape of the biofilm (Nielsen et al., 2000; Wolfaardt et al., 1994). It has been found that quorum sensing plays an important role in biofilm production by different bacterial species, including *B. cenocepacia* and *P. aeruginosa*, in agreement with the fact that bacteria are intimately connected in biofilms (Davies et al., 1998; Hentzer et al., 2002; Allesen-Holm et al., 2006; Yang et al., 2007).

12.6 Role of biofilms in bioremediation

Biofilm-mediated remediation is a cost-effective and environmentally acceptable method of removing pollutants from the environment. Biofilms are effective in bioremediation because they absorb, immobilize, and decompose a variety of contaminants (Mitra and Mukhopadhyay, 2016). The variable local concentrations of various nutrients and oxygen inside the biofilm matrix, as well as the microbial division of labor, cause differential gene expression within biofilms. The variable gene expression could be crucial for the degradation of a wide array of contaminants via a variety of metabolic pathways. The xenobiotics found in soil and water, resulting in responses such as chemotaxis, swarming, swimming, twitching motility, and quorum sensing, aid in synchronized movement of bacteria toward pollutant and better biodegradation (Lacal et al., 2013). Because of their large microbial biomass and high competence to immobilize a variety of compounds by biomineralization, bioaccumulation, and biosorption, microbes that produce polymers and biofilms on the surface of hydrocarbons are more suited for bioremediation (Barkay and Schaefer, 2001). Large volumes of dilute aqueous solutions (industrial and municipal wastewaters) are often treated with biofilm-based reactors. The methods used by primary biofilm reactors are classified as biofilm airlift suspension (BAS), biofilm fluidized bed (BFB), expanded granular sludge blanket (EGSB), internal circulation (IC), and upflow sludge blanket (USB). An upward liquid flow keeps the particles fluidized in USB, BFB, and EGSB reactors (Nicolella et al., 2000).

Efficient bioremediation of polluted environments can be limited by the bioavailability of organic contaminants. Bioavailability and thus the breakdown of hydrophobic substances are enhanced by employing chemotactic bacteria having physiological features such as biosurfactant synthesis and chemotaxis (Pandey and Jain, 2002; Paul et al., 2005). When cells come into intimate contact with a surface, the mechanisms of biofilm formation and surfactant synthesis kick in, resulting in increased bioavailability and biodegradation. In comparison to hydrophobic contaminants that are adsorbed in the nonaqueous phase liquid (NAPL) associated with contaminated soils, the rate of biodegradation is increased when the target contaminants are dissolved in an aqueous medium (Stelmack, 1999; Law and Aitken, 2003).

12.7 Role of biofilms in remediation of polycyclic aromatic hydrocarbons

Polycyclic aromatic hydrocarbons (PAHs), which accumulate in adipose tissues and have possible carcinogenic or mutational consequences, can be consumed by benthic creatures and transported up the food chain. In a study, the researchers found that biofilms made these compounds more soluble and allowed them to be transported from refractory crystals into cells for biotransformation (Johnsen and Karlson, 2004). Bacteria's primary means of overcoming the mass transfer limits of resistant PAHs is biofilm formation. Later studies have established that mixed biofilms generated with activated sludge inoculum had enhanced biodegradation efficiency on mixed PAH substrates than those formed with a surfactant (Rodriguez and Bishop, 2008). In such studies, the most relevant cause for greater biodegradation was not the enhanced solubility of PAHs due to surfactants, but the cometabolic pathways that allowed for the simultaneous degradation of numerous PAHs. It has been shown that bacteria specialized in adherence to PAHs aid in PAH breakdown (Bastiaens et al., 2000; Johnsen et al., 2005). In a microenvironment with huge biomass and extracellular polymeric substances

(EPSs) of bacterial biofilms, effective bioavailability and degradation of PAHs were observed (Mangwani et al., 2014a,b, 2015). Few bacteria, such as *Mycobacterium frederiksbergense* (Wick et al., 2001, 2002), *Alcaligenes*, and *Pseudomonas* (Eriksson et al., 2002), have been observed to produce biofilms on PAH surfaces and to secrete EPSs to enhance their solubility and therefore their bioavailability. Marine biofilm EPSs, in particular, are a rich source of active biomolecules that can have a major impact on PAH interactions with cells. As a result, biofilms with a higher EPS content are better at degrading hydrophobic substances (Mangwani et al., 2014b). Different structural aspects of biofilms have been found to have a substantial impact on bacteria's ability to degrade PAHs (Mangwani et al., 2016). In addition to this, EPSs impact the solubility of numerous hydrophobic substances (Gutierrez et al., 2013). Mitra and Mukhopadhyay (2016) found that this feature of biofilms is particularly beneficial for the bioremediation of contaminants in soil.

12.8 Role of biofilms in remediation of petroleum

Petroleum and its compounds are one of the most important threats to the ocean's ecology (Scoma et al., 2017). The increased use of petroleum in industries has led to an increase in its buildup in the ocean environment, posing a threat to marine life. Oil spills have been shown to alter the generation of hydrocarbon-degrading microorganisms in soil (Song and Bartha, 1990). Hydrocarbon-degrading bacteria with a fast growth rate can also be used to remediate oil spills in the marine environment (Atlas and Hazen, 2011; Ron and Rosenberg, 2014). Several Gammaproteobacteria, such as *Alcanivorax*, *Arthrobacter*, *Bacillus*, *Cycloclasticus Pseudomonas*, and *Rhodococcus* spp., have been used in this setting (Harayama et al., 2004). In a study, four high-performance hydrocarbon-degrading bacterial strains isolated from different environments were used to test the bioremediation efficiency of membrane-bacterial systems in crude oil: two marine hydrocarbonoclastic Gammaproteobacteria, *Alcanivorax borkumensis* SK2 (Yakimov et al., 1998) and *Oleibacter marinus* (Catania et al., 2015), and one soil, long-chain n-alkane-degrading actinobacterium *Gordonia* sp. SoCg (Lo Piccolo et al., 2011). The ability of these formulations to degrade crude oil was tested and compared to that of planktonic bacterial cells. Biofilm-associated cells have been shown to have stronger resistance and survive at low temperatures and also maintain their degradative capacity for a longer period of time than planktonic cells (Alessandrello et al., 2017). Biofilms could thus supply a continuous stream of microorganisms with high degrading capabilities, boosting bioremediation efficiency (Lear, 2016). The search for new microbial strains capable of biofilm-mediated oil breakdown is ongoing (Dasgupta et al., 2013). A microbial consortium consisting of *Acinetobacter radioresistens* and *B. subtilis* with a surfactant-generating strain has been successfully demonstrated to break down petroleum more effectively than the microbial consortia solely comprising of degraders (Mnif et al., 2015). Water-insoluble fertilizers, such as uric acid, can help hydrocarbon degraders get the nitrogen they need to help oil biodegradation in the open environment (Koren et al., 2003). Although a variety of mechanical, physical, and biological remediation procedures have been used to eliminate oil from the environment, biodegradation by microorganisms has been identified as a crucial player in cleanup activities (Mapelli et al., 2017) and is considered as the best option.

12.9 Role of biofilms in remediation of chlorinated aromatic compounds

Chlorinated aromatic compounds are also refractory chemicals that can swiftly travel through soils and are found in a variety of industrial effluents. They are known as one of the most common pollutants of soil and groundwater and can be cancerous even at minor doses (Puhakka et al., 1995; Kargi and Ekker, 2005). Kargi and Ekker (2005) employed a rotating perforated tube biofilm reactor with mixed microbial biomass of activated sludge culture to extract 2,4-dichlorophenol (DCP). The reactor was supplemented with DCP-degrading *P. putida* from synthetic wastewater; about 100% of the DCP was destroyed. This was also demonstrated for diclofop methyl and methyl 2-[4-(2,4-dichlorophenoxy)phenoxy] pyruvate, a two-ring chlorinated herbicide that accumulated in biofilms through adsorption on microbial exopolymers. When starving, the biofilm community metabolized the accumulated diclofop methyl more efficiently (Wolfaardt, 1995).

It has been proved that improved strains could improve biofilm-mediated bioremediation by manipulating metabolic pathways and enzymes involved in degradation or in raising the copy number of degradative genes. Biofilm production would benefit strains having biodegradation and chemotactic characteristics; e.g., a chemotactic strain modified to contain catabolic genes would be particularly efficient in biodegradation. Perumbakkam et al. (2006) engineered *Acinetobacter* sp. BD413 biofilms with atrazine-degrading genes, resulting in the development of an atrazine-degrading biofilm-mediated mechanism and concluded that bioremediation can also be aided by the development of specialized mixed communities. In addition, coadhesion and synergistic interaction with biofilm-forming species may be a viable method of maintaining other strains and propagating them (Castonguay et al., 2006; Kuboniwa et al., 2006). To create efficient bioremediation techniques, it is essential to combine genetically engineered microorganisms with optimization of physical and chemical factors and substrate concentration in bioreactors (Cheng et al., 2006).

12.10 Role of biofilms in remediation of heavy metals and radionuclides

Remediation of heavy metals and radionuclides is one of the important applications of biofilms. Biological mechanisms, unlike organic pollutants, are incapable of completely eliminating heavy metal toxins from the environment, necessitating a heavy metal cleanup. Heavy metals occur naturally as dilute components of geochemical cycles, and thus it is reasonable to believe that microbial communities can interact with and use them. However, when these heavy metals occur as contaminants, their concentrations are normally greater than their natural levels. Metal and radionuclide bioremediation heavily relies on the variety and distribution of microorganisms that reside in contaminated areas as well as on the genes that encode phenotypes important for metal-microbe interactions. To achieve heavy metal bioremediation while minimizing hazards, immobilization, concentration, and partitioning into an environmental compartment are required (Barkay and Schaefer, 2001; Lloyd, 2003).

Sulfate-reducing bacterial biofilms produced in culture and subjected to a medium containing 20–200 mM copper were observed to accumulate copper sulfide (White and Gadd, 2000). A simultaneous increase in a biofilm's exopolysaccharide (EPS) content implied that EPSs and biofilms play a role in trapping metal precipitates. Another study indicated that polycrystalline $NaUO_2PO_4$ was congregated around the cell wall of *Citrobacter* sp. N14 due to the adsorption to lipopolysaccharides, thereby assisting in the bioprecipitation of the uranium salt (Macaskie et al., 2000). Labrenz et al. (2000) discovered the synthesis of sphalerite (ZnS) by a natural biofilm that is a member of the aerotolerant *Desulfobacterota*. A biofilm of single bacterial sp. or consortia of *Cellulosimicrobium* sp. (Naeem et al., 2013) and *Halomonas aquamarina* TA-04 (Focardi et al., 2012) grown on various matrices can eliminate more than 90% of 500 g/mL of chromate. Grumbein et al. (2014) discovered that many heavy metal ions, including Al (III), Cu(II), Fe(II),

Fe(III), and Zn(II), can be accumulated by the biofilm produced by *B. subtilis*. The metal adsorption properties of biofilms are employed in biotechnology for detoxification of groundwater contaminated with uranium (Cao et al., 2011).

During the last decade, technologies that take advantage of the metal adsorptive characteristics of biomaterials have been heavily developed. Costley and Wallis (2001) used a rotating biological contactor (RBC) with alternate sorption and desorption cycles to remove 84% of Cd^{2+}, Cu^{2+}, and Zn^{2+} from synthetic wastewater, suggesting the biofilm's potential to repopulate after desorption cycles. Although the species richness of the RBC biofilm was not evaluated in this work, its capability to deal with such environmental stresses and flourish on numerous metallic species showed that it was made up of varied and diverse communities. Biofilms of mixed culture of microorganisms were able to hold more mercury and diversity in packed-bed bioreactors for the wastewater treatment contaminated with mercury as compared to monoculture biofilms in the presence of rapidly changing mercury concentrations (von Canstein et al., 2002). These findings revealed that diversified biofilms were more effective in metal bioremediation, making biofilm-based solutions relevant to research and implementation.

More microbial diversity can lead to the formation of a web of ecological and metabolic niches, permitting biofilm-linked microbes to thrive in the phase of fast-changing environmental stress, resulting in improved performance whether in a bioreactor or in the wild. Biofilms of metal-resistant bacteria were discovered in a moving bed sand filter designed to handle industrial effluents comprising high concentrations of heavy metals. The biofilms not only acted as biosorption materials but also decreased the pH to allow for the bioprecipitation of several heavy metals (Diels et al., 2003). Jong and Parry (2003) used a mixed-species biofilm of sulfate-reducing bacteria to take advantage of the oxidation-reduction abilities to precipitate the metal sulfides of As, Cu, Fe, Ni, and Zn. This treatment removed 98% of Zn, Cu, and Ni as well as 78 and 82% of As and Fe, respectively. The total biosorption potential of biofilms was restricted by the resistance of some bacteria in the community to metals, such as resistance to Cr (VI) in *Stenotrophomonas* sp. (Morel et al., 2009). Furthermore, certain isolates were able to tolerate larger concentrations of Zn and Mn heavy metal ions (2000 mg/L); thus, the advantage of their effectiveness allows for the establishment of a biofilm capable of removing the greatest quantity of heavy metals (Pani et al., 2017).

All of these technologies benefited from biofilms' intrinsic resistance to heavy metals and their ability to absorb them. This outcome is mostly owing to the existence of EPSs produced by the microbial biofilm community itself, as opposed to planktonic cells, which do not produce EPSs (Teitzel and Parsek, 2003). In a review by Pal and Paul (2008), the benefits of EPSs for heavy metal bioremediation were detailed for specific bacteria. The importance of EPSs is becoming more widely acknowledged, and, with the present array of genetic engineering techniques, new inventive ways to use biofilm-produced EPSs via modified bacterial strains may emerge in the future. As a result of their physiological response during the absorption of inorganic or organic solutes and water, biofilms can impact the fate of other substances in their proximity (Flemming, 1995), which could be useful for remediation. Valls and Lorenzo (2002) have outlined several situations in which genetic engineering can be used to combine or improve a cellular characteristic that is already present in some strains. Molecular techniques enable the creation of enhanced strains with specialized metal-binding capabilities by expressing metal-chelating peptides and proteins, improving metal precipitation processes, and introducing metal transformation activities into robust environmental strains (Singh et al., 2006).

12.11 Other applications

Many different bioremediation applications use biofilms. Biofilms are the first organisms in aquatic systems to interact with nutrients and contaminants, and, therefore, they can be employed as an initial environmental monitor in aquatic habitats (Peacock et al., 2004). Various biofilm indicator qualities, such as biomass, species composition, pigment production, photosynthesis, and enzymatic activity, can be examined for monitoring environmental pollution. The species found in river biofilms change depending on the season and the quantity of pollution (Brummer et al., 2000). Environmental contaminations, such as those by Zn and Cd, can affect species diversity inside biofilms, and so estimating microbial biofilm species diversity can also signal pollution (Arini et al., 2012). The alteration in the biomass of normal algal biofilm communities after exposure to herbicides and heavy metals has been studied and could be utilized as a pollution indicator (Navarro et al., 2002). Furthermore, because pigment composition among members of the biofilm community can change after exposure to harmful substances, pigment composition analysis and patterns can be used as biomarkers (Dorigo et al., 2004; Dewez et al., 2008). Chemical attractants desorb or disperse from contact due to chemotaxis, according to Law and Aitken (2003). This leads to enhanced degradation rates of the chemical attractant. They also help in removing acid from acid mine drains, which is a potentially difficult long-term environmental challenge associated with mining. The flow of oxygen limits the oxidation of sulfide particles in tailings, open-pit mines, and waste rock piles. Bacterial biofilms considerably increase the rates of cleanup under such settings by increasing oxygen availability (Bang et al., 2000). The microorganisms in biofilms are beneficial for micropollutant bioremediation because they may absorb or degrade a wide range of severe contaminants via a variety of catabolic routes (Edwards and Kjellerup, 2013). Titanium dioxide nanoparticles can be found in practically all cosmetics, paints, and a variety of other household items. As a result, a large number of these nanoparticles end up in wastewater discharged from residential areas all over the world. Because the titania component is photocatalytically activated, titanium dioxide nanoparticles can pose a serious threat to the microenvironment. Mathur et al. (2017) used biofilms to monitor the uptake and removal of these particles to determine their toxicity. The remediation of nanoparticles was relatively convenient due to the presence of surfactants and humic substances in the wastewater.

12.12 Emerging scope for biofilms

Wyss Institute researchers developed Biofilm-Integrated Nanofiber Display (BIND) by genetically attaching peptide domains to the amyloid protein CsgA and molecular programming of the bacterial extracellular matrix material (a dominant proteinaceous component in *E. coli* biofilms). BIND is a technique for performing a variety of tasks on the biofilm matrix like substrate adhesion, nanoparticle biotemplating, protein covalent immobilization, or a combination of these functions. BIND is a nanobiotechnological platform for generating durable materials with programmable functionality, showcasing the value of biofilms as large-scale designable biomaterials (Botyanszki et al., 2015). Although cell-free biocatalysis has numerous advantages, it is limited in its practical application due to the high costs of enzyme modification, purification, and immobilization on solid substrates (Maurya and Raj, 2020).

To lower the degree of variation in biofilm structures, a novel biofilm application built a 3D printer integrated with genetic engineering. An *E. coli* biofilm is suspended in an alginate solution and hardened on a calcium-containing substrate to create this 3D printer. Curli fibers are produced by *E. coli* during biofilm formation, which aids in adhesion to the printing substrate. Even after treatment with biofilm-dissolving chemicals, the fiber pattern remained arrayed in the spatial pattern, forming a biofilm-reproducing structure. The 3D printer may be used in industries as a practical piece of apparatus for processes like metal-ion sequestration and water purification or as models for biofilms that produce curli (Schmieden et al., 2018). The ability of a biofilm to adhere to a surface is critical for its functioning. Therefore, Smith et al. (2018) created a stage based on the *P. fluorescens* Pf0 1's Lap system to achieve this goal. This system controls the release of cell surface-associated cargo proteins, which modulate adhesion of bacterial cells to the surface (Smith et al., 2018).

12.13 Limitations to the use of biofilms in bioremediation

Apart from so many applications, biofilm-based bioremediation has some drawbacks. When compared to chemical treatments for pollutant breakdown, bioremediation is comparatively slow. Bioremediation has several drawbacks, one of which is its lack of reliability, which limits its use in specific conditions, particularly when there is a lot of contamination. Some chemicals, particularly man-made plastics and certain halogenated aromatic compounds, are resistant to biodegradation. Another drawback of bioremediation is the possibility of metabolic hazardous compounds forming during microbial breakdown. Bioremediation is an ideal option only when the level of contamination is relatively low and immediate restoration is not required or chemical treatment is not at all an option. When critical nutrients for microbial development are scarce, bioremediation may be delayed or unsuccessful, especially when there is high-level pollution. The efficiency of bioremediation is determined by the bioavailability of pollutants to bacteria, and toxins that are not encased in other materials such as clay are more receptive to biodegradation. When the reactor capacity is dependent on oxygen diffusion, biofilm reactors should not be used with rapidly developing microorganisms. Because of these drawbacks, bioremediation can be compared to other available strategies for its effectiveness in removing contaminants from the environment (Heitzer and Sayler, 1993). Apart from bioremediation, microbial biofilms have been used to produce a range of useful chemicals, including lactic acid, table olive oil, and cellulose. The ability to generate power using microbial fuel cells has advanced biofilm research. During the decomposition of inorganic and organic substances found in industrial wastewater, chemical energy is generated, which is then used to power microbial fuel cells.

12.14 Conclusions

Biofilm-forming bacteria have recently been shown to play a crucial role in bioremediation due to their greater ability to adapt to various severe environmental conditions and survival. Due to the oxidation of organic pollutants, microbial biofilms are used as microbial fuel cells for the production of power. Biofilms are also used to make biosurfactants, bioemulsifiers, and biocapsules, among other things. Biofilms formed on implanted medical equipment, on the other hand, has a considerable impact on public health. This problem is currently under control because of the introduction of modern technologies to eliminate and control biofilm-related infections. However, in order to avoid food deterioration or pathogenic microflora, new and more effective antibiofilm solutions are needed to manage biofilm-associated infections, particularly in the food industry. Biofilm-mediated bioremediation requires adequate technical skills and an interdisciplinary approach from many fields such as soil science, civil engineering, and environmental microbiology. Developing a reliable biofilm-mediated remediation technique for large-scale applications remains a challenge. To improve treatment efficiency, a thorough understanding of biofilm formation and its unique mechanisms is required.

References

Ajao, A.T., Adebayo, G.B., Yakubu, S.E., 2011. Bioremediation of textile industrial effluent using mixed culture of *Pseudomonas aeruginosa* and *Bacillus ubtilis* immobilized on agar-agar in a bioreactor. J. Microbiol. Biotechnol. Res. 1 (3), 50–56.

Akpor, O.B., Muchie, M., 2010. Remediation of heavy metals in drinking water and wastewater treatment systems: processes and application. Int. J. Phys. Sci. 5 (12), 1807–1817.

Alessandrello, M.J., Juárez Tomás, M.S., Raimondo, E.E., Vullo, D.L., Ferrero, M.A., 2017. Petroleum oil removal by immobilized bacterial cells on polyurethane foam under different temperature conditions. Mar. Pollut. Bull. 122, 156–160.

Allesen-Holm, M., Barken, K.B., Yang, L., Klausen, M., Webb, J.S., Kjelleberg, S., Molin, S., Givskov, M., Tolker-Nielsen, T., 2006. A characterization of DNA release in Pseudomonas aeruginosa cultures and biofilms. Mol. Microbiol. 59, 1114–1128.

Alori, E.T., Babalola, O.O., 2018. Microbial inoculant for improving crop quality and human health. Front. Microbiol. 9, 1–12. https://doi.org/10.3389/fmicb.2018.02213. 2213.

Andersson, S., Kuttuva Rajarao, G., Land, C.J., Dalhammar, G.J., 2008. Biofilm formation and interactions of bacterial strains found in wastewater treatment systems. FEMS Microbiol. Lett. 283, 83–90.

Arini, A., Feurtet-Mazel, A., Maury-Brachet, R., Pokrovsky, O.S., Coste, M., Delmas, F., 2012. Recovery potential of periphytic biofilms translocated in artificial streams after industrial contamination (Cd and Zn). Ecotoxicology 21, 1403–1414.

Atlas, R.M., Hazen, T.C., 2011. Oil biodegradation and bioremediation: a tale of the two worst spills in U.S. history. Environ. Sci. Technol. 45, 6709–6715.

Bang, S.W., Clark, D.S., Keasling, J.D., 2000. Engineering hydrogen sulfide production and cadmium removal by expression of the thiosulfate reductase gene (phsABC) from Salmonella enterica serovar typhimurium in Escherichia coli. Appl. Environ. Microbiol. 66 (9), 3939–3944.

Barkay, T., Schaefer, J., 2001. Metal and radionuclide bioremediation: issues, considerations and potentials. Curr. Opin. Microbiol. 4, 318–323.

Bastiaens, L., Springael, D., Wattiau, P., Harms, H., deWachter, R., Verachtert, H., Diels, L., 2000. Isolation of adherent polycyclic aromatic hydrocarbon-degrading bacteria using PAH-sorbing carriers. Appl. Environ. Microbiol. 66, 1834–1843.

Boelee, N., Temmink, H., Janssen, M., Buisman, C., Wijffels, R.J., 2014. Balancing the organic load and light supply in symbiotic microalgal bacterial biofilm reactors treating synthetic municipal wastewater. Ecol. Eng. 64, 213–221.

Boles, B.R., Thoendel, M., Singh, P.K., 2004. Self-generated diversity produces "insurance effects" in biofilm communities. Proc. Natl. Acad. Sci. U. S. A. 101 (47), 16630–16635. https://doi.org/10.1073/pnas.0407460101.

Botyanszki, Z., Tay, P.K.R., Nguyen, P.Q., Nussbaumer, M.G., Joshi, N.S.J., 2015. Engineered catalytic biofilms: site-specific enzyme immobilization onto *E. coli* curli nanofibers. Biotechnol. Bioeng. 112, 2016–2024.

Branda, S.S., Vik, S., Friedman, L., Kolter, R., 2005. Biofilms: the matrix revisited. Trends Microbiol. 13, 20–26.

Brummer, I.H., Fehr, W., Wagner-Dobler, I., 2000. Biofilm community structure in polluted rivers: abundance of dominant phylogenetic groups over a complete annual cycle. Appl. Environ. Microbiol. 66, 3078–3082.

Burmølle, M., Ren, D., Bjarnsholt, T., Sørensen, S.J., 2014. Interactions in multispecies biofilms: do they actually matter? Trends Microbiol. 22, 84–91.

Butler, C., Boltz, J., 2014. Biofilm processes and control in water and wastewater treatment. In: Comprehensive Water Quality and Purification.

Cairney, T., 1993. Contaminated Land. Blackie, London, p. 4.

Cao, B., Ahmed, B., Kennedy, D.W., Wang, Z., Shi, L., Marshall, M.J., Fredrickson, J.K., Isern, N.G., Majors, P.D., Beyenal, H., 2011. Contribution of extracellular polymeric substances from *Shewanella* sp. HRCR-1 biofilms to U (VI) immobilization. Environ. Sci. Technol. 45, 5483–5490.

Castonguay, M.H., van der Schaaf, S., Koester, W., Krooneman, J., van der Meer, W., Harmsen, H., Landini, P., 2006. Biofilm formation by *Escherichia coli* is stimulated by synergistic interactions and co-adhesion mechanisms with adherence-proficient bacteria. Res. Microbiol. 157, 471–478.

Catania, V., Santisi, S., Signa, G., Vizzini, S., Mazzola, A., Cappello, S., Yakimov, M.M., Quatrini, P., 2015. Intrinsic bioremediation potential of a chronically polluted marine coastal area. Mar. Pollut. Bull. 99, 138–149.

Chen, Y., Chai, Y., Guo, J.H., Losick, R., 2012. Evidence for cyclic Di-GMP-mediated signaling in *Bacillus subtilis*. J. Bacteriol. 194, 5080–5090.

Cheng, C., Phipps, D., Alkhaddar, R., 2006. Thermophilic aerobic wastewater treatment of waste metal working fluids. Water Environ. J. 20, 227–232.

Colagiorgi, A., Bruini, I., Di Ciccio, P.A., Zanardi, E., Ghidini, S., Ianieri, A., 2017. *Listeria monocytogenes* biofilms in the wonderland of food industry. Pathogens 6, 41.

Costley, S.C., Wallis, F.M., 2001. Bioremediation of heavy metals in a synthetic wastewater using a rotating biological contactor. Water Res. 35, 3715–3723.

Crouzet, M., Le Senechal, C., Brözel, V.S., Costaglioli, P., Barthe, C., Bonneu, M., Garbay, B., Vilain, S., 2014. Exploring early steps in biofilm formation: set-up of an experimental system for molecular studies. BMC Microbiol. 14, 253.

Dasgupta, D., Ghosh, R., Sengupta, T.K., 2013. Biofilm-mediated enhanced crude oil degradation by newly isolated pseudomonas species. ISRN Biotechnol. 2013, 250749.

Davies, D.G., Parsek, M.R., Pearson, J.P., Iglewski, B.H., Costerton, J.W., Greenberg, E.P., 1998. The involvement of cell-to-cell signals in the development of a bacterial biofilm. Science 280, 295–298.

de Lipthay, J.R., Tuxen, N., Johnsen, K., Hansen, L.H., Albrechtsen, H.J., Bjerg, P.L., Aamand, J., 2003. In situ exposure to low herbicide concentrations affects microbial population composition and catabolic gene frequency in an aerobic shallow aquifer. Appl. Environ. Microbiol. 69, 461–467.

Dewez, D., Didur, O., Vincent-Héroux, J., Popovic, R., 2008. Validation of photosynthetic-fluorescence parameters as biomarkers for isoproturon toxic effect on alga *Scenedesmus obliquus*. Environ. Pollut. 151, 93–100.

Diels, L., Spaans, P.H., Van Roy, S., Hooyberghs, L., Ryngaert, A., Wouters, H., Walter, E., Winters, J., Macaskie, L., Finlay, J., Pernfuss, B., Woebking, H., Pümpel, T., Tsezos, M., 2003. Heavy metals removal by sand filters inoculated with metal sorbing and precipitating bacteria. Hydrometallurgy 71, 235–241.

do Valle Gomes, M.Z., Nitschke, M., 2012. Evaluation of rhamnolipid and surfactin to reduce the adhesion and remove biofilms of individual and mixed cultures of food pathogenic bacteria. Food Control 25, 441–447.

Dorigo, U., Bourrain, X., Bérard, A., Leboulanger, C., 2004. Seasonal changes in the sensitivity of river microalgae to atrazine and isoproturon along a contamination gradient. Sci. Total Environ. 318, 101–114.

Douterelo, I., Fish, K., Boxall, J.J., 2018. Succession of bacterial and fungal communities within biofilms of a chlorinated drinking water distribution system. Water Res. 141, 74–85.

Dupres, V., Alsteens, D., Andre, G., Dufrêne, Y.F.J., 2010. Microbial nanoscopy: a closer look at microbial cell surfaces. Trends Microbiol. 18, 397–405.

Edwards, S.J., Kjellerup, B.V., 2013. Applications of biofilms in bioremediation and biotransformation of persistent organic pollutants, pharmaceuticals/personal care products, and heavy metals. Appl. Microbiol. Biotechnol. 97, 9909–9921.

Eriksson, M., Dalhammar, G., Mohn, W.W., 2002. Bacterial growth and biofilm production on pyrene. FEMS Microbiol. Ecol. 40, 21–27.

Ezeonu, C.S., Tagbo, R., Anike, E.N., Oje, O.A., Onwurah, I.N.E., 2012. Biotechnological tools for environmental sustainability: prospects and challenges for environments in Nigeria – a standard review. Biotechnol. Res. Int. 2012, 450802.

Fazli, M., O'Connell, A., Nilsson, M., Niehaus, K., Dow, J.M., Givskov, M., Ryan, R.P., Tolker-Nielsen, T., 2011. The CRP/FNR family protein Bcam1349 is a c-di-GMP effector that regulates biofilm formation in the respiratory pathogen *Burkholderia cenocepacia*. Mol. Microbiol. 82, 327–341.

Flemming, H.C., 1995. Sorption sites in biofilms. Water Sci. Technol. 32, 27–33.

Flemming, H.C., Wingender, J., 2001. Relevance of microbial extracellular polymeric substances (EPSs) – Part II: Technical aspects. Water Sci. Technol. 43, 9–16.

Flemming, H.C., Wingender, J., 2010. The biofilm matrix. Nat. Rev. Microbiol. 8, 623–633.

Focardi, S., Pepi, M., Landi, G., Gasperini, S., Ruta, M., Di Biasio, P., Focardi, S.E., 2012. Hexavalent chromium reduction by whole cells and cell free extract of the moderate halophilic bacterial strain *Halomonas* sp. TA-04. Int. Biodeterior. Biodegrad. 66, 63–70.

Garcia, B., Latasa, C., Solano, C., Garcia-del Portillo, F., Gamazo, C., Lasa, I., 2004. Role of the GGDEF protein family in Salmonella cellulose biosynthesis and biofilm formation. Mol. Microbiol. 54, 264–277.

Ghanbari, A., Dehghany, J., Schwebs, T., Müsken, M., Häussler, S., Meyer-Hermann, M., 2016. Inoculation density and nutrient level determine the formation of mushroom-shaped structures in *Pseudomonas aeruginosa* biofilms. Sci. Rep. 6, 32097.

Gieg, L.M., Fowler, S.J., Berdugo-Clavijo, C., 2014. Syntrophic biodegradation of hydrocarbon contaminants. Curr. Opin. Biotechnol. 27, 21–29.

Gjermansen, M., Ragas, P., Sternberg, C., Molin, S., Tolker-Nielsen, T., 2005. Characterization of starvation-induced dispersion in *Pseudomonas putida* biofilms. Environ. Microbiol. 7, 894–906.

Gjermansen, M., Ragas, P., Tolker-Nielsen, T., 2006. Proteins with GGDEF and EAL domains regulate *Pseudomonas putida* biofilm formation and dispersal. FEMS Microbiol. Lett. 265, 215–224.

Gjermansen, M., Nilsson, M., Yang, L., Tolker-Nielsen, T., 2010. Characterization of starvation-induced dispersion in *Pseudomonas putida* biofilms: genetic elements and molecular mechanisms. Mol. Microbiol. 75, 815–826.

Goymer, P., Kahn, S.G., Malone, J.G., Gehrig, S.M., Spiers, A.J., Rainey, P.B., 2006. Adaptive divergence in experimental populations of *Pseudomonas fluorescens*. II. Role of the GGDEF regulator WspR in evolution and development of the wrinkly spreader phenotype. Genetics 173, 515–526.

Grumbein, S., Opitz, M., Lieleg, O., 2014. Selected metal ions protect *Bacillus subtilis* biofilms from erosion. Metallomics 6, 1441–1450.

Guezennec, A.-G., Michel, C., Joulian, C., Dictor, M.C., Battaglia-Brunet, F., 2012. Treatment of Arsenic Contaminated Mining Water Using Biofilms. Interfaces Against Pollution.

Gutierrez, T., Berry, D., Yang, T., Mishamandani, S., McKay, L., Teske, A., Aitken, M.D., 2013. Role of bacterial exopolysaccharides (EPS) in the fate of the oil released during the Deepwater horizon oil spill. PLoS One 8, 67717.

Harayama, S., Kasai, Y., Hara, A., 2004. Microbial communities in oil-contaminated seawater. Curr. Opin. Biotechnol. 15, 205–214.

Heitzer, A., Sayler, G.S., 1993. Monitoring the efficacy of bioremediation. Trends Biotechnol. 11, 334–343.

Hengge, R., 2009. Principles of c-di-GMP signalling in bacteria. Nat. Rev. Microbiol. 7, 263–273.

Hentzer, M., Riedel, K., Rasmussen, T.B., Heydorn, A., Andersen, J.B., Parsek, M.R., Rice, S.A., Eberl, L., Molin, S., Hoiby, N., Kjelleberg, S., Givskov, M., 2002. Inhibition of quorum sensing in *Pseudomonas aeruginosa* biofilm bacteria by a halogenated furanone compound. Microbiology 148, 87–102.

Hickman, J.W., Tifrea, D.F., Harwood, C.S., 2005. A chemosensory system that regulates biofilm formation through modulation of cyclic diguanylate levels. Proc. Natl. Acad. Sci. USA 102, 14422–14427.

Horemans, B., Breugelmans, P., Hofkens, J., Smolders, E., Springael, D., 2013. Environmental dissolved organic matter governs biofilm formation and subsequent linuron degradation activity of a linuron-degrading bacterial consortium. Appl. Environ. Microbiol. 79, 4534–4542.

Horn, H., Morgenroth, E., 2006. Transport of oxygen, sodium chloride, and sodium nitrate in biofilms. Chem. Eng. Sci. 61, 1347–1356.

Ibrahim, S.S., Bakery, H.H., El-shawarby, R.M., Abuosalem, M.E., Nabila, M.A., 2014. Incidence of some pollutants in Ismailia canal and using different methods for removal. Benha Vet. Med. J. 27 (1), 62–77.

Jackson, K.D., Starkey, M., Kremer, S., Parsek, M.R., Wozniak, D.J., 2004. Identification of psl, a locus encoding a potential exopolysaccharidethat is essential for *Pseudomonas aeruginosa* PAO1 biofilm formation. J. Bacteriol. 186, 4466–4475.

Jamal, M., Ahmad, W., Andleeb, S., Jalil, F., Imran, M., Nawaz, M.A., Hussain, T., Ali, M., Rafiq, M., Kamil, M.A., 2018. Bacterial biofilm and associated infections. J. Chin. Med. Assoc. 81, 7–11.

Johnsen, A.R., Karlson, U., 2004. Evaluation of bacterial strategies to promote the bioavailability of polycyclic aromatic hydrocarbons. Appl. Microbiol. Biotechnol. 63, 452–459.

Johnsen, A.R., Wick, L.Y., Harms, H., 2005. Principles of microbial PAH-degradation in soil. Environ. Pollut. 133, 71–84.

Jong, T., Parry, D.L., 2003. Removal of sulfate and heavy metals by sulphate reducing bacteria in short-term bench scale upflow anaerobic packed bed reactor runs. Water Res. 37, 3379–3389.

Jung, J.H., Choi, N.Y., Lee, S.Y., 2013. Biofilm formation and exopolysaccharide (EPS) production by *Cronobacter sakazakii* depending on environmental conditions. Food Microbiol. 34, 70–80.

Kargi, F., Ekker, S., 2005. Removal of 2,4-dichlorophenol and toxicity from synthetic wastewater in a rotating perforated tube biofilm reactor. Process Biochem. 40, 2105–2111.

Kensa, V.M., 2011. Bioremediation – an overview. J. Ind. Pollut. Control. 27 (2), 161–168.

Kirillina, O., Fetherston, J.D., Bobrov, A.G., Abney, J., Perry, R.D., 2004. HmsP, a putative phosphodiesterase, and HmsT, a putative diguanylatecyclase, control Hms-dependent biofilm formation in *Yersinia pestis*. Mol. Microbiol. 54, 75–88.

Klausen, M., Heydorn, A., Ragas, P., Lambertsen, L., Aaes-Jorgensen, A., Molin, S., Tolker-Nielsen, T., 2003. Biofilm formation by *Pseudomonas aeruginosa* wild type, flagella and type IV pili mutants. Mol. Microbiol. 48, 1511–1524.

Koren, O., Knezevic, V., Ron, E.Z., Rosenberg, E., 2003. Petroleum pollution bioremediation using water-insoluble uric acid as the nitrogen source. Appl. Environ. Microbiol. 69, 6337–6339.

Kreft, J.U., Wimpenny, J.W., 2001. Effect of EPS on biofilm structure and function as revealed by an individual-based model of biofilm growth. Water Sci. Technol. 43, 135–141.

Kuboniwa, M., Tribble, G.D., James, C.E., Kilic, A.O., Tao, L., Herzberg, M.C., Shizukuishi, S., Lamont, R.J., 2006. *Streptococcus gordonii* utilizes several distinct gene functions to recruit *Porphyromonas gingivalis* into a mixed community. Mol. Microbiol. 60, 121–139.

Kulasakara, H., Lee, V., Brencic, A., Liberati, N., Urbach, J., Miyata, S., Lee, D.G., Neely, A.N., Hyodo, M., Hayakawa, Y., Ausubel, F.M., Lory, S., 2006. Analysis of *Pseudomonas aeruginosa* diguanylatecyclases and phosphodiesterases reveals a role for bis-(3′-5′)-cyclic-GMP in virulence. Proc. Natl. Acad. Sci. USA 103, 2839–2844.

Labrenz, M., Druschel, G.K., Thomsen-Ebert, T., Gilbert, B., Welch, S.A., Kemner, K.M., Logan, G.A., Summons, R.E., De Stasio, G., Bond, P.L., Lai, B., Kelly, S.D., Banfield, J. F., 2000. Formation of sphalerite (ZnS) deposits in natural biofilms of sulfate-reducing bacteria. Science 290, 1744–1747.

Lacal, J., Reyes-Darias, J.A., García-Fontana, C., Ramos, J.L., Krell, T., 2013. Tactic responses to pollutants and their potential to increase biodegradation efficiency. J. Appl. Microbiol. 114, 923–933.

Law, A.M.J., Aitken, M.D., 2003. Bacterial chemotaxis to naphthalene desorbing from a nonaqueous liquid. Appl. Environ. Microbiol. 69, 5968–5973.

Lear, G., 2016. Biofilms in Bioremediation. Caister Academic Press.

Li, S., Zhang, N., Zhang, Z., Luo, J., Shen, B., Zhang, R., Shen, Q., 2013. Antagonist *Bacillus* subtiilis HJ5 controls Verticillium wilt of cotton by root colonization and biofilm formation. Biol. Fertil. Soils 49, 295–303.

Lim, B., Beyhan, S., Meir, J., Yildiz, F.H., 2006. Cyclic-diGMP signal transduction systems in *Vibrio cholerae*: modulation of rugosity and biofilm formation. Mol. Microbiol. 60, 331–348.

Lloyd, J.R., 2003. Microbial reduction of metals and radionuclides. FEMS Microbiol. Rev. 27, 411–425.

Lo Piccolo, L., De Pasquale, C., Fodale, R., Puglia, A.M., Quatrini, P., 2011. Involvement of an alkane hydroxylase system of *Gordonia* sp. strain SoCg in degradation of solid nalkanes. Appl. Environ. Microbiol. 77, 1204–1213.

Macaskie, L.E., Bonthrone, K.M., Yong, P., Goddard, D.T., 2000. Enzymatically mediated bioprecipitation of uranium by *Citrobacter* sp.: a concerted role for exocellular lipopolysaccharide and associated phosphatase in biomineral formation. Microbiology (Reading) 146, 1855–1867.

Mailafia, S., Agbede, S., 2016. Evaluation of bacterial and fungal isolates of biofilm of water distribution systems and receptacles in Abuja, Nigeria. Eur. J. Exp. Biol. 6, 12–19.

Mangwani, N., Shukla, S.K., Kumari, S., Rao, T.S., Das, S., 2014a. Characterization of *Stenotrophomonas acidaminiphila* NCW702 biofilm for implication in the degradation of polycyclic aromatic hydrocarbons. J. Appl. Microbiol. 117, 1012–1024.

Mangwani, N., Shukla, S.K., Rao, T.S., Das, S., 2014b. Calciummediated modulation of *Pseudomonas mendocina* NR802 biofilm influences the phenanthrene degradation. Colloids Surf. B 114, 301–309.

Mangwani, N., Kumari, S., Das, S., 2015. Involvement of quorum sensing genes in biofilm development and degradation of polycyclic aromatic hydrocarbons by a marine bacterium *Pseudomonas aeruginosa* N6P6. Appl. Microbiol. Biotechnol. 99, 10283–10297.

Mangwani, N., Shukla, S.K., Kumari, S., Das, S., Rao, T.S., 2016. Effect of biofilm parameters and extracellular polymeric substance composition on polycyclic aromatic hydrocarbon degradation. RSC Adv. 6, 57540–57551.

Mapelli, F., Scoma, A., Michoud, G., Aulenta, F., Boon, N., Borin, S., Kalogerakis, N., Daffonchio, D., 2017. Biotechnologies for marine oil spill cleanup: indissoluble ties with microorganisms. Trends Biotechnol. 35, 860–870.

Massie, J.P., Reynolds, E.L., Koestler, B.J., Cong, J.P., Agostoni, M., Waters, C.M., 2012. Quantification of high-specificity cyclic diguanylate signaling. Proc. Natl. Acad. Sci. USA 109, 12746–12751.

Mathur, A., Parashar, A., Chandrasekaran, N., Mukherjee, A., 2017. Nano-TiO_2 enhances biofilm formation in a bacterial isolate from activated sludge of a waste water treatment plant. Int. Biodeterior. Biodegrad. 116, 17–25.

Matsukawa, M., Greenberg, E.P., 2004. Putative exopolysaccharide synthesis genes influence *Pseudomonas aeruginosa* biofilm development. J. Bacteriol. 186, 4449–4456.

Maurya, A., Raj, A., 2020. Recent advances in the application of biofilm in bioremediation of industrial wastewater and organic pollutants. In: Microorganisms for Sustainable Environment and Health, pp. 81–118.

Miqueleto, A.P., Dolosic, C.C., Pozzi, E., Foresti, E., Zaiat, M., 2010. Influence of carbon sources and C/N ratio on EPS production in anaerobic sequencing batch biofilm reactors for wastewater treatment. Bioresour. Technol. 101, 1324–1330.

Mitra, A., Mukhopadhyay, S., 2016. Biofilm mediated decontamination of pollutants from the environment. AIMS Bioeng. 3, 44–59.

Mnif, I., Mnif, S., Sahnoun, R., Maktouf, S., Ayedi, Y., Ellouze-Chaabouni, S., Ghribi, D., 2015. Biodegradation of diesel oil by a novel microbial consortium: comparison between co-inoculation with biosurfactant-producing strain and exogenously added biosurfactants. Environ. Sci. Pollut. Res. Int. 22, 14852–14861.

Monds, R.D., Newell, P.D., Gross, R.H., O'Toole, G.A., 2007. Phosphate dependent modulation of c-di-GMP levels regulates *Pseudomonas fluorescens* Pf0-1 biofilm formation by controlling secretion of the adhesion LapA. Mol. Microbiol. 63, 656–679.

More, T.T., Yadav, J.S.S., Yan, S., Tyagi, R.D., Surampalli, R.Y., 2014. Extracellular polymeric substances of bacteria and their potential environmental applications. J. Environ. Manag. 144, 1–25.

Morel, M.A., Ubalde, M.C., Olivera-Bravo, S., Callejas, C., Gill, P.R., CastroSowinski, S., 2009. Cellular and biochemical response to Cr(VI) in *Stenotrophomonas* sp. FEMS Microbiol. Lett. 291, 162–168.

Naeem, A., Batool, R., Jamil, N., 2013. Cr (VI) reduction by *Cellulosimicrobium* sp. isolated from tannery effluent. Biol. Chem. 37, 315–322.

Navarro, E., Guasch, H., Sabater, S., 2002. Use of microbenthic algal communities in ecotoxicological tests for the assessment of water quality: the Ter river case study. J. Appl. Phycol. 14, 41–48.

Nicolella, C., van Loosdrecht, M.C., Heijnen, S.J., 2000. Particle based biofilm reactor technology. Trends Biotechnol. 18, 312–320.

Nielsen, A.T., Tolker-Nielsen, T., Barken, K.B., Molin, S., 2000. Role of commensal relationships on the spatial structure of a surface-attached microbial consortium. Environ. Microbiol. 2, 59–68.

Nilsson, M., Chiang, W.C., Fazli, M., Gjermansen, M., Givskov, M., TolkerNielsen, T., 2011. Influence of putative exopolysaccharide genes on *Pseudomonas putida* KT2440 biofilm stability. Environ. Microbiol. 13, 1357–1369.

O'Connor, J.R., Kuwada, N.J., Huangyutitham, V., Wiggins, P.A., Harwood, C.S., 2012. Surface sensing and lateral subcellular localization of WspA, the receptor in a chemosensory-like system leading to c-di-GMP production. Mol. Microbiol. 86, 720–729.

Pal, A., Paul, A.K., 2008. Microbial extracellular polymeric substances: central elements in heavy metal bioremediation. Indian J. Microbiol. 48, 49–64.

Pandey, G., Jain, R.K., 2002. Bacterial chemotaxis toward environmental pollutants: role in bioremediation. Appl. Environ. Microbiol. 68, 5789–5795.

Pani, T., Das, A., Osborne, J.W., 2017. Bioremoval of zinc and manganese by bacterial biofilm: a bioreactorbased approach. J. Photochem. Photobiol. B 175, 211–218.

Paul, D., Pandey, G., Pandey, J., Jain, R.K., 2005. Accessing microbial diversity for bioremediation and environmental restoration. Trends Biotechnol. 23, 135–142.

Peacock, A.D., Chang, Y.J., Istok, J.D., Krumholz, L., Geyer, R., Kinsall, B., Watson, D., Sublette, K.L., White, D.C., 2004. Utilization of microbial biofilms as monitors of bioremediation. Microb. Ecol. 47, 284–292.

Pérez-Mendoza, D., Coulthurst, S.J., Sanjuán, J., Salmond, G.P., 2011. N-Acetylglucosamine-dependent biofilm formation in *Pectobacterium atrosepticum* is cryptic and activated by elevated c-di-GMP levels. Microbiology 157, 3340–3348.

Perumbakkam, S., Hess, T.F., Crawford, R.L., 2006. A bioremediation approach using natural transformation in pure-culture and mixed-population biofilms. Biodegradation 17 (6), 545–557.

Petrie, L., North, N.N., Dollhopf, S.L., Balkwill, D.L., Kostka, J.E., 2003. Enumeration and characterization of iron(III)-reducing microbial communities from acidic subsurface sediments contaminated with uranium(VI). Appl. Environ. Microbiol. 69, 7467–7479.

Pool, J.R., Kruse, N.A., Vis, M.L., 2013. Assessment of mine drainage remediated streams using diatom assemblages and biofilm enzyme activities. Hydrobiologia 709, 101–116.

Pratt, L.A., Kolter, R., 1999. Genetic analyses of bacterial biofilm formation. Curr. Opin. Microbiol. 2, 598–603.

Puhakka, J.A., Melin, E.S., Järvinen, K.T., Koro, P.M., Rintala, J.A., Hartikainen, P., Shieh, W.K., Ferguson, J.F., 1995. Fluidized bed biofilms for chlorophenol mineralization. Water Sci. Technol. 31, 227–235.

Purcell, E.B., McKee, R.W., McBride, S.M., Waters, C.M., Tamayo, R., 2012. Cyclic diguanylate inversely regulates motility and aggregation in *Clostridium difficile*. J. Bacteriol. 194, 3307–3316.

Qi, L., Christopher, G.F., 2019. Role of flagella, Type IV Pili, biosurfactants, and extracellular polymeric substance polysaccharides on the formation of pellicles by *Pseudomonas aeruginosa*. Langmuir 35, 5294–5304.

Rajendran, R., Sherry, L., Nile, C.J., Sherriff, A., Johnson, E., Hanson, M., Williams, C., Munro, C.A., Jones, B.J., Ramage, G., 2016. Biofilm formation is a risk factor for mortality in patients with *Candida albicans* bloodstream infection—Scotland, 2012 2013. Clin. Microbiol. Infect. 22, 87–93.

Rodriguez, S., Bishop, P., 2008. Enhancing the biodegradation of polycyclic aromatic hydrocarbons: effects of nonionic surfactant addition on biofilm function and structure. J. Environ. Eng. 134, 505–512.

Ron, E.Z., Rosenberg, E., 2014. Enhanced bioremediation of oil spills in the sea. Curr. Opin. Biotechnol. 27, 191–194.

Ross, P., Weinhouse, H., Aloni, Y., Michaeli, D., Weinberger-Ohana, P., Mayer, R., Braun, S., de Vroom, E., van der Marel, G.A., van Boom, J.H., Benziman, M., 1987. Regulation of cellulose synthesis in *Acetobacter xylinum* by cyclic diguanylic acid. Nature 325, 279–281.

Ryan, R.P., Lucey, J., O'Donovan, K., McCarthy, Y., Yang, L., TolkerNielsen, T., Dow, J. M., 2009. HD-GYP domain proteins regulate biofilm formation and virulence in *Pseudomonas aeruginosa*. Environ. Microbiol. 11, 1126–1136.

Schmieden, D.T., Basalo Vázquez, S.J., Sangüesa, H.C., van der Does, M., Idema, T., Meyer, A.S., 2018. Printing of patterned, engineered *E. coli* biofilms with a low-cost 3D printer. ACS Synth. Biol. 7, 1328–11337.

Scoma, A., Yakimov, M.M., Daffonchio, D., Boon, N., 2017. Self-healing capacity of deep-sea ecosystems affected by petroleum hydrocarbons. EMBO Rep. 18, 868–872.

Simm, R., Morr, M., Kader, A., Nimtz, M., Romling, U., 2004. GGDEF and EAL domains inversely regulate cyclic di-GMP levels and transition from sessility to motility. Mol. Microbiol. 53, 1123–1134.

Singh, R., Paul, D., Jain, R.K., 2006. Biofilms: implications in bioremediation. Trends Microbiol. 14 (9), 389–397.

Smith, T.J., Sondermann, H., O'Toole, G.A., 2018. Co-opting the lap system of *Pseudomonas fluorescens* to reversibly customize bacterial cell surfaces. ACS Synth. Biol. 7, 2612–2617.

Song, H.G., Bartha, R., 1990. Effects of jet fuel spills on the microbial community of soil. Appl. Environ. Microbiol. 56, 646–651.

Srinivasan, V., Bhavan, P.S., Krishnakumar, J., 2014. Bioremediation of textile dye effluent by *Bacillus* and *Pseudomonas* Spp. Int. J. Sci. Environ. Technol. 3 (6), 2215–2224.

Stelmack, P.L., 1999. Bacterial adhesion to soil contaminants in the presence of surfactants. Appl. Environ. Microbiol. 65, 163–168.

Sutherland, I.W., 2001. The biofilm matrix – an immobilized but dynamic microbial environment. Trends Microbiol. 9, 222–227.

Tang, X., Liu, S., Zhang, Z., Zhuang, G., 2015. Identification of the release and effects of AHLs in anammox culture for bacteria communication. Chem. Eng. J. 273, 184–191.

Tarangini, K., 2009. Biosorption of Heavy Metals, Using Individual and Mixed Cultures of *Bacillus subtilis* and *Pseudomonas aeruginosa*. Master of Technology Thesis, Chemical Engineering (Biochemical Engineering and Biotechnology), National Institute of Technology, Rourkela, Orissa, India.

Teitzel, G.M., Parsek, M.R., 2003. Heavy metal resistance of biofilm and planktonic *Pseudomonas aeruginosa*. Appl. Environ. Microbiol. 69, 2313–2320.

Tischler, A.D., Camilli, A., 2004. Cyclic diguanylate (c-di-GMP) regulates *Vibrio cholerae* biofilm formation. Mol. Microbiol. 53, 857–869.

Tolker-Nielsen, T., Brinch, U.C., Ragas, P.C., Andersen, J.B., Jacobsen, C.S., Molin, S., 2000. Development and dynamics of *Pseudomonas* sp. biofilms. J. Bacteriol. 182, 6482–6489.

Turki, Y., Mehri, I., Lajnef, R., Rejab, A.B., Khessairi, A., Cherif, H., Ouzari, H., Hassen, A., 2017. Biofilms in bioremediation and wastewater treatment: characterization of bacterial community structure and diversity during seasons in municipal wastewater treatment process. Environ. Sci. Pollut. Res. Int. 24, 3519–3530.

Valentini, M., Filloux, A., 2016. Biofilms and cyclic di-GMP (c-di-GMP) signaling: lessons from *Pseudomonas aeruginosa* and other bacteria. J. Biol. Chem. 291, 12547–12555.

Valls, M., Lorenzo, V.D., 2002. Exploiting the genetic and biochemical capacities of bacteria for the remediation of heavy metal pollution. FEMS Microbiol. Rev. 26, 327–338.

von Canstein, H., Kelly, S., Li, Y., Wagner-Döbler, I., 2002. Species diversity improves the efficiency of mercury-reducing biofilms under changing environmental conditions. Appl. Environ. Microbiol. 68, 2829–2837.

Vu, B., Chen, M., Crawford, R.J., Ivanova, E.P., 2009. Bacterial extracellular polysaccharides involved in biofilm formation. Molecules 14, 2535–2554. https://doi.org/10.3390/molecules14072535.

White, C., Gadd, G.M., 2000. Copper accumulation by sulfate-reducing bacterial biofilms. FEMS Microbiol. Lett. 183, 313–318.

Wick, L.Y., Colangelo, T., Harms, H., 2001. Kinetics of mass transfer-limited bacterial growth on solid PAHs. Environ. Sci. Technol. 35, 354–361.

Wick, L.Y., De Munain, A., Springael, D., Harms, H., 2002. Responses of *Mycobacterium* sp. LB501T to the low bioavailability of solid anthracene. Appl. Microbiol. Biotechnol. 58, 378–385.

Williams, K.H., Wilkins, M.J., N'Guessan, A.L., Arey, B., Dodova, E., Dohnalkova, A., Holmes, D., Lovley, D.R., Long, P.E., 2013. Field evidence of selenium bioreduction in a uranium-contaminated aquifer. Environ. Microbiol. Rep. 5, 444–452.

Wolfaardt, G.M., 1995. Bioaccumulation of the herbicide diclofop in extracellular polymers and its utilization by a biofilm community during starvation. Appl. Environ. Microbiol. 61, 152–158.

Wolfaardt, G.M., Lawrence, J.R., Robarts, R.D., Caldwell, S.J., Caldwell, D.E., 1994. Multicellular organization in a degradative biofilm community. Appl. Environ. Microbiol. 60, 434–446.

Wozniak, D.J., Wyckoff, T.J., Starkey, M., Keyser, R., Azadi, P., O'Toole, G.A., Parsek, M. R., 2003. Alginate is not a significant component of the extracellular polysaccharide matrix of PA14 and PAO1 *Pseudomonas aeruginosa* biofilms. Proc. Natl. Acad. Sci. USA 100, 7907–7912.

Yakimov, M.M., Golyshin, P.N., Lang, S., Moore, E.R., Abraham, W.R., Lünsdorf, H., Timmis, K.N., 1998. *Alcanivorax borkumensis* gen. nov., sp. nov., a new, hydrocarbon-degrading and surfactant-producing marine bacterium. Int. J. Syst. Bacteriol. 48, 339–348.

Yang, L., Barken, K.B., Skindersoe, M.E., Christensen, A.B., Givskov, M., Tolker-Nielsen, T., 2007. Effects of iron on DNA release and biofilm development by *Pseudomonas aeruginosa*. Microbiology 153, 1318–1328.

CHAPTER 13

Role of microbial enzymes in bioremediation: Emerging opportunities and limitations

Akanksha Pandey[a], Sachin Kumar[a], Navneet[a], Sandeep Kumar[a], Minakshi Rajput[b], Monika Singh[b], Chitranshu Pandey[c], and Nikunj Bhardwaj[d]

[a]Department of Botany and Microbiology, Gurukula Kangri (Deemed to be University), Haridwar, Uttarakhand, India
[b]Department of Biotechnology, School of Applied and Life Sciences, Uttaranchal University, Dehradun, India
[c]Department of Biotechnology, MRD Life Science, Lucknow, India
[d]Department of Zoology, Maa Shakumbhari University, Saharanpur, India

13.1 Introduction

Pollution caused by the deleterious effects of undesirable, unwanted matter unintentionally or deliberately released into the environment is a major global concern. With the emergence of the industrial and agricultural revolution, intensive use of inorganic and organic pollutants has increased, mostly through anthropogenic activities and natural processes to some extent (Malla et al., 2018; Rodríguez et al., 2020). Inorganic pollutants mainly include heavy metals, dyes, and radioactive compounds, whereas petroleum oil and hydrocarbons, polycyclic aromatic hydrocarbons (PAHs), fertilizers, pesticides, dimethyl phthalate, tributyltin, explosives, etc., contribute to organic pollution (Mondal et al., 2019). These hazardous and recalcitrant contaminants harm the aquatic and terrestrial environments, necessitating pollutant monitoring and management (Bharagava et al., 2020). For the treatment of chief wastes generated through industrial, domestic, and agricultural malpractices, various conventional, commercial technologies such as landfilling, biofiltration, burning or incineration, adsorption, membrane separation, ultraviolet (UV) decomposition, etc., are available (Megharaj and Naidu, 2017). Besides their availability for pollutant removal, these traditional approaches meet the challenges of specificity, cost adequacy, and generation of secondary nonbiodegradable recalcitrant products; hence, there is a need for a sustainable, environmental-friendly bioremediation approach (Perelo, 2010; Karigar and Rao, 2011; Bala et al., 2022).

Microbial Inoculants
https://doi.org/10.1016/B978-0-323-99043-1.00007-4

Bioremediation is a multidisciplinary, integrated, sustainable remediation technology that uses the biological activities of algae, bacteria, archaea, fungi, green plants, or their enzymes to tackle or clean up harmful contaminants from polluted sites. Microbial remediation, phytoremediation, or a combination of these two might be a possible strategy as the pollution problem must be rectified without carrying over to the near future (Prakash et al., 2013; Sharma et al., 2018).

13.2 Enzymes

Enzymes are important biocatalysts capable of converting a specific substrate into desired products by reducing the activation energy of a reaction in a favorable environment. Enzymes have so far been extensively used in analytical chemistry and medicine; now, efforts are being taken to use these biological catalysts for environmental safety and waste management. To decrease the release of hazardous by-products and eliminate the risks associated with the use of harmful chemicals, enzymatic strategies are likely to supersede several conventional industrial methodologies (Singh et al., 2021).

Enzymes work in three different ways for environmental purposes by acting as the main catalyst in the transformation/detoxification of waste products, as bioindicators of ecosystem health, and as biosensors to pinpoint and assess the levels of contamination before, during, and after the restoration stages (Rao et al., 2014).

13.3 Microbial enzymes

There are more than 1000 different types of enzymes, including hydrolases, lyases, oxidoreductases, transferases, peroxidases, etc., in a single microbial cell (Ahuja et al., 2004). Microbial enzymes play a significant part in the breakdown of hazardous substances in the environment and carry out reactions in an intracellular/extracellular manner or in a free/immobilized form (Karigar and Rao, 2011). As microbe-mediated processes are slow and generate a large amount of hazardous secondary by-products, enzyme-assisted biotransformation in remediation is more environment-friendly with the benefit of target-degrading potential and better action on target chemicals (Ghosh et al., 2017; Kumar and Bharadvaja, 2019). Microbial biomass in contamination sites faces many constraints, including the risk of microbial predators or antagonists and competition with the existing soil microbiota (Rao et al., 2010). Furthermore, maintenance and transport of microbial

Table 1 Advantages and disadvantages of microbial enzymes.

Advantages	Disadvantages
1. Better flexibility in terms of operational conditions as a microbial population needs more specific parameters like temperature, salinity, pH, contaminant concentration, etc.	**1.** Difficulties in identification and selection of the microbe source and plant
2. The enzymatic process is rapid, easier to control, accessible, and highly specific	**2.** The time-consuming and capital-intensive process for the production, extraction, and purification of enzymes on an industrial scale is tough
3. Enzyme-mediated biotransformation decreases the amount of toxic by-product formation	**3.** Limited number, stability, and catalytic activity of enzymes across a wide range of contaminants and environmental conditions
4. Enzyme-based biosensors and nanozymes provide a more advanced analytical tool for qualitative and quantitative assessment of compounds of interest	**4.** Enzymatic treatment performance or efficiency varies from the experimental laboratory to the on-site field

culture on a large scale are more expensive, and even elevated concentrations of toxicants or their products may cause irreversible damage or metabolic inactivation of microbial cells (Ruggaber and Talley, 2006; Bhatt et al., 2019). Therefore, to overcome such limitations of microbe-mediated processes, enzymes have been isolated from microbial cells, which enhance the enzymatic function relative to the entire cell. Enzymes work more effectively on pollutants than do microbial cells due to their smaller size promoting a greater interaction with pollutants, faster mobility, and more focused, rapid, efficient degradation (Saravanan et al., 2021). Some advantages and disadvantages of microbial enzymes are reported in Table 1.

13.4 Bioremediation and enzymes

There are different methods of bioremediation that have been used to lower the amount of harmful contaminants in the environment (Bala et al., 2022). Depending on the type of biological agent used, bioremediation may be microbial remediation (use of microbes to transform toxic pollutants into nontoxic products) or phytoremediation (use of green plants to decontaminate the polluted site). Furthermore, remediation of pollutants at their actual contaminant site is related to in situ remediation, whereas removal of waste materials via transfer and treatment in special landfills and bioreactors is linked to ex situ remediation. The latter approach is rapid and more effective and needs solid-phase (land farming, composting, and soil biopiles) or slurry-phase (large tank bioreactors) treatment (Azubuike et al., 2016; Arora et al., 2022).

The outcomes of bioremediation using pure and partially purified enzymes rely more on enzymatic activity rather than on the growth of a specific microorganism in a polluted area. To overcome some of the limitations of enzymatic bioremediation described earlier, immobilization techniques have been used by researchers to improve the process efficiency, which also lowers the costs associated with enzyme production and purification (Ruggaber and Talley, 2006; Singh et al., 2021). Industrial-scale enzyme degradation necessitates a greater quantity of pure enzymes, yet this quantity may be constrained by the cost of producing and purifying enzymes. This can be accomplished by the use of genetic engineering, which produces the desired quantities of enzymes with higher specificity at a relatively smaller cost than the traditional enzyme production procedure (Sharma et al., 2018).

Several enzymes have been used to break down environmental contaminants, and each enzyme has a specific mechanism. Enzymes isolated from a sole microbe are preferable in the remediation process due to their superior catalytic efficiency when compared to those from a combination of enzymes. Various microbial enzymes, including both extracellular and intracellular, reported in different research studies from bacteria and fungi are presented in Fig. 1.

13.5 Enzyme-assisted microbial remediation for hazardous pollutants

The utilization of microbes and their catabolic enzymes to remediate hazardous xenobiotic compounds from the natural environment through complete degradation/transformation into safe products like carbon dioxide and water is a key premise for the bioremediation approach. Several bacteria and fungi have been reported to remove toxic heavy metals, to decolorize synthetic dyes and biological stains from textile effluents, and to degrade palm oil, polychlorinated biphenyls (PCBs), etc., from industrial wastewater along with degrading *N*-methyl carbamate pesticides, polycyclic aromatic hydrocarbons (PAHs), phenolic compounds, and organophosphates from

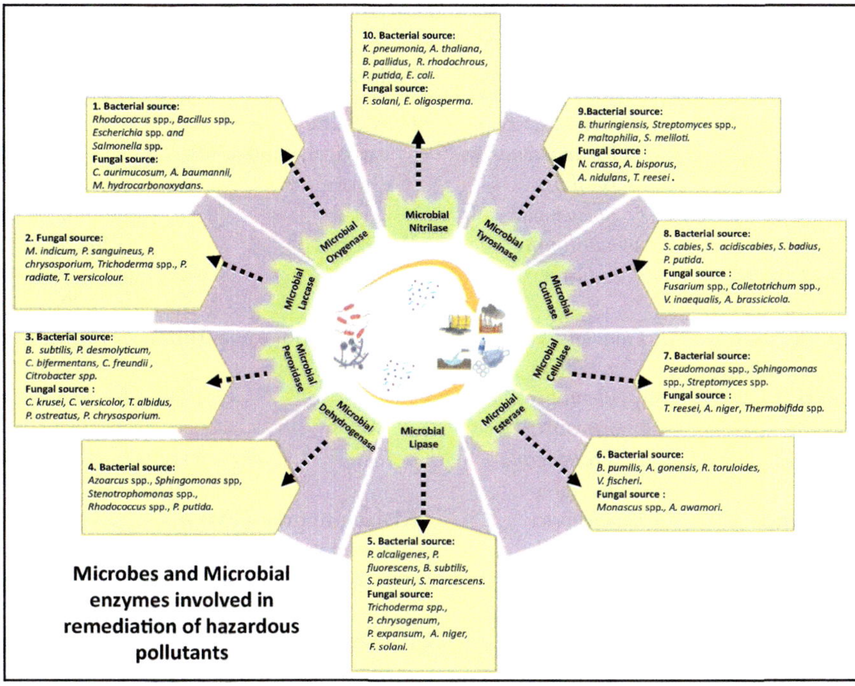

Fig. 1 Some important degrading microbial enzymes with their microbial sources.

soil and water. The phenomena of biodegradation are divided into convergent and divergent modes of degradation based on the complexity of the degradation pathways. In the convergent mode, a small number of aromatic ring cleavage substrates, such as catechol, gent sate, protocatechuate, and their derivatives, are transformed from structurally varying aromatic molecules. The hydroxylation of the aromatic nucleus by peripheral enzymes, especially oxygenases and dehydrogenases, was discovered to convert structurally different substrates into one of these central intermediates. The presence of azoreductase, laccase, peroxidase, tyrosinase, and nicotinamide adenine dinucleotide phosphate-dichlorophenolindophenol (NADP–DCIP) reductase enzymes affects a bacterium's ability to degrade substances (Singh et al., 2022). Several enzymes used in the bioremediation of different harmful chemical compounds are enlisted in Table 2.

13.5.1 Microbial oxygenases

Microbial oxygenase is related to the oxidoreductase family of enzymes that performs oxidation reactions by incorporating oxygen atoms into the

Table 2 Enzymes used in the bioremediation of different harmful chemical compounds.

Enzymes	Producing microorganisms	Pollutants	References
Carbamate hydrolase	*Achromobacter* sp., *Pseudomonas* sp., *Bacillus*, *Shewanella*	Carbofuran, red SNR dye, carbaryl, blue SNR, chromium	Durve et al. (2012)
Ascomycetes laccase	*Monocillium indicum*	Peroxides	Shekher et al. (2011)
Lignin peroxidases (LiPs)	*Coriolopsis polyzona*, *Pleurotus ostreatus*, *Trichoderma versicolor*	PCBs, biopolymers (kraft, lignin)	Pointing (2001)
Manganese peroxidases (MnPs)	*Enterobacter*, *Klebsiella pneumoniae* GM-04	Malachite green (MG), bisperse blue-284	Roy et al. (2020)
Carboxylesterases	*Pseudomonas aeruginosa* PA1	Red SNR dye blue SNR, malathion, and parathion	Durve et al. (2012)
Peroxidases	*Phanerochaete chrysosporium*	TNT (2,4,6-trinitrotoluene)	Cameron et al. (2000)
Laccase and esterase	*Bacillus subtilis*	Cypermethrin (insecticide)	Gangola et al. (2018)
Lipase	*Bacillus pumilus*	Palm oil degradation	Bhandari et al. (2021)
Dehydrogenase	*Sphingobacterium rhizophila*	Catabolism of 2,4-xylenol	Chen et al. (2014)
Carbamate hydrolase	*Pseudomonas* sp., *Achromobacter* sp.	Carbofuran, carbaryl	Mulbry and Eaton (1991)
Horseradish peroxidase	*Armoracia*, *Rusticana*	Chromium	Thatoi et al. (2014)
Oxidoreductases	*Aspergillus ochraceus*, *Bjerkandera adusta*, *Phlebia*, *Irpex lacteus*	Congo red, malachite green	Singh et al. (2022)

Table 2 Enzymes used in the bioremediation of different harmful chemical compounds—cont'd

Enzymes	Producing microorganisms	Pollutants	References
Protease	*B. subtilis*	Degradation of casein and feather	Bhandari et al. (2021)
Tyrosinase	*Agaricus bisporus*	Fluoranthene, anthracene, phenanthrene, mono- and diphenols, pyrene, bisphenol A	Rao et al. (2014)
Esterase	*Synechocystis*	Dimethyl phthalate	Sharma et al. (2019)
Phytase	*Aspergillus niger* NCIM 563	Chlorpyrifos	Saravanan et al. (2021)
Monooxygenase	*Bacillus* sp., *Staphylococcus aureus*	Bemacron yellow HP-2R (BY), bemacron blue RS 01 (BB), bemacron red RS (BR)	Aktar et al. (2019)
Azoreductase	*Shewanella marisflavi*	Xylidine ponceau 2R (dye)	Xu et al. (2016)
Reductase	*Alishewanella* sp. CBL-2	Sumifex tourqi (blue dye)	Ajaz et al. (2018)
Laccase	*Geobacillus thermocatenulatus*	Congo red and bromophenol blue	Bhandari et al. (2021)
Chloroperoxidase	*Caldariomyces fumago*	Anthracene, acenaphthene, azulene, benzothiophene, fluoranthene, dibenzothiophene, fluorene, 2-methyl anthracene, 9-methyl anthracene, pyrene, phenanthrene, thianthrene, pentachlorophenol, chlorophenols	Rao et al. (2014)
Manganese peroxidases	*B. subtilis* X3, *Bacillus firmus*	Lead, mercury, copper, and chromium	Pande et al. (2022)
Cellulases	*Trichoderma reesei*, *Thermobifida*	Hydrogen sulfide, sodium sulfide, methyl mercaptan	Sharma et al. (2019)

reduced substrate and is responsible for the oxidative biodegradation of aromatic compounds present in organic pollutants. Depending on the number of oxygen molecules transferred, two types of oxygenases are available: (a) monooxygenases, also known as mixed-function oxidases that catalyze the transfer of a single oxygen molecule to the substrate and (b) dioxygenases, also known as oxygen transferases, which introduce two oxygen molecules into a substrate. In the disintegration of highly chlorinated chemicals, which frequently turn into important environmental contaminants, microbial oxygenase enzymes play a significant role. Aerobes have a large number of oxygenases, which catalyze the reductive activation of O_2 and the incorporation of oxygen atoms into substrates. In this study, the catalytic properties of four enzymes that are often implicated in the breakdown of pesticides flavin-dependent monooxygenases, heme iron-dependent cytochrome P450 monooxygenases, Rieske nonheme iron-dependent oxygenases, and ring-cleavage dioxygenases are outlined (Hayaishi et al., 1955). Among monooxygenases, methane monooxygenase is the enzyme with the finest characterization. This enzyme is involved in the hydrolysis of aromatic and heterocyclic hydrocarbons as well as substituted methanes, alkanes, cycloalkanes, alkenes, and haloalkenes. As part of nature's strategy for breaking down aromatic molecules in the environment, catechol dioxygenases are now used. They are soil bacteria that play a role in converting aromatic precursors into aliphatic products (Karigar and Rao, 2011). Methane monooxygenases occur in two forms: first is found in the cytoplasmic membrane and the other is located in the cytoplasm. Bacteria of *Methylocella palustris* produce soluble MMOs that can break down a variety of contaminants, including hydrocarbons and aliphatic and aromatic chemicals (Singh and Singh, 2017). For instance, naphthalene dioxygenase from *Pseudomonas putida* is isolated and involved in the breakdown of naphthalene (Di Gennaro et al., 1997).

Microbial source: Few important sources of monooxygenase enzymes are *Saccharomyces*, *Streptomyces*, *Basidiomycete*, *Dehalococcoides*, *Rhodococcus*, *Bacillus*, *Escherichia*, and *Salmonella* (Mousavi et al., 2021). The dioxygenase enzyme was reported to be found in *Corynebacterium aurimucosum*, *Acinetobacter baumannii*, *Microbacterium hydrocarbonoxydans*, and *M. hydrocarbonoxydans* (Muthukamalam et al., 2017).

13.5.2 Microbial laccases

Microbial laccase is a copper-containing oxidase that catalyzes the cleavage of polyphenolic and many aromatic compounds by simultaneously reducing molecular oxygen to water with the generation of free radicals. With simultaneous electron removal from the reducing phenols, aromatic amines, and heterocyclic compounds, laccase catalyzes the complete four-electron reduction of dioxygen to water. Besides this, lignins can be metabolized by laccases into phenoxy radicals, allowing them to participate in copolymerization or homopolymerization processes. Numerous bacteria create intracellular and extracellular laccases that may catalyze the oxidation of ortho- and para-diphenols and aminophenols, polyphenols, polyamines, lignins, and aryl diamines as well as certain inorganic ions. Laccases attack the methoxy groups in phenolic and methoxy-phenolic acids and oxidize and decarboxylate them (demethylation) (Karigar and Rao, 2011). The first *Ascomycetes* laccase with peroxidase activity was identified as *Monocillium indicum*. While *Pycnoporus sanguineus* generates laccase as phenol oxidase, *Pseudupeneus cinnabarinus* produces laccase as a ligninolytic enzyme. While laccase has been linked to delignification, sporulation, pigment generation, fruiting body development, and plant disease in fungi, it has a role in the lignification of plants like *Podospora anserina* and *Neurospora crassa* (Shekher et al., 2011). On porous glass beads, laccase from the fungus *Trametes versicolor* was immobilized. It demonstrates its effectiveness as a potent enzyme in the bioremediation of a variety of contaminants, including phenolic compounds, aromatic heterocyclic compounds, and aromatic compounds containing amines. By removing electrons from the organic substrate, laccase can lower the dioxygen pollution molecules in the water (Chakroun et al., 2010).

Microbial source: Laccase production was reported in *Magnaporthe grisea*, *Gaeumannomyces- graminis*, *Ophiostoma novo-ulmi*, *Melanocarpus albomyces*, *Marginella*, *M. indicum*, *Streptomyces cyaneus*, *Streptomyces lavendulae*, *Marinomonas mediterranea*, *Theiophora terrestris*, *Phanerochaete chrysosporium*, and *Lenzites*, *betulina* and in white-rot fungi such as *T. versicolor*, *Pleurotus ostreatus*, *Phlebia radiata*, *Trichoderma* species, such as *T. harzianum*, *T. atroviride*, and *T. longibrachiatum*, *Pycnoporus cinnabarinus*, and *Marasmius quercophilus* (Chakroun et al., 2010).

13.5.3 Microbial peroxidases

Only one type of heme-containing peroxidase is present in mammals, whereas the other group is present in fungi, bacteria, and plants. Microbial peroxidase is a heme or heme-free protein containing the class of oxidoreductases that are known to use hydrogen peroxide (H_2O_2) or other organic hydroperoxides to catalyze the oxidation of several inorganic and organic substrates. Peroxidase has a molecular weight of 11.45 kDa. The peroxidase–catalase superfamily is a major group of microbial-associated peroxidases and has been extensively used in the removal of toxic pollutants, in sewage treatment, dyes, and lignin degradation. These enzymes are highly efficient in the degradation of lignin and many other polyphenolic compounds. A set of specialized enzymes called peroxidases, such as glutathione peroxidase, NAD peroxidase, and fatty acid peroxidase and some nonspecific enzymes from other sources are also referred to as peroxidases. Heme-containing enzymes may or may not be present in peroxidases (Sharma et al., 2018). The most common sources of biodegrading peroxidases are *Bacillus subtilis*, *Streptomyces* spp., *Pantoea agglomerans*, *Pseudomonas aeruginosa*, *Staphylococcus aureus*, *Providencia* spp., *P. desmolyticum*, *Clostridium bifermentans*, *Candida krusei*, *Citrobacter freundii*, *Coriolus versicolor* (L.), *Tyromyces albidus* (Schaeff.), *Trametes gallica*, *Pleurotus ostreatus*, *Phanerochaete chrysosporium*, *Thanatephorus cucumeris*, and *Auricularia auricula-judae* (Bansal and Kanwar, 2013).

Two of the most common peroxidases involved in the biodegradation of toxic pollutants are manganese peroxidases and lignin peroxidases.

Manganese peroxidases

Manganese peroxidases (MnPs) are extracellular enzymes that include heme and are generated by fungi that break down lignin. They may oxidize Mn^{2+} into Mn^{3+} through a multistep mechanism. The enzyme MnPTra-48424 was isolated and purified from *Trametes* sp. (a white-rot fungus) having one heme group bearing manganese-binding sites and several acidic amino acid moieties. This enzyme strongly decolorized various dyes, including indigo, azo, anthraquinone, and triphenylmethane, whereas other dyes, including methyl green and indigo carmine, were combined with organic solvents and heavy metal ions. In *Peniophora incarnata*, manganese-dependent peroxidase was found, whereas degraded anthracene was encoded by a manganese-dependent peroxidase (MnP) gene pimp1. MnP gene pimp1 was also reported in yeast *Saccharomyces cerevisiae* that helps to speed-up the rate of bioremediation (Lee et al., 2016).

Microbial source: Some manganese peroxidase-producing fungal strains are *Agaricus bisporus*, *Phanerochaete chrysosporium*, *Trametes versicolor*, *Schizophyllum commune*, *Anthracophyllum discolor*, *Galactomyces reessii*, *Pleurotus ostreatus*, *Penicillium chrysogenum*, and *Ceriporiopsis subvermispora*. Along with these fungi, some MnP-producing bacteria are also found, including *Bacillus amyloliquefaciens*, *Klebsiella pneumoniae*, *B. cereus*, *Salmonella enterica*, *Enterobacter aerogenes*, *E. cloaceae*, *Lactobacillus kefiri*, *Bacillus aryabhattai*, *Lysinibacillus* sp., *B. subtilis*, *B. pumilus*, and *Paenibacillus* sp. (Kumar and Arora, 2022).

Lignin peroxidases

Lignin peroxidases (LiPs) are heme-containing monomeric enzymes and secondary metabolites that are released by bacteria, fungi, and other organisms (Xu et al., 2014). In terms of specificity and thermostability, bacterial peroxidases are more effective in degrading lignin than are fungal peroxidases. Lignin, a component of plant cell walls, is typically broken down by LiPs. LiP enzymes may oxidize aromatic compounds with redox potentials greater than 1.4V (Piontek et al., 2001).

Microbial source: *Ganoderma lucidum*, *Bjerkandera adusta*, *Phanerochaete chrysosporium*, *Aspergillus* sp., *Phanerochaete* sp., *Phlebia radiata*, *Trametes versicolor*, *Pleurotus eryngii*, *P. ostreatus*, *Streptomyces coelicolor*, *Thermobifida fusca*, *Rhodococcus jostii*, *Acinetobacter calcoaceticus*, and *Streptomyces viridosporus* (Janusz et al., 2017).

13.5.4 Microbial lipases

Microbial lipases belong to the family of hydrolytic enzymes that are associated with triacylglycerol degradation into glycerol and free fatty acids, performing important reactions, including hydrolysis, alcoholysis, thiolysis, aminolysis, esterification, etc. Hydrolytic lipases can degrade lipid content from microbes, plants, and animals. The hydrolysis, esterification, and alcoholysis processes are all catalyzed by special lipase (triacylglycerol acyl hydrolase) enzymes. Long-chain acylglycerols (C8) are hydrolyzed by lipases (EC3.1.1.3). The lipid molecules produced by microbes, animals, and plants can all be decomposed by hydrolytic lipases. In comparison to traditional lipid degradation mechanisms, microorganisms can effectively biodegrade lipid waste in modest circumstances, creating lipolytic enzymes and promoting environmental sustainability. *Thermomyces lanuginosus* was the source of the first industrially produced, commercially useful recombinant lipase, known as lipolase (Sharma et al., 2018). The medium was optimized to increase the production of microbial lipases using various statistical techniques. A new crude oil-degrading lipase from the bacterium *P. aeruginosa* was produced using statistical tools, and it was used to bioremediate crude oils (Verma et al., 2012).

Microbial source: Bacterial strains such as *Pseudomonas alcaligenes*, *P. aeruginosa*, *Pseudomonas fragi*, *P. fluorescens*, *B. subtilis*, *Bacillus nealsonii*, *P. mendocina*, *Acinetobacter radioresistens*, *Staphylococcus pasteuri*, *Staphylococcus warneri*,

Staphylococcus xylosus, *Marinobacter lipolyticus*, *Geobacillus stearothermophilus*, *Natronococcus* sp., *Pseudomonas plantarii*, *Chromobacterium viscosum*, *Bacillus thermocatenulatus*, *Serratia marcescens*, *B. prodigiosus*, *B. fluorescens*, *B. pyocyaneus*, and *Pseudomonas cepacia* and some species of fungi such as *Penicillium expansum*, *Trichoderma*, *P. chrysogenum*, *Aspergillus niger*, *Fusarium solani*, *Yarrowia lipolytica*, *A. oryzae*, *Rhizomucor miehei*, *Geotrichum candidum*, *Candida antarctica*, *C. rugosa*, *Candida lipolytica*, *P. camemberti*, *Tachytes lanuginosus*, *P. roqueforti*, *Meyerozyma guilliermondii*, *A. flavus*, *C. tropicalis*, *P. abeanum*, and *R. nodosus* produce lipases in higher quantities (Chandra et al., 2020).

13.5.5 Microbial esterases

Esterase, a type of lipolytic enzyme, can hydrolyze a diverse range of chemical compounds containing ester bonds. Short-chain fatty esters ($\leq$C8) can be hydrolyzed by esterase (EC 3.1.1.1). The polypeptide chain of esterase frequently contains a highly conserved catalytic triad made up of the three amino acids serine, aspartic acid, and histidine; of the triad's serine residues, the active-site serine is most likely to carry out a nucleophilic attack on its substrate during catalytic hydrolysis. Eight different families (Families I–VIII) have traditionally been used to classify microbial esterases, all of which share the same folding pattern (Shin et al., 2021). Esterases have excellent chemo- and/or enantioselectivity and can catalyze three different types of reactions: esterification, interesterification, and transesterification (Sharma et al., 2017). Insecticides, herbicides, and other pollutants, including organophosphates and chlorinated organic compounds, are broken down by the esterase enzymes carboxylesterases and phosphotriesterases. Pesticides, insecticides, and fungicides sprayed on crops have all been degraded using carboxylesterases. The carboxylesterase E2 from strain *P. aeruginosa* PA1 was visible on the outer membrane of *E. coli* for the absorption of mercury in the contaminated location. The amidohydrolase metalloenzyme family includes the bacterial phosphodiesterase subgroup. The hydrolysis of OP trimesters is predominantly catalyzed by phosphotriesterases. The comprehensive characterization of two closely related bacterial phosphotriesterases, OpdA from *Agrobacterium radiobacter*, OPH from *P. diminuta* (Horne et al., 2002).

Microbial source: *Micropolyspora faeni*, *Monascus* sp., *B. pumilus*, *Anoxybacillus gonensis* A4, recombinant *Escherichia coli*, *B. subtilis*, *P. fluorescens*, *A. awamori*, *Rhodosporidium toruloides*, *Vibrio fischeri*, *Aureobasidium pullulans*, *Comamonas acidovorans*, *Clostridium botulinum*, *Acinetobacter* sp., and

Rhodopseudomonas palustris (Sayali et al., 2013). Some reported bacterial and fungal sources of microbial phosphotriesterases are *Pseudomonas pseudoalcaligenes, P. diminuta, A. radiobacter, Alteromonas haloplanktis, S. cerevisiae, Pseudomonas monteilii, G. stearothermophilus, Flavobacterium* sp., *Alteromonas* sp., and *Geobacillus kaustophilus* (Latip et al., 2019).

13.5.6 Microbial cellulases

Cellulases are important enzymes in the textile manufacturing, biofuel production, and paper pulp industries to degrade complex carbohydrate polymers like cellulose into simple sugars. Different forms of celluloses in microbial cells, whether intracellular or extracellular, can help in the hydrolysis of lignocellulosic content present in the soil. A combination of endoglucanase and exoglucanase enzymes work in synergy to degrade paper ink in the pulp industry during the paper recycling process. According to a recent market report on "Global Industrial Enzymes" by Freedonia in January 2018, the demand for industrial enzymes is anticipated to grow by 4.0% every year to \$5.0 billion in 2021 (Jayasekara and Ratnayake, 2019). Scientists are extremely interested in alkaline cellulases produced by marine microorganisms because they are more suited for gritty industrial operations in a variety of sectors, including waste management, pulp and paper manufacturing, food processing, and textile and laundry (Karigar and Rao, 2011). When exposed to 10 ppm of Cr, the bacterium *B. amyloliquefaciens*, which was isolated from leather tanning industrial wastes, produced nearly $20\,U\,mL^{-1}$ of cellulases (VI) (Aslam et al., 2019).

Microbial source: *Trichoderma reesei*, *Streptomyces*, and *Thermobifida* and genera of bacteria, such as *Pseudomonas* and *Sphingomonas* (dye degradation). *Humicola insolens*, *Thermomonospora* sp. (Biostone washing) *A. niger*, (Biopolishing and biofinishing), *Humicola* (*H. insolens* and *H. griseathermoidea*), and *Bacillus* sp. (production of cellulases) (Aslam et al., 2019; Hayashida and Yoshioka, 1980).

13.5.7 Microbial cutinases

Cutinases are serine α/β-hydrolases that are associated with the cleavage of ester bonds and play an important role in the degradation of apoplastic materials cutin and suberin. Cutinase has been modified to degrade polyethylene terephthalate. Cutin's fatty acids typically have one to three hydroxyl groups and are in the n-C16 to n-C18 range. Although ether connections and peroxide bridges have also been proposed, ester bonds dominate in cutins. One of the initial stages of infection is the enzymatic hydrolysis of cutin. As effectively as pancreatic lipases, cutinases can hydrolyze a wide range of synthetic esters and exhibit action on both the short and long chains of emulsified triacylglycerols. *Fusarium solani f.* sp. *pisi* secreted extracellular cutinase in response to cutin hydrolysate. Saturation was reached at 80 mg/mL of cutin, and the rate of cutinase synthesis was dependent on the quantity of cutin hydrolysate added to the medium containing glucose. Yeast esterase cannot decompose malathion as quickly as can fungal cutinase with excellent stability. To make polyvinyl chloride flexible, phthalate esters are primarily employed in the manufacturing process. Paints, lacquers, and cosmetics also use them to a lesser extent (Dutta et al., 2009). Dipropyl phthalate (DPrP) and dimethyl phthalate (DMP), two of the most prevalent phthalates, were identified by the United States Environmental Protection Agency as priority pollutants. It has been noted that cutinase reduced DPrP (Kim et al., 2005).

Microbial source: *S. scabies*, *Streptomyces acidiscabies*, *S. badius*, *P. putida*, *P. mendocina*, *P. aeruginosa*, *Thermomonospora fusca*, *Thermoactinomyces vulgaris*, *Botrytis cinerea*. *F. Solani pisi*, *F. oxysporum*, *F. roseum culmorum*, *F. roseum sambucinum*, *Colletotrichum kahawae*, *C. gloeosporioides*, *C. capsica*, *Monilinia fructicola*, *Venturia inaequalis*, *Alternaria brassicicola*, *Helminthosporium sativum*, *Phytophthora cactorum*, and *Ulocladium consortiale* (Dutta et al., 2009).

13.5.8 Microbial dehydrogenases

Dehydrogenases are oxidoreductase enzymes that accelerate the transport of hydrogen from the substrate to several electron carriers like NAD, FMN, and NADP, which are found in a wide variety of species, including humans, bacteria, yeast, plants, and animals. Microbial alcohol dehydrogenase is an enzyme that catalyzes the reversible conversion of alcohol to aldehyde or ketone. It can be divided into two types: (a) NAD^+- or $NAD(P)^+$-dependent dehydrogenases and (b) NAD^+- or $NAD(P)^+$-independent enzymes that use cofactors such as heme, cofactor F420, and pyrroloquinoline quinone. Similar to this, aldehyde dehydrogenase catalyzes the oxidation of aldehyde to carboxylic acid under the control of $NAD(P)^+$. Dehydrogenases are one of the most significant enzymes in the soil environment and are used as a measure of the total soil microbial activity. Utilizing flavin adenine dinucleotide as a cofactor, 3-ketosteroid Δ^1-dehydrogenase catalyzes the 1(2)-dehydrogenation of 3-ketosteroid substrates. By initiating the opening of the steroid nucleus, the enzyme plays a key role in microbial steroid breakdown under both aerobic and anaerobic conditions (Rohman and Dijkstra, 2021).

Microbial source: *Sphingomonas* sp., *Stenotrophomonas maltophilia*, *Azoarcus evansii*, *Rhodococcus* sp., *P. putida*, *Amycolatopsis tucumanensis*, *Aromatoleum aromaticum*, and *Azoarcus* sp. (Bhandari et al., 2021).

13.5.9 Microbial tyrosinases

Tyrosinases belong to the oxidoreductase family and are copper-containing multifunctional metalloenzymes performing the main function of providing dark pigmentation to the skin of animals. Tyrosinase performs a hydroxylation reaction of monophenol to o-diphenol, followed by its oxidation to form o-quinones, which further aggregate to form dark blackish pigments. It is also responsible for the conversion of L-tyrosine to L-DOPA, required for melanin synthesis. Tyrosinase is more helpful in the removal of toxic phenols (chlorophenols, catechols, etc.) and dyes (Direct Orange 39, Direct Blue 71, Reactive Black 5, etc.) from the environment through their oxidation. This enzyme is primarily employed in phenol biosensors, in the breakdown of phenols and amines, in the elimination of xenobiotic substances, in catechol oxidation, and in phenolic compound polymerization (Rao et al., 2010).

Microbial source: *Streptomyces glaucescens*, *N. crassa*, *A. bisporus*, *Aspergillus nidulans*, *T. reesei*, *Mycothermus thermophilus*, *Thermomicrobium roseum*, *Trametes sanguinea*, *Bacillus thuringiensis*, *Streptomyces polyantibioticus*, *Pseudomonas maltophilia*, and *Sinorhizobium meliloti* (El-Sayed et al., 2021).

13.5.10 Microbial nitrilases

The hydrolysis of nitrile (RCN) molecules into the appropriate carboxylic acid and ammonia is catalyzed by nitrilase enzymes (nitrilases). These enzymes have been found in the genomes of mammals, yeast, and plants, and they have been identified and described in plants, bacteria, and fungi. More than 30 nitrilases have been described since the first nitrilase activity was found in plants in 1958 and in bacteria in 1964. The nonpeptide carbon-nitrogen bonds are hydrolyzed by enzymes that belong to the nitrilase superfamily, commonly known as CN hydrolases. As a result of these enzymes hydrolyzing a nitrile compound's CN group, the matching carboxylic acid is synthesized and ammonia is released. Nitrilases are classified into one of three categories according to substrate specificity: aliphatic nitrilases, which predominantly work on aliphatic nitriles such cyanoalanine, glutaronitrile, and acrylonitrile; aromatic and heterocyclic nitrilases, which specifically target benzonitrile and cyanopyridine; and arylacetonitrilases, which mainly act on indole-3-acetonitrile (IAN), phenylacetonitrile, and phenylpropionitrile (Howden and Preston, 2009).

Microbial source: *K. pneumoniae* sp. *Ozaenae, B. pallidus, Rhodococcus rhodochrous, P. putida, E. coli, Fusarium solani, Alcaligenes faecalis, Pyrococcus abyssi, Exophiala oligosperma, Acidovorax facilis, Streptomyces* sp., and *B. thuringiensis* (Verma and Sangave, 2015).

13.6 Present scenario for microbial remediation of enzymes

13.6.1 Immobilization of Enzymes

Enzyme immobilization is the process of anchoring enzyme molecules to or within an appropriate strong stable support to confine them to a different phase other than the substrate and product (Rao et al., 2014). The immobilizing support, also called carrier matrices, chosen for immobilization should have the following characteristics: inertness, stability, enhanced enzyme specificity, reduction of product inhibition, and high renderability (Khan and Alzohairy, 2010). Immobilization on different carrier matrices can occur through adsorption on ion exchange matrices/porous carbon, etc., covalent binding to agarose/cellulose, etc., entrapment through cellulose acetate/acrylamide, etc., or membrane confinement (Rao et al., 2014). This enzyme-based advanced tool provides long-term, repeated usability and operational stability to the enzymes (Sharma et al., 2018).

13.6.2 Nanotechnology-based nanozymes

An efficient, productive, and novel concept of nanozymes has been developed as a result of advancements in both nanotechnology and enzyme technology to remediate pollutants in the environment (Kumar and Bharadvaja, 2019). Nanozymes are next-generation artificial enzymes that mimic the catalytic activity of natural enzymes with similar mechanisms and kinetic profiles (Gao and Yan, 2016). A nanozyme-based technique in bioremediation is believed to become a more pronounced approach due to high stability, cost-effectiveness, reusability, and faster and more efficient functioning (Zhang et al., 2021). Nanozymes can simultaneously recognize and eliminate contaminants. Based on mimicking behavior, Wong et al. (2021) grouped nanozymes into four categories: type I nanozymes (that include metalloenzymes of metal oxides or sulfides), type II nanozymes (that include nonmetallic or functional nanozymes of graphene oxide, carbon dots, and nanotubes), type III nanozymes (that include metal-carbon nanocomposites), and type IV nanozymes (that include structural mimics). There are several efforts being taken for designing nanozymes with optimized catalytic functioning.

13.6.3 Enzyme engineering and related genetic approach

To dispose of pollutants from the environment, bioremediation using microorganisms is a desirable alternative to traditional methods. Over the past few years, advances in molecular technology and genetic and microbiological studies have allowed us to provide the driving force needed to develop genetically engineered microbes and enzymes related to bioremediation (Chen et al., 1999). In recent years, several genetic engineering techniques have been designed, which have been effective in introducing relevant biodegrading features in enzymes. The initial excitement in enzyme engineering should continue to open the door for the rational or irrational design of enzymes with unique biodegradable properties (Singh et al., 2021). Biocatalyst designing based on modification of amino acid sequences to increase enzymatic efficiency, production, and stability toward various stresses is the desired goal of enzyme engineering and recombinant DNA technology in enzyme-mediated remediation (Sharma et al., 2018; Saravanan et al., 2021). Genetic engineering generates recombinant enzymes with a higher production in the host and an enhanced degradation rate for xenobiotic compounds. Sutherland et al. (2004) studies revealed the enzyme discovery and product development along with therapeutical aspects of microbial remediating enzymes and also given three stages of enzyme development for the biodegradation process: (1) identification of the main enzyme from the native place; (2) isolation of the enzyme coding region of the gene; and, finally, (3) scale-up pollutant remediating performance of enzymes (Fig. 2).

13.6.4 Limitations of microbial enzymes in pollutant removal

Although enzyme-mediated bioremediation is a highly promising technology, it has several shortcomings, including the nonapplicability of enzymes to a wide range of pollutants, undesirable results of enzymes on active polluted sites, intensive process of isolation and purification of remediating enzymes, etc. Some specific enzymes like microbial hydrolases can only be used in the bioremediation of nonbiodegradable materials, possibly as

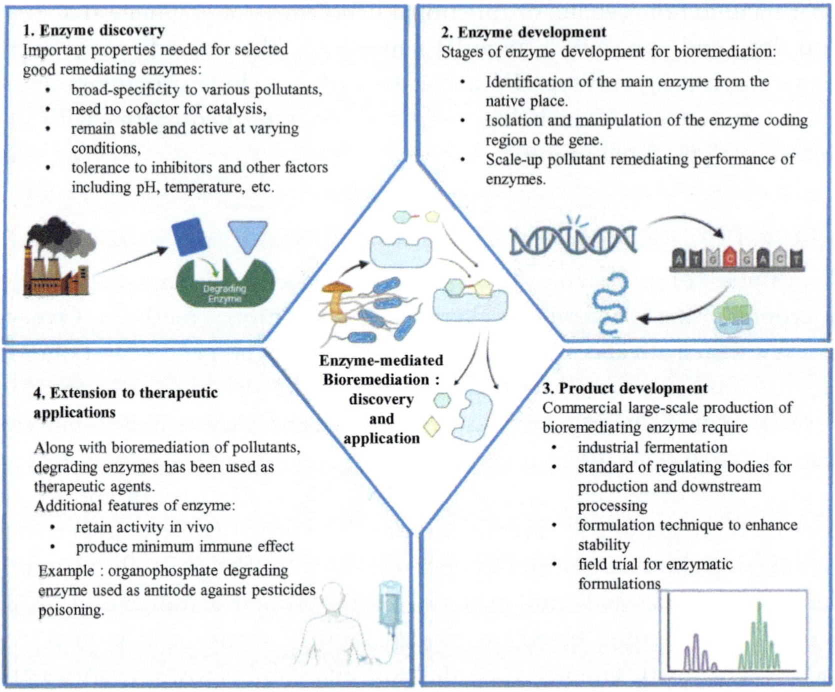

Fig. 2 Enzyme discovery and product development along with the therapeutical aspects of microbial remediating enzymes.

a result of their unique characteristics. Using lipases or esterases for biodegradation often takes more time than other types of treatments (Sharma et al., 2019). Furthermore, a result of a mutation in enzymes encoding genes of microbial lipases or in esterases hydrolyzing harmful toxicants can lead to the generation of more toxic products than the original molecule. Therefore, further study is needed to undertake bioremediation employing microbial enzymes to create a healthy and green planet. Fig. 3 shows some limitations of the enzyme-mediated bioremediation process of environmental pollutants.

Extracellular enzymes cannot proliferate and increase their number like microbes to consume a significant amount of substrate material. Since enzymes cannot reproduce, an increase in their abundance must come from somewhere else in the system. Sometimes, it has been observed that enzymes become nonfunctional after interacting with some pollutants due to loss in their reactive state (Gianfreda and Rao, 2004). Enzyme kinetics must be

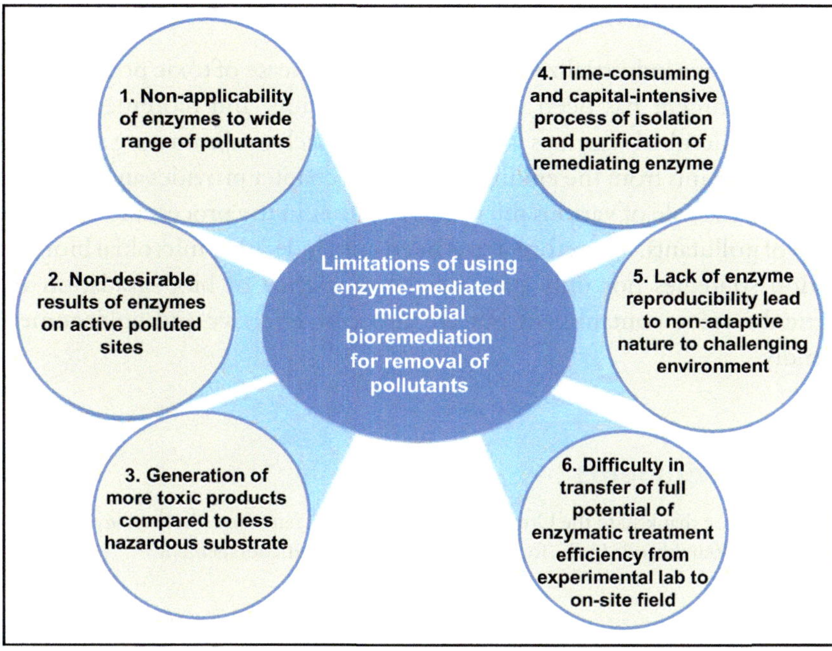

Fig. 3 Several limitations and drawbacks of enzyme-mediated bioremediation to clean up environmental pollutants.

designed in accordance with site-specific variables since the enzymatic concentration depends on the reaction rate. Enzymes are not able to cope under dynamic and mutagenic environments as microbes adapt themselves to challenging environments. Enzymatic tolerance toward changing environments is retained only within certain limits of their adaptability. Extracellular enzymes are expensive to use in bioremediation, which is the major disadvantage (Durán and Esposito, 2000). Because enzymatic purity is directly concerned with their specialized functions and eliminates negative consequences, enzymes used in bioremediation must be produced in its pure form, which comes at an extremely high cost. Enzyme production in its raw form is somewhat less expensive, but it frequently has undesirable outcomes (Vishnoi and Dixit, 2019). The use of less expensive substrates for the growth of enzyme-synthesizing microbial biomass can reduce the manufacturing costs of bioremediation. The enhancement of bacterial and fungal biomass levels utilized in the production of enzymes is the focus of future technological innovations (Gianfreda and Rao, 2004; Vishnoi and Dixit, 2019).

13.7 Conclusions

Due to massive industrialization processes, the release of toxic pollutants into the environment has been immensely increasing, and bioremediation by means of microbial enzymes is a valuable approach to eliminating and detoxifying pollutants from the environment. This chapter provides an insight into the immense role of various microbial enzymes in the process of bioremediation of pollutants. Therefore, it has been concluded that microbial bioremediation strategies not only increase the efficiency of biodegradation of a particular toxic contaminant but are also cost-effective and environment-friendly.

Acknowledgments

The authors are thankful to the Uttarakhand Council of Science and Technology (UCOST), Dehradun, Uttarakhand, India, for financial support and encouragement.

References

Ahuja, S.K., Ferreira, G.M., Moreira, A.R., 2004. Utilization of enzymes for environmental applications. Crit. Rev. Biotechnol. 24 (2–3), 125–154.

Ajaz, M., Elahi, A., Rehman, A., 2018. Degradation of azo dye by the bacterium, *Alishewanella* sp. CBL-2 isolated from industrial effluent and its potential use in decontamination of wastewater. J. Water Reuse Desalin. 8 (4), 507–515.

Aktar, K., Zerin, T., Banik, A., 2019. Biodegradation of textile dyes by bacteria isolated from textile industry effluents. Stamford J. Microbiol. 9 (1), 5–8.

Arora, S., Saxena, S., Sutaria, D., Sethi, J., 2022. Bioremediation: an ecofriendly approach for the treatment of oil spills. In: Advances in Oil-Water Separation. Elsevier, pp. 353–373.

Aslam, S., Hussain, A., Qazi, J.I., 2019. Production of cellulase by *Bacillus amyloliquefaciens*-ASK11 under high chromium stress. Waste Biomass Valoriz. 10 (1), 53–61.

Azubuike, C.C., Chikere, C.B., Okpokwasili, G.C., 2016. Bioremediation techniques—classification based on site of application: principles, advantages, limitations and prospects. World J. Microbiol. Biotechnol. 32 (11), 1–18.

Bala, S., Garg, D., Thirumalesh, B.V., Sharma, M., Sridhar, K., Inbaraj, B.S., Tripathi, M., 2022. Recent strategies for bioremediation of emerging pollutants: a review for a green and sustainable environment. Toxics 10 (8), 484.

Bansal, N., Kanwar, S.S., 2013. Peroxidase(s) in environment protection. Sci. World J. 2013, 714639.

Bhandari, S., Poudel, D.K., Marahatha, R., Dawadi, S., Khadayat, K., Phuyal, S., Shrestha, S., Gaire, S., Basnet, K., Khadka, U., Parajuli, N., 2021. Microbial enzymes used in bioremediation. J. Chem. 2021, 1–17.

Bharagava, R.N., Saxena, G., Mulla, S.I., 2020. Introduction to industrial wastes containing organic and inorganic pollutants and bioremediation approaches for environmental management. In: Bioremediation of Industrial Waste for Environmental. Springer, Singapore, pp. 1–18.

Bhatt, P., Pathak, R., Bhatt, K., 2019. System biology, simulation, and network analysis of enzymes in waste removal from the environment. In: Smart Bioremediation. Academic Press, pp. 347–358.

Cameron, M.D., Timofeevski, S., Aust, S.D., 2000. Enzymology of *Phanerochaete chrysosporium* with respect to the degradation of recalcitrant compounds and xenobiotics. Appl. Microbiol. Biotechnol. 54 (6), 751–758.

Chakroun, H., Mechichi, T., Martinez, M.J., Dhouib, A., Sayadi, S., 2010. Purification and characterization of a novel laccase from the ascomycete *Trichoderma atroviride*: application on bioremediation of phenolic compounds. Process Biochem. 45 (4), 507–513.

Chandra, P., Singh, R., Arora, P.K., 2020. Microbial lipases and their industrial applications: a comprehensive review. Microb. Cell Factories 19 (1), 1–42.

Chen, W., Brühlmann, F., Richins, R.D., Mulchandani, A., 1999. Engineering of improved microbes and enzymes for bioremediation. Curr. Opin. Biotechnol. 10 (2), 137–141.

Chen, Y.F., Chao, H., Zhou, N.Y., 2014. The catabolism of 2, 4-xylenol and p-cresol share the enzymes for the oxidation of para-methyl group in *Pseudomonas putida* NCIMB 9866. Appl. Microbiol. Biotechnol. 98 (3), 1349–1356.

Di Gennaro, P., Galli, E., Albini, G., Pelizzoni, F., Sello, G., Bestetti, G., 1997. Production of substituted naphthalene dihydrodiols by engineered *Escherichia coli* containing the cloned naphthalene 1, 2-dioxygenase gene from *Pseudomonas fluorescens* N3. Res. Microbiol. 148 (4), 355–364.

Durán, N., Esposito, E., 2000. Potential applications of oxidative enzymes and phenoloxidase-like compounds in wastewater and soil treatment: a review. Appl. Catal. B Environ. 28 (2), 83–99.

Durve, A.A., Gupta, A.R., Naphade, S.R., 2012. Decolourisation of textile dyes and biological stains by bacterial strains isolated from industrial effluents. Adv. Appl. Sci. Res. 3 (5), 2660–2671.

Dutta, K., Sen, S., Veeranki, V.D., 2009. Production, characterization and applications of microbial cutinases. Process Biochem. 44 (2), 127–134.

El-Sayed, A.S., Maamoun, H.S., Rabie, G.H., Shaker, I., Alaidaroos, B.A., Ali, M.G., 2021. Microbial tyrosinase: biochemical, molecular properties and pharmaceutical applications. Biomed. Pharmacol. J. 14 (3), 1281–1295.

Gangola, S., Sharma, A., Bhatt, P., Khati, P., Chaudhary, P., 2018. Presence of esterase and laccase in *Bacillus subtilis* facilitates biodegradation and detoxification of cypermethrin. Sci. Rep. 8 (1), 1–11.

Gao, L., Yan, X., 2016. Nanozymes: an emerging field bridging nanotechnology and biology. Sci. China Life Sci. 59 (4), 400–402.

Ghosh, A., Dastidar, M.G., Sreekrishnan, T.R., 2017. Bioremediation of chromium complex dyes and treatment of sludge generated during the process. Int. Biodeterior. Biodegrad. 119, 448–460.

Gianfreda, L., Rao, M.A., 2004. Potential of extra cellular enzymes in remediation of polluted soils: a review. Enzym. Microb. Technol. 35 (4), 339–354.

Hayaishi, O., Katagiri, M., Rothberg, S., 1955. Mechanism of the pyrocatechase reaction. J. Am. Chem. Soc. 77 (20), 5450–5451.

Hayashida, S., Yoshioka, H., 1980. Production and purification of thermostable cellulases from *Humicola insolens* YH-8. Agric. Biol. Chem. 44 (8), 1721–1728.

Horne, I., Sutherland, T.D., Harcourt, R.L., Russell, R.J., Oakeshott, J.G., 2002. Identification of an opd (organophosphate degradation) gene in an *Agrobacterium* isolate. Appl. Environ. Microbiol. 68 (7), 3371–3376.

Howden, A.J., Preston, G.M., 2009. Nitrilase enzymes and their role in plant–microbe interactions. Microb. Biotechnol. 2 (4), 441–451.

Janusz, G., Pawlik, A., Sulej, J., Świderska-Burek, U., Jarosz-Wilkołazka, A., Paszczyński, A., 2017. Lignin degradation: microorganisms, enzymes involved, genomes analysis and evolution. FEMS Microbiol. Rev. 41 (6), 941–962.

Jayasekara, S., Ratnayake, R., 2019. Microbial cellulases: an overview and applications. In: Cellulose. IntechOpen, pp. 1–21.

Karigar, C.S., Rao, S.S., 2011. Role of microbial enzymes in the bioremediation of pollutants: a review. Enzyme Res. 2011, 805187.

Khan, A.A., Alzohairy, M.A., 2010. Recent advances and applications of immobilized enzyme technologies: a review. Res. J. Biol. Sci. 5 (8), 565–575.

Kim, Y.H., Ahn, J.Y., Moon, S.H., Lee, J., 2005. Biodegradation and detoxification of organophosphate insecticide, malathion by *Fusarium oxysporum* f. sp. pisi cutinase. Chemosphere 60 (10), 1349–1355.

Kumar, A., Arora, P.K., 2022. Biotechnological applications of manganese peroxidases for sustainable management. Front. Environ. Sci. 10, 875157.

Kumar, L., Bharadvaja, N., 2019. Enzymatic bioremediation: a smart tool to fight environmental pollutants. In: Smart Bioremediation Technologies. Academic Press, pp. 99–118.

Latip, W., Knight, V.F., Abdul Halim, N., Ong, K.K., Mohd Kassim, N.A., Wan Yunus, W.M.Z., Mohamad Ali, M.S., 2019. Microbial phosphotriesterase: structure, function, and biotechnological applications. Catalysts 9 (8), 671.

Lee, A.H., Kang, C.M., Lee, Y.M., Lee, H., Yun, C.W., Kim, G.H., Kim, J.J., 2016. Heterologous expression of a new manganese-dependent peroxidase gene from *Peniophora incarnata* KUC8836 and its ability to remove anthracene in *Saccharomyces cerevisiae.* J. Biosci. Bioeng. 122 (6), 716–721.

Malla, M.A., Dubey, A., Yadav, S., Kumar, A., Hashem, A., AbdAllah, E.F., 2018. Understanding and designing the strategies for the microbe-mediated remediation of environmental contaminants using omics approaches. Front. Microbiol. 9, 1132.

Megharaj, M., Naidu, R., 2017. Soil and brownfield bioremediation. Microb. Biotechnol. 10 (5), 1244–1249.

Mondal, M., Halder, G., Oinam, G., Indrama, T., Tiwari, O.N., 2019. Bioremediation of organic and inorganic pollutants using microalgae. In: New and Future Developments in Microbial Biotechnology and Bioengineering. Elsevier, pp. 223–235.

Mousavi, S.M., Hashemi, S.A., Iman Moezzi, S.M., Ravan, N., Gholami, A., Lai, C., Behbudi, G., 2021. Recent advances in enzymes for the bioremediation of pollutants. Biochem. Res. Int. 2021, 5599204.

Mulbry, W.W., Eaton, R.W., 1991. Purification and characterization of the N-methylcarbamate hydrolase from *Pseudomonas* strain CRL-OK. Appl. Environ. Microbiol. 57 (12), 3679–3682.

Muthukamalam, S., Sivagangavathi, S., Dhrishya, D., Sudha Rani, S., 2017. Characterization of dioxygenases and biosurfactants produced by crude oil degrading soil bacteria. Braz. J. Microbiol. 48, 637–647.

Pande, V., Pandey, S.C., Sati, D., Bhatt, P., Samant, M., 2022. Microbial interventions in bioremediation of heavy metal contaminants in agroecosystem. Front. Microbiol. 13, 824084.

Perelo, L.W., 2010. In situ and bioremediation of organic pollutants in aquatic sediments. J. Hazard. Mater. 177 (1–3), 81–89.

Piontek, K., Smith, A.T., Blodig, W., 2001. Lignin peroxidase structure and function. Biochem. Soc. Trans. 29 (2), 111–116.

Pointing, S.B., 2001. Feasibility of bioremediation by white rot fungi. Appl. Microbiol. Biotechnol. 57, 20–33.

Prakash, D., Gabani, P., Chandel, A.K., Ronen, Z., Singh, O.V., 2013. Bioremediation: a genuine technology to remediate radionuclides from the environment. Microb. Biotechnol. 6 (4), 349–360.

Rao, M.A., Scelza, R., Scotti, R., Gianfreda, L., 2010. Role of enzymes in the remediation of polluted environments. J. Soil Sci. Plant Nutr. 10 (3), 333–353.

Rao, M.A., Scelza, R., Acevedo, F., Diez, M.C., Gianfreda, L., 2014. Enzymes as useful tools for environmental purposes. Chemosphere 107, 145–162.

Rodríguez, A., Castrejón-Godínez, M.L., Salazar-Bustamante, E., Gama-Martínez, Y., Sánchez-Salinas, E., Mussali-Galante, P., Tovar-Sánchez, E., Ortiz-Hernández, M., 2020. Omics approaches to pesticide biodegradation. Curr. Microbiol. 77 (4), 545–563.

Rohman, A., Dijkstra, B.W., 2021. Application of microbial 3-ketosteroid Δ1-dehydrogenases in biotechnology. Biotechnol. Adv. 49, 107751.

Roy, D.C., Biswas, S.K., Sheam, M.M., Hasan, M.R., Saha, A.K., Roy, A.K., Tang, S.S., 2020. Bioremediation of malachite green dye by two bacterial strains isolated from textile effluents. Curr. Res. Microb. Sci. 1, 37–43.

Ruggaber, T.P., Talley, J.W., 2006. Enhancing bioremediation with enzymatic processes: a review. Pract. Period. Hazard. Toxic Radioact. Waste Manage. 10 (2), 73–85.

Saravanan, A., Kumar, P.S., Vo, D.V.N., Jeevanantham, S., Karishma, S., Yaashikaa, P.R., 2021. A review on catalytic-enzyme degradation of toxic environmental pollutants: microbial enzymes. J. Hazard. Mater. 419, 126451.

Sayali, K., Sadichha, P., Surekha, S., 2013. Microbial esterases: an overview. Int. J. Curr. Microbiol. Appl. Sci. 2 (7), 135–146.

Sharma, T., Sharma, A., Kanwar, S.S., 2017. An overview on esterases: structure, classification, sources and their application. Recent Adv. Biotechnol. 2, 216–229.

Sharma, B., Dangi, A.K., Shukla, P., 2018. Contemporary enzyme-based technologies for bioremediation: a review. J. Environ. Manag. 210, 10–22.

Sharma, A., Sharma, T., Sharma, T., Sharma, S., Kanwar, S.S., 2019. Role of microbial hydrolases in bioremediation. In: Microbes and Enzymes in Soil Health and Bioremediation, Microorganisms for Sustainability. Springer Nature, pp. 149–164.

Shekher, R., Sehgal, S., Kamthania, M., Kumar, A., 2011. Laccase: microbial sources, production, purification, and potential biotechnological applications. Enzyme Res. 2011, 217861.

Shin, W.R., Um, H.J., Kim, Y.C., Kim, S.C., Cho, B.K., Ahn, J.Y., Kim, Y.H., 2021. Biochemical characterization and molecular docking analysis of novel esterases from *Sphingobium chungbukense* DJ77. Int. J. Biol. Macromol. 168, 403–411.

Singh, J.S., Singh, D.P., 2017. Methanotrophs: an emerging bioremediation tool with unique broad spectrum methanotrophs. In: Agro-Environmental Sustainability. 2. Springer, pp. 1–18.

Singh, S.K., Singh, M.K., Singh, V.K., Modi, A., Jaiswal, P., Rashmi, K., Kumar, A., 2021. Microbial enzymes and their exploitation in remediation of environmental contaminants. In: Microbe Mediated Remediation of Environmental Contaminants. Woodhead Publishing, pp. 59–71.

Singh, R., Pandey, A., Sagar, N.A., Kumar, N., 2022. Advances in dye removal technologies from wastewater by microbes. In: Synergistic Approaches for Bioremediation of Environmental Pollutants: Recent Advances and Challenges. Academic Press, pp. 185–197.

Sutherland, T.D., Horne, I., Weir, K.M., Coppin, C.W., Williams, M.R., Selleck, M., Oakeshott, J.G., 2004. Enzymatic bioremediation: from enzyme discovery to applications. Clin. Exp. Pharmacol. Physiol. 31 (11), 817–821.

Thatoi, H., Das, S., Mishra, J., Rath, B.P., Das, N., 2014. Bacterial chromate reductase, a potential enzyme for bioremediation of hexavalent chromium: a review. J. Environ. Manag. 146, 383–399.

Verma, V., Sangave, P., 2015. Screening, isolation and identification of nitrilase producing bacteria from soil. Int. J. Pharm. Sci. Res. 6 (5), 1950–1997.

Verma, S., Saxena, J., Prasanna, R., Sharma, V., Nain, L., 2012. Medium optimization for a novel crude-oil degrading lipase from *Pseudomonas aeruginosa* SL-72 using statistical approaches for bioremediation of crude-oil. Biocatal. Agric. Biotechnol. 1 (4), 321–329.

Vishnoi, N., Dixit, S., 2019. Bioremediation: new prospects for environmental cleaning by fungal enzymes. In: Recent Advancement in White Biotechnology Through Fungi. Springer, Cham, pp. 17–52.

Wong, E.L., Vuong, K.Q., Chow, E., 2021. Nanozymes for environmental pollutant monitoring and remediation. Sensors 21 (2), 408.

Xu, P., Liu, L., Zeng, G., Huang, D., Lai, C., Zhao, M., He, Y., 2014. Heavy metal-induced glutathione accumulation and its role in heavy metal detoxification in *Phanerochaete chrysosporium*. Appl. Microbiol. Biotechnol. 98 (14), 6409–6418.

Xu, F., Mou, Z., Geng, J., Zhang, X., Li, C.Z., 2016. Azo dye decolorization by a halotolerant exoelectrogenic decolorizer isolated from marine sediment. Chemosphere 158, 30–36.

Zhang, R., Yan, X., Fan, K., 2021. Nanozymes inspired by natural enzymes. Acc. Mater. Res. 2 (7), 534–547.

CHAPTER 14

Applications of microbial formulations in the pharmaceutical industry

Amita Sharma[a] **and Prabhjot Kaur**[b]
[a]Department of Botany, Kurukshetra University, Thanesar, Haryana, India
[b]Department of Nephrology, Postgraduate Institute of Medical Education and Research (PGIMER), Chandigarh, Punjab, India

14.1 Introduction

A microbial formulation refers to a combination of one or more microbial strains/species conjugated with a safe, economical, and easy-to-use material as the carrier. Formulation is the industrial or laboratory process of unifying microbes with a bacterial strain carrier such that the carrier will deliver the right amount of physiologically active cells at the right time, thus ultimately enhancing the effectiveness of the microbial action. A carrier (a) should be nearly sterilized, (b) have good water-holding capacity, (c) should be physically or chemically uniform, (d) have a high content of organic matter, and (e) should be suitable for a wide range of microbial strains and pathogens. Selecting the type of formulation is a crucial step that will determine the success or failure of the desired application. The material of choice should be capable of delivering the appropriate amount of viable and physiologically active cells and be economically affordable and easy to use.

A wide variety of carrier materials are used for making formulations, viz., cellulose powder, charcoal, farmyard manure, peat, talc, vermicompost, bagasse, etc. As bioformulations have to withstand different stresses during production and storage, microbes must possess high survival rates and sustainable capabilities to show efficiency for extended durations (Ferreira and Castro, 2005; Zayed, 2016; Rani and Kumar, 2019). Microbial formulations have a wide range of applications in a myriad of commercial sectors like food preservation, agriculture, and phytoremediation, as biocontrol agents, and in medicine (Singh et al., 2016; Ohadi et al., 2020; Radaic et al., 2020). Numerous microbial products or microbes are utilized as formulations with valuable pharmacological functions. Utilization of microbial

Microbial Inoculants
https://doi.org/10.1016/B978-0-323-99043-1.00015-3

products like biosurfactants, bacteriocins, and polymers in formulations needs intensive evaluation and is catching the eye of the scientific community very well. Probiotic formulations are also widely explored to gain insights into their therapeutic applications.

14.2 Probiotic formulations

Probiotics use is considered a modern-day approach to curb and cure infections. Probiotic therapy is based on reaping health benefits by applying useful bacteria against the harmful bacteria of the digestive tract. In the past decade, modifications of the gut microflora with use of probiotic strains therapeutically have been proven to be remarkably beneficial. Improvement of the gut microbiota is considered as a promising approach in the context of end-stage renal disease (ESRD), as in some cases positive outcomes were documented (Di Cerbo et al., 2013). Probiotics are normally sold as nutritional/food supplements and pharmaceutical products. Studies have shed light on the role of probiotics in the maintenance of health, prevention/treatment of diseases, improvement of immunity, and reduction in the risk of developing diseases in future (Iannitti and Palmieri, 2010; Sreeja and Prajapati, 2013). Probiotics play a role in improving cognitive health and feminine health, handling lactose intolerance, managing cholesterol levels, and treating Crohn's disease and diarrhea by restoring the intestinal microbiota (Iannitti and Palmieri, 2010; Reiff and Kelly, 2010; Dinan et al., 2013; Govender et al., 2014; Reid et al., 2015). Probiotic strains act by different mechanisms like antagonizing gastrointestinal pathogens, creating competition for nutrients and receptor sites, producing hydrogen peroxide, and exhibiting antimicrobial activity to stimulate the immune system (Bendali et al., 2011; Nicoleta et al., 2015). Studies have revealed the evidence-based efficacy of probiotics in the treatment of diarrhea due to antibiotics and *Clostridium difficile* (Rondanelli et al., 2017). Furthermore, probiotic supplementation has been proven to be efficient in the eradication of *Helicobacter pylori* (Zagari et al., 2018).

However, the status of probiotics as drugs is still unclear. The availability of more evidences and further clinical trials will determine the therapeutic status of probiotics. Stringent regulations are needed for the establishment of probiotics as drugs since the efficacy of probiotic strains is restricted to few microbial species. The plausible role of probiotics in restoring the gastrointestinal microbiota cannot be undermined, but a subset of the population might not get the benefits of probiotics due to their clinical conditions like gut disorders, diabetes mellitus, malignancy, etc. Furthermore, probiotics may turn into harmful pathogens and take advantage of the weak immunity of the at-risk population. These facts call for a strict regulation and use of probiotic interventions with utmost care in these high-risk categories of population, including immunocompromised, allergic, critically ill, elderly, and pregnant patients (Kothari et al., 2019).

Recently, studies have highlighted the concept of tailoring probiotic microbial strains by genetic engineering to gain insights into the therapeutic potential of these strains and to resolve the doubts pertaining to use of these probiotics as drugs (Wells., 2011; Mazhar et al., 2020). Genetically modified (GM) probiotic strains have several advantages over conventional probiotic strains, such as targeted action, lower delivery cost, increased shelf life, better tolerance to stress, etc. GM probiotics have proven their efficacy as anticancer therapeutics because of their precise targeting of tumor cells (Heap et al., 2014). Thus, recombinant probiotic microbes are emerging as promising candidates for the prevention and treatment of human clinical conditions, but it is indispensable to assess the risk and safety of probiotic strains under both in vitro and in vivo conditions (Mazhar et al., 2020; Torres et al., 2016).

14.3 Biosurfactants

Biosurfactants are mostly anionic or neutral agents with surface-active properties. They may be cationic in nature. Biosurfactants are chemicals derived from natural forms, primarily obtained from fungi (yeast or filamentous fungi) and bacteria. They are secondary metabolites that can be obtained from microbes by precipitation, extraction, and distillation processes (Banat et al., 2010; Varvaresou and Iakovou, 2015; Santos et al., 2016). They are diverse compounds owing to their different microbial origins, cultivation conditions, and substrates for growth. They are potential candidates to be utilized in the cosmetic industry, as they are natural surfactants with significant advantages over their synthetic counterparts. They possess unique properties, viz., biodegradability, low toxicity, emulsification, detergency function, wetting, human skin compatibility, and stability under extreme conditions, including pH, salinity, and temperature. In addition to these, they can be synthesized from renewable and cheaper resources. In the recent decade, they have been gaining immense attention for their wide range of commercial applications in the oil, cosmetic, textile, food, agricultural, pharmaceutical industries, etc.

Biosurfactants can be categorized into low-molecular-weight compounds and high-molecular-weight compounds. They are further classified based on their chemical composition. Lipopeptides, glycopeptides, glycolipids, and phospholipids are classes of low-molecular-weight compounds, whereas high-molecular-weight compounds include lipoproteins, protein-polysaccharide-fatty acid complexes and protein-lipopolysaccharide-fatty acid complexes. Glycolipids are considered the most promising class of biosurfactants in the food, cosmetic, and pharmaceutical industries (Lukic et al., 2016; Kubicki et al., 2019). In recent years, the demand of lipopeptides has also been increasing due to their antitumor and antimicrobial properties (Meena and Kanwar, 2015; Patel et al., 2015). Biosurfactants possess antiviral, antifungal, antibacterial, and antibiofilm properties (Naughton et al., 2019).

Biosurfactants are used in capsule and tablet formulations owing to their wetting effect, thus enhancing both wetting and drug disintegration so as to improvise dissolution. Biosurfactants are potential candidates for topical formulations like gels, lotions, creams, etc., for better emulsification, thereby enhancing penetration (Ghafourian et al., 2015). A novel gel formulation from *Nigella sativa* L. seeds was developed to determine its potential as an effective antimicrobial agent against some bacteria-causing acne. Topical gel formulations containing ethyl acetate extract of *Nigella* sp. seeds inhibited the growth of *Staphylococcus aureus* and *Pseudomonas acnes* with significant efficacy (Nawarathne et al., 2019). Moreover, biosurfactants improve the performance or efficacy of pharmaceutical products besides enhancing the chemical and physical characteristics of pharmaceutical formulations (Koehl et al., 2020).

Recently, biosurfactants have been proving their potential in the development of drug delivery systems (DDSs). To overcome the challenge of improving the oral bioavailability of hydrophobic drugs, an microemulsion drug delivery system (MDSS) is considered a promising strategy. Microemulsions formulated with biosurfactants have several benefits over their chemical counterparts, including reduced toxicity and gastric irritation (Rodrigues, 2015). Biosurfactants show brilliant self-assembling, better emulsification, active stabilization, hydrophobic or hydrophilic linking, and safe delivery in MDSSs (Ohadi et al., 2020).

14.4 Bacteriocins

Bacteriocins are proteinaceous substances with antimicrobial activity at specified concentrations against pathogenic microorganisms. They are multifunctional proteins produced by bacteria and some members of archaea. They are categorized into three major classes based on their structural and physiochemical attributes: class I (lantibiotics), class II, and class III bacteriocins. Bacteriocins produced by *Pediococcus acidilactici* BA28 are used in formulations of topical care therapy with positive outcomes in the treatment of peptic ulcers (Kaur et al., 2012). Bacteriocins have great biotechnological potential and have recently propelled their way into a wide range of commercial sectors like the food industry, pharmaceutical industry, agriculture, etc. Bacteriocins act against harmful microbes by affecting gene expression and targeting the cell envelope, thus ultimately inhibiting their growth (Negash and Tsehai, 2020).

The past 2 decades have witnessed the emergence of a number of multidrug-resistant pathogens, leading to public health problems globally. As per the declaration of the Centers for Disease Control and Prevention (CDC), the human race has moved into the postantibiotic era and the World Health Organization (WHO) has forewarned about the dire consequences of an antibiotic resistance crisis (Michael et al., 2014; Ovchinnikov et al., 2020).

Bacteriocins have unique attributes that make them potential candidates to be therapeutically used as novel antimicrobials in human and veterinary health. Their beneficial attributes include: (a) less cytotoxicity, (b) potency against pathogens under both in vitro and in vivo conditions, (c) prospective in situ production by probiotics, (d) less environmental persistence, and (e) possibility of tailoring bacteriocin genes by genetic engineering (Hols et al., 2019; Schofs et al., 2020). They have established their applications in oral care, skin care, woman care, anticancer treatment, etc. Bacteriocins are addressed as designer drugs owing to their specificity to target pathogens.

Staphylococcus is one of the common pathogens causing community-acquired and nosocomial infections. An antibacterial peptide nisin, synthesized by *Lactococcus lactis*, has been approved by the FDA for therapeutic purposes to be used in dairy cattle as a disinfectant. This formulation is marketed as wipe out as it was found to be effective in the prevention of infections caused by pathogens like *S. aureus*, *Streptococcus dysgalactiae*, and *Streptococcus uberis* (Ahmad et al., 2017). Kranjec et al. (2020) successfully indicated that a combination of the bacteriocins garvicin KS and micrococcin P1 has the therapeutic potential to act against biofilms produced by methicillin-resistant *S. aureus* (MRSA) under in vitro conditions. Additionally, this bacteriocin-based antimicrobial formulation restored the sensitivity of infectious strains against beta-lactam antibiotics. Ovchinnikov et al. (2020) developed a therapeutic three-component formulation of the bacteriocins garvicin KS and micrococcin P1 with penicillin G, to treat skin infections of a murine model caused by MRSA. Collectively, these studies suggest that bacteriocins are promising candidates with immense potential to be used as therapeutic tools to combat biofilm-related MRSA infections.

Bacteriocins represent a better alternative to the use of antibiotics and to tackle the global challenge of multidrug resistance. They have a GRAS status and therefore are gaining immense consideration as they are safe for human consumption. However, large-scale utilization of bacteriocin formulations need to overcome some limitations, including a narrow spectrum of action, low yield, high-dose requirement, and sensitivity to proteolytic enzymes. Recent developments in nanotechnology have shown a prospective approach to enhance bacteriocin properties and overcome the hurdles in the path of their industrial-scale utilization (Sidhu and Nehra, 2017; Sulthana and Archer, 2021). The association of bacteriocins with nanoparticles and encapsulation of bacteriocins enhance the antimicrobial range of bacteriocins, thus resulting in increased survival, protection from proteolytic enzymes, and effectiveness against pathogens (Fig. 1). Thus, the combination of nanotechnology and biotechnology is the need of the hour to encourage commercial production of bacteriocins and to deal with multidrug resistance.

In a study, a combination of silver nanoparticles with polymyxin B enhanced its activity against gram-negative bacterial pathogens. Polymyxin B in conjugation with nanoparticles showed efficient removal of toxins, thereby causing inhibition of biofilm formation on surgical blades

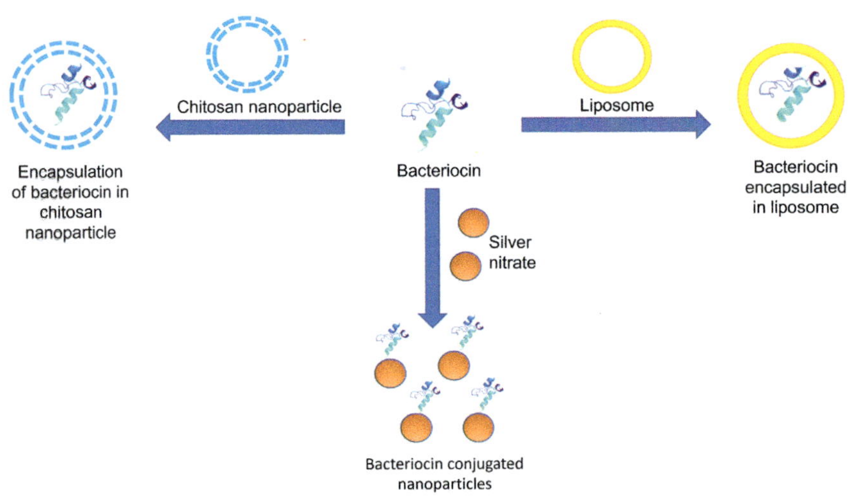

Fig. 1 Bacteriocin conjugation/encapsulation with nanoparticles.

(Baptista et al., 2018). In a similar study, synergy between silver nanoparticles and bacteriocin isolated from *Lactobacillus paracasei* against MDR bacteria like *Pseudomonas aeruginosa*, *Staphylococcus aureus*, *Proteus vulgaris*, *Escherichia coli*, etc. was reported (Gomaa, 2019). Thus, bacteriocins have the capability of revolutionizing the biomedical industry as a potential replacement for current antibiotics, which are currently threatening mankind due to proliferation of multidrug-resistant microbial strains. Nanoformulated bacteriocins provide an attractive option with enhanced stability and a wide spectrum of antimicrobial action. Nanoscale drug delivery systems have the potential to overcome drawbacks of conventional bacteriocin action, and they facilitate their action by improving pharmacodynamics and pharmacokinetics. Different bacteriocins that are associated with drug delivery systems include nisin, bacteriocin-like substance P34, plantaricin 432, enterocin, pediocin PA-1, etc. (Radaic et al., 2020).

In several studies, bacteriocins are delivered by encapsulation with liposomes. Liposomes are biodegradable, spherical vesicles comprising one or more phospholipid bilayers. Nanosized liposomes, i.e., nanoliposomes, represent an excellent vehicle for stability and targeted delivery of several compounds, including bacteriocins (Sidhu and Nehra, 2017).

14.5 Therapeutic microbes

A live biotherapeutic is a live microorganism that is capable of preventing or curing human diseases. Therapeutic microbes or live biotherapeutic products (LBPs) are emerging as a new therapeutic domain to prevent and treat gastrointestinal diseases. They interact with the mucosal surfaces of the gut, thus targeting the microbiota or molecules produced by pathogens that are responsible for their clinical state (Fig. 2). Many strains have been identified to be used as therapeutic interventions like *E. coli*, *Bacillus* sp., *Weissella* sp., etc. (O'Toole et al., 2017). *Lactobacillus* and *Lactococcus* species have been genetically altered to come up with a novel platform for delivery of desired therapeutic compounds (Plavec and Berlec, 2019). Live APIs (active pharmaceutical ingredients) obtained from microbes have huge therapeutic potential as they contain all the machinery of live cells, thus allowing molecular interactions, self-renewal, and adaptive responses. Since APIs are live cells, they show different pharmacological processes like distribution, metabolism, absorption, excretion, etc. They are exposed to a number of external stresses during production, storage, and drug delivery. Precise targeting of APIs to the disease site greatly heightens the efficiency of microbial

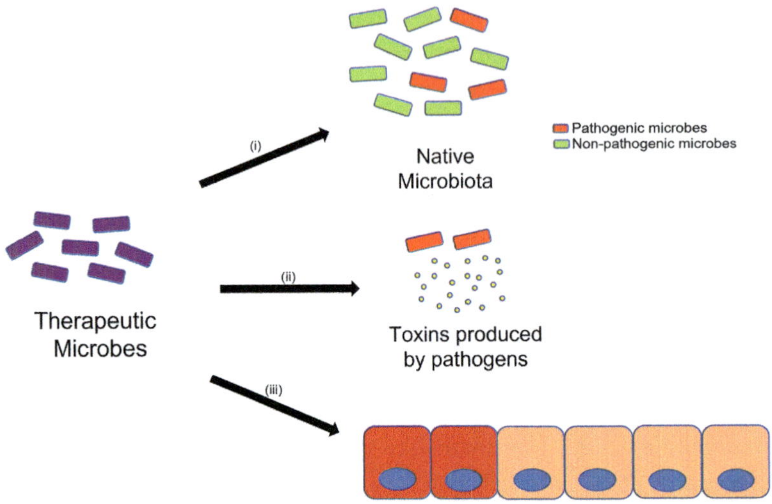

Fig. 2 Therapeutic microbes deal with (i) the native microbiota, (ii) molecules produced by pathogens, and (iii) infected cells/tissues.

therapeutics. Target delivery can be achieved by utilizing different materials that are mucus-penetrating, mucoadhesive, colon-targeting, etc. (Mimee et al., 2016; Kyzioł et al., 2020).

In a study, encapsulation of the LBP model *Lactobacillus casei* ATCC 393 in alginate beads (cross-linked with calcium chloride) was systemically evaluated for improved characteristics. It was suggested that encapsulation did not interfere with microbial action and it further enhanced several properties like survival, acid resistance, sustained release, etc. (Jimenez et al., 2019; Qiu et al., 2021). The manipulation of live microbial cells to treat or prevent human diseases can be successfully achieved through synthetic biology. Microbes offer unique features for creation of synthetic circuits like easy manipulation, robustness and simple metabolism, etc. Microbial diversity represents a vast repertoire that serves as an inexhaustible resource with novel properties for generating synthetic genetic circuits. Microbial therapeutics are a cutting-edge technology for addressing various unresolved health issues. Fighting and preventing infectious diseases, elimination of tumor cells, treatment of metabolic disorders, and delivery of therapeutically significant molecules are some specific achievements of engineered microbes, establishing the fact that live microbial therapeutics represent next-generation therapeutics (Riglar and Silver, 2018; Lim and Song, 2019; Cubillos-Ruiz et al., 2021).

14.6 Conclusions

Microbe-based formulations are gaining considerable attention as they represent a safe and effective alternative to current antibiotics and other medicinal approaches. Probiotics are considered as magic bullets owing to their vast applications in medical care, but they are still marketed as supplements instead of drugs. The reason for this is that all probiotics cannot be given the GRAS status, and, therefore, strict regulation must be exercised to launch probiotics as drugs in the market. Tailoring of conventional microbes by genetic engineering has come up with modern tactics to address shortcomings, and these engineered microbes show precise targeting and many other positive aspects. Biocontainment of the designed bacteria is one of the major concerns to prevent escape of recombinant bacteria. Furthermore, gene transfer from a recombinant strain to a wild-type strain is another challenge that needs to be addressed. Increased characterization of a vast repertoire of microbial genes with integration of modern techniques will enable us to target urgent and unmet needs. Continued research related to microbial therapeutics will facilitate advancements in the biomedical field and in the development of tools that can conduct precise investigations.

References

Ahmad, V., Khan, M.S., Jamal, Q., Alzohairy, M.A., Al Karaawi, M.A., Siddiqui, M.U., 2017. Antimicrobial potential of bacteriocins: in therapy, agriculture and food preservation. Int. J. Antimicrob. Agents 49 (1), 1–11.

Banat, I.M., Franzetti, A., Gandolfi, I., Bestetti, G., Martinotti, M.G., Fracchia, L., Smyth, T. J., Marchant, R., 2010. Microbial biosurfactants production, applications and future potential. Appl. Microbiol. Biotechnol. 87 (2), 427–444.

Baptista, P.V., McCusker, M.P., Carvalho, A., Ferreira, D.A., Mohan, N.M., Martins, M., Fernandes, A.R., 2018. Nano-strategies to fight multidrug resistant bacteria—"a battle of the titans". Front. Microbiol. 9, 1441.

Bendali, F., Durand, A., Hébraud, M., Sadoun, D., 2011. Lactobacillus paracasei subsp paracasei: an Algerian isolate with antibacterial activity against enteric pathogens and probiotic fitness. J. Food Nutr. Res. 50 (3), 139–149.

Cubillos-Ruiz, A., Guo, T., Sokolovska, A., Miller, P.F., Collins, J.J., Lu, T.K., Lora, J.M., 2021. Engineering living therapeutics with synthetic biology. Nat. Rev. Drug Discov. 20 (12), 941–960.

Di Cerbo, A., Pezzuto, F., Palmieri, L., Rottigni, V., Iannitti, T., Palmieri, B., 2013. Clinical and experimental use of probiotic formulations for management of end-stage renal disease: an update. Int. Urol. Nephrol. 45 (6), 1569–1576.

Dinan, T.G., Stanton, C., Cryan, J.F., 2013. Psychobiotics: a novel class of psychotropic. Biol. Psychiatry 74 (10), 720–726.

Ferreira, E.M., Castro, I.V., 2005. Residues of the Cork industry as carriers for the production of legume inoculants. Silva Lusitana 13, 159–167.

Ghafourian, T., Nokhodchi, A., Kaialy, W., 2015. Surfactants as penetration enhancers for dermal and transdermal drug delivery. In: Percutaneous Penetration Enhancers Chemical Methods in Penetration Enhancement. Springer, Berlin/Heidelberg, Germany, pp. 207–230.

Gomaa, E.Z., 2019. Synergistic antibacterial efficiency of bacteriocin and silver nanoparticles produced by probiotic *Lactobacillus paracasei* against multidrug resistant bacteria. Int. J. Pept. Res. Ther. 25 (3), 1113–1125.

Govender, M., Choonara, Y.E., Kumar, P., du Toit, L.C., van Vuuren, S., Pillay, V., 2014. A review of the advancements in probiotic delivery: conventional vs. non-conventional formulations for intestinal flora supplementation. AAPS PharmSciTech 15 (1), 29–43.

Heap, J.T., Theys, J., Ehsaan, M., Kubiak, A.M., Dubois, L., Paesmans, K., Van Mellaert, L., Knox, R., Kuehne, S.A., Lambin, P., Minton, N.P., 2014. Spores of Clostridium engineered for clinical efficacy and safety cause regression and cure of tumors in vivo. Oncotarget 5 (7), 1761–1769.

Hols, P., Ledesma-García, L., Gabant, P., Mignolet, J., 2019. Mobilization of microbiota commensals and their bacteriocins for therapeutics. Trends Microbiol. 27 (8), 690–702.

Iannitti, T., Palmieri, B., 2010. Therapeutical use of probiotic formulations in clinical practice. Clin. Nutr. (Edinburgh, Scotland) 29 (6), 701–725.

Jimenez, M., Langer, R., Traverso, G., 2019. Microbial therapeutics: new opportunities for drug delivery. J. Exp. Med. 216 (5), 1005–1009.

Kaur, B., Garg, N., Sachdev, A., Kumar, B., Mittu, B., Chauhan, A., 2012. Isolation and molecular characterization of anti-Helicobacter pylori. Bacteriocin producing Pediococcus acidilactici BA28. Open Access Sci. Rep. 1, 1–6.

Koehl, N.J., Holm, R., Kuentz, M., Jannin, V., Griffin, B.T., 2020. Exploring the impact of surfactant type and digestion: highly digestible surfactants improve oral bioavailability of Nilotinib. Mol. Pharm. 17 (9), 3202–3213.

Kothari, D., Patel, S., Kim, S.K., 2019. Probiotic supplements might not be universally-effective and safe: a review. Biomed. Pharmacother. 111, 537–547.

Kranjec, C., Ovchinnikov, K.V., Grønseth, T., Ebineshan, K., Srikantam, A., Diep, D.B., 2020. A bacteriocin-based antimicrobial formulation to effectively disrupt the cell viability of methicillin-resistant Staphylococcus aureus (MRSA) biofilms. NPJ Biofilms Microbiomes 6 (1), 58.

Kubicki, S., Bollinger, A., Katzke, N., Jaeger, K.E., Loeschcke, A., Thies, S., 2019. Marine biosurfactants: biosynthesis, structural diversity and biotechnological applications. Mar. Drugs 17 (7), 408.

Kyzioł, A., Khan, W., Sebastian, V., Kyzioł, K., 2020. Tackling microbial infections and increasing resistance involving formulations based on antimicrobial polymers. Chem. Eng. J. 385.

Lim, D., Song, M., 2019. Development of bacteria as diagnostics and therapeutics by genetic engineering. J. Microbiol. (Seoul, Korea) 57 (8), 637–643.

Lukic, M., Pantelic, I., Savic, S., 2016. An overview of novel surfactants for formulation of cosmetics with certain emphasis on acidic active substances. Tenside Surfactant Deterg. 53 (1), 7–19.

Mazhar, S.F., Afzal, M., Almatroudi, A., Munir, S., Ashfaq, U.A., Rasool, M., Raza, H., Munir, H.M.W., Rajoka, M.S.R., Khurshid, M., 2020. The prospects for the therapeutic implications of genetically engineered probiotics. J. Food Qual. 2020. 9676452, 11 pp.

Meena, K.R., Kanwar, S.S., 2015. Lipopeptides as the antifungal and antibacterial agents: applications in food safety and therapeutics. Biomed. Res. Int. 2015, 473050.

Michael, C.A., Dominey-Howes, D., Labbate, M., 2014. The antimicrobial resistance crisis: causes, consequences, and management. Front. Public Health 2, 145.

Mimee, M., Citorik, R.J., Lu, T.K., 2016. Microbiome therapeutics—advances and challenges. Adv. Drug Deliv. Rev. 105 (Pt A), 44–54.

Naughton, P.J., Marchant, R., Naughton, V., Banat, I.M., 2019. Microbial biosurfactants: current trends and applications in agricultural and biomedical industries. J. Appl. Microbiol. 127 (1), 12–28.

Nawarathne, N.W., Wijesekera, K., Wijayaratne, W., Napagoda, M., 2019. Development of novel topical cosmeceutical formulations from *Nigella sativa* L. with antimicrobial activity against acne-causing microorganisms. SciWorldJ 2019, 5985207.

Negash, A.W., Tsehai, B.A., 2020. Current applications of bacteriocin. Int. J. Microbiol. 2020, 1–7.

Nicoleta, A., Monica, B., Gabriela, B., 2015. Probiotics and therapeutic effect in clinical practice—review. Roman. Biotechnol. Lett. 20 (1), 10162–10175.

Ohadi, M., Shahravan, A., Dehghannoudeh, N., Eslaminejad, T., Banat, I.M., Dehghannoudeh, G., 2020. Potential use of microbial surfactant in microemulsion drug delivery system: a systematic review. Drug Des. Devel. Ther. 14, 541–550.

O'Toole, P.W., Marchesi, J.R., Hill, C., 2017. Next-generation probiotics: the spectrum from probiotics to live biotherapeutics. Nat. Microbiol. 2, 17057.

Ovchinnikov, K.V., Kranjec, C., Thorstensen, T., Carlsen, H., Diep, D.B., 2020. Successful development of bacteriocins into therapeutic formulation for treatment of MRSA skin infection in a murine model. Antimicrob. Agents Chemother. 64 (12). e00829-20.

Patel, S., Ahmed, S., Eswari, J.S., 2015. Therapeutic cyclic lipopeptides mining from microbes: latest strides and hurdles. World J. Microbiol. Biotechnol. 31 (8), 1177–1193.

Plavec, T.V., Berlec, A., 2019. Engineering of lactic acid bacteria for delivery of therapeutic proteins and peptides. Appl. Microbiol. Biotechnol. 103 (5), 2053–2066.

Qiu, K., Huang, Y., Anselmo, A.C., 2021. Polymer and crosslinker content influences performance of encapsulated live biotherapeutic products. Cell. Mol. Bioeng. 14 (5), 487–499.

Radaic, A., de Jesus, M.B., Kapila, Y.L., 2020. Bacterial anti-microbial peptides and nano-sized drug delivery systems: the state of the art toward improved bacteriocins. J. Control. Release 321, 100–118.

Rani, U., Kumar, V., 2019. Microbial bioformulations: Present and future aspects. In: Prasad, R., Kumar, V., Kumar, M., Choudhary, D. (Eds.), Nanobiotechnology in Bioformulations. Nanotechnology in the Life Sciences. Springer, Cham.

Reid, G., Brigidi, P., Burton, J.P., Contractor, N., Duncan, S., Fargier, E., Hill, C., Lebeer, S., Martín, R., McBain, A.J., Mor, G., O'Neill, C., Rodríguez, J.M., Swann, J., van Hemert, S., Ansell, J., 2015. Microbes central to human reproduction. Am. J. Reprod. Immunol. (New York, NY: 1989) 73 (1), 1–11.

Reiff, C., Kelly, D., 2010. Inflammatory bowel disease, gut bacteria and probiotic therapy. Int. J. Med. Microbiol. 300 (1), 25–33.

Riglar, D.T., Silver, P.A., 2018. Engineering bacteria for diagnostic and therapeutic applications. Nat. Rev. Microbiol. 16 (4), 214–225.

Rodrigues, L.R., 2015. Microbial surfactants: fundamentals and applicability in the formulation of nano-sized drug delivery vectors. J. Colloid Interface Sci. 449, 304–316.

Rondanelli, M., Faliva, M.A., Perna, S., Giacosa, A., Peroni, G., Castellazzi, A.M., 2017. Using probiotics in clinical practice: where are we now? A review of existing meta-analyses. Gut Microbes 8 (6), 521–543.

Santos, D.K., Rufino, R.D., Luna, J.M., Santos, V.A., Sarubbo, L.A., 2016. Biosurfactants: multifunctional biomolecules of the 21st century. Int. J. Mol. Sci. 17 (3), 401.

Schofs, L., Sparo, M.D., Sánchez Bruni, S.F., 2020. Gram-positive bacteriocins: usage as antimicrobial agents in veterinary medicine. Vet. Res. Commun. 44 (3–4), 89–100.

Sidhu, P.K., Nehra, K., 2017. Bacteriocin-nanoconjugates as emerging compounds for enhancing antimicrobial activity of bacteriocins. J. King Saud Univ. Sci. 31 (4), 758–767.

Singh, J.S., Koushal, S., Kumar, A., Vimal, S.R., Gupta, V.K., 2016. Book review: microbial inoculants in sustainable agricultural productivity—vol. II: functional application. Front. Microbiol. 7, 2105.

Sreeja, V., Prajapati, J.B., 2013. Probiotic formulations: application and status as pharmaceuticals—a review. Probiotics Antimicrob. Proteins 5 (2), 81–91.

Sulthana, R., Archer, A.C., 2021. Bacteriocin nanoconjugates: boon to medical and food industry. J. Appl. Microbiol. 131 (3), 1056–1071.

Torres, L., Krüger, A., Csibra, E., Gianni, E., Pinheiro, V.B., 2016. Synthetic biology approaches to biological containment: pre-emptively tackling potential risks. Essays Biochem. 60 (4), 393–410.

Varvaresou, A., Iakovou, K., 2015. Biosurfactants in cosmetics and biopharmaceuticals. Lett. Appl. Microbiol. 61 (3), 214–223.

Wells., 2011. Mucosal vaccination and therapy with genetically modified lactic acid bacteria. Annu. Rev. Food Sci. Technol. 2 (1), 423–445.

Zagari, R.M., Romiti, A., Ierardi, E., Gravina, A.G., Panarese, A., Grande, G., Savarino, E., Maconi, G., Stasi, E., Eusebi, L.H., Farinati, F., Conigliaro, R., Bazzoli, F., Romano, M., 2018. The "three-in-one" formulation of bismuth quadruple therapy for Helicobacter pylori eradication with or without probiotics supplementation: efficacy and safety in daily clinical practice. Helicobacter 23 (4), e12502.

Zayed, M.S., 2016. Advances in formulation development technologies. In: Singh, D., Singh, H., Prabha, R. (Eds.), Microbial Inoculants in Sustainable Agricultural Productivity. Springer, New Delhi.

CHAPTER 15

An evaluation of the lacunae in current techniques using microbial inoculants for enhanced bioremediation and nutrient recovery

N. Jennifer Michellin Kiruba and Agnieszka Saeid
Faculty of Chemistry, Department of Engineering and Technology of Chemical Processes, Wroclaw University of Science and Technology, Wrocław, Poland

15.1 Background

15.1.1 Waste biorefinery and circular bioeconomy

The development of technologies involving bioremediation or any other clean energy solutions inevitably converges with certain terms, such as circular bioeconomy, waste streams, waste biorefinery, and waste stream valorization. The common principle behind bioremediation is the application of a living organism, such as a plant or microorganism, to break down or remove toxic compounds, turning them into nontoxic derivatives, to clean up a polluted environment. While the whole technology was proposed as an optional alternative for other physical and chemical techniques applied to remove contaminants, nowadays it has turned into a necessity. The Organization for Economic Co-operation and Development (OECD) produced a scenario report in 2007 via their International Futures Programme (IFP) on the bioeconomy, which considered the probable trajectory of economic values and associated biological products and services and labeled the set of activities surrounding it, including invention, production, research and development, and industrial processes. The focus of this report was dependent on the economic trends observed in the healthcare sector and the influence of biotechnology on it. Later, the product market was enlarged to accommodate diverse sectors such as agriculture, environment, and energy to drive up the factors of sustainability and ecoefficiency in the face of burgeoning population.

https://doi.org/10.1016/B978-0-323-99043-1.00008-6

Growing population leads to increasing solid wastes and wastewater, which in turn requires more waste management mobilization and technology, and, on the other hand, leading up to resource scarcity. According to the estimates of Worldmeter (2022), the total human population living as of February 2022 was approximately 7.9 billion. In Europe alone, more than 5 ton of wastes were generated per inhabitant, whereas the complete waste generated between 2010 and 2018 increased by 5%, which was more than 114 million tons (Namlis and Komilis, 2019). Issues such as environmental stability, resource scarcity, and food security abound globally, as the population is expected to reach more than 10 billion by 2050. Managing wastes and integrating them into a circular economy rather than a linear economy is a challenge and a sustainable option, especially due to the ever-growing populace and urbanization. A circular economy can be described as a closed-loop system or a cradle-cradle concept, where resources are repetitively recycled without being released back into the environment as toxic substances. The transition of the current economic activities to a circular bioeconomy could be considered as the next possible Industrial Revolution, like that of the change-over from coal to petrochemicals. The European Commission proposed the concept of "bioeconomy" for formulating policies and defined it as the emerging role of biotechnology in contributing to the world economy in the form of knowledge and applications in the field of primary production, healthcare, and renewable biomass for environmental sustainability (Tait et al., 2007).

In a circular bioeconomy, biomass from biological sources is fed in a cascading economic system via waste streams, where it is efficiently utilized for the sustainable generation of high value-added products such as bioenergy, biomaterials, food, and feed. The economic and sustainable approach is dependent on the concepts of recycle, reuse, and remanufacture of innovative biological products. One of the major advantages of this approach is its use of waste valorization from various sources such as agricultural, food, and marine, among others. A highly efficient waste management system can be developed from the concept of biorefining, which is one of the foremost strategies in the circular bioeconomy. The collusion of the two concepts results in a waste biorefinery wherein industrial bioprocesses are developed to produce highly marketable and sustainable biomaterials, bioactive compounds and metabolites, and biofuels that greatly reduce carbon footprint, from biomass resources from waste streams. This strategy greatly mitigates resource scarcity, and safeguards the environment from further exploitation, providing an economy that is low in carbon emissions (Leong et al., 2021).

Several types of waste biorefineries have been studied regarding biomass from waste streams, such as food processing industries like breweries and bakeries, the paper and pulp industry, byproducts from agroindustrial processes, fisheries, aquaculture, agriculture, forestry, and sludge from wastewaters. Castilla-Archilla et al. (2019) classifies biorefineries into three different generations, where the first generation utilizes corn, maize, sugarcane, and other crops as raw materials for the generation of bioproducts such as ethanol, polyhydroxyalkanoates, and biopolymers through fermentation of feedstock without any pretreatment of biomass. The second generation biorefineries use agro-industrial wastes as feedstock, which has the distinct advantage of being a sustainable alternative to incineration of the wastes generating more carbon emissions. The feedstock procured from different industries could require varying degrees of pretreatment, depending upon the biodegradability of the substrate, such as lignocellulosic biomass and other simpler food waste. The third generation or tertiary biorefineries are unique in their own right, as they exploit the metabolism of cyanobacteria or microalgae for the generation of biofuels and other bioproducts, though this mode of production is extremely sustainable as cyanobacteria and microalgae act as carbon sinks and generate biomass on their own from sunlight with meager amounts of other sources of energy, in addition to their rapid periods of growth. But the economic feasibility of this biorefinery is subject to many questions and requires us to delve much more into their technology and design as well as their industrial implementation.

15.1.2 Waste streams

Typically, industrial processes in any biorefinery begin with the acquisition of some form of biomass, which is converted to a renewable energy for sustainability. Primary resources for biomass are often agricultural residues, which are easily available forms of parts of plants, such as seeds, leaves, stems, stalks, roots, and secondary sources include processed or nonprocessed dung, shells, husks, and bagasse. Residue biomass from agriculture is also a tertiary resource, which includes food and sewage waste (Bentsen, 2019). To properly define waste biorefineries, it is necessary to determine the type of biomass that is available for processing, depending on the industries and their respective waste streams. Bourguignon (2015) in his Briefing for the European Parliamentary Research Service, defines waste streams as the course of transfer of each particular type of waste from its origin in an industrial or domestic source, to disposal or other means such as recycle or reuse. The waste streams are further divided into two primary categories, which are material-based streams that include many waste types such as glass, metals, plastics, rubber, paper, wood, textiles, and specifically all types of biowaste, and product-based streams that typically include e-waste, batteries, accumulators, end-of-life vehicles, construction, mining, demolition waste, and packaging waste.

15.2 Current waste stream treatment techniques

15.2.1 Construction and demolition

The impact of construction and demolition waste (CDW) in terms of socioeconomic as well as environmental sustainability cannot be ignored, even though these wastes are a vital source of job opportunities and gross domestic product (GDP). The sector reckons a significant share of the GDP of Europe as well as Latin American at 6.3% and 5.7%, respectively (Ruiz et al., 2020). In Europe, the sector is alone responsible for more than 35% of total waste produced compared to all other commercial sectors, and generates nearly 900 million tons of waste, which pales in comparison to the 1.13 billion tons produced by China.

Compared to all other sectors, CDW is more adherent to the linear model of the economy, and often ends up in illegal dumpsites or landfills devoid of any environmental prevention measures, resulting in a massive carbon footprint in the form of water pollution, greenhouse gas emissions, and resource depletion (Sáez and Osmani, 2019). Adopting the concept of the circular economy, economic strategic frameworks have been proposed and accepted for sustainable material recovery and reduce CDW generation. One of the most influential economic frameworks implemented so far has been the Construction Waste Disposal Charging Scheme in Hong Kong, which works on the principle of making the polluters pay for it, wherein high landfill charges and associated waste disposal schemes are brought in to make the producer adopt recovery or reduction practices over disposal. CDW can be targeted by strategic changes to project design by focusing on waste prevention, by assembling a reuse and recycle market for recoverable resource material.

Sustainability in construction can also be achieved by using a tactical design for deconstruction or disassembly, which will in return enhance the market for recyclable or prefabricated construction materials. While applicability of this design is not quite at an implementable stage for all locations, if executed on a large scale, it is capable of reaping enormous economic profits due to its environmental implications. Similarly, embracing the usage of prefabricated construction primary and secondary resources has been shown to effectively reduce labor-exhaustive processes in the sector, thereby lessening the output of concrete and wood waste streams. In Europe, a CDW management protocol is commonly adopted that involves a waste management report (WMR) and a site waste management plan (SWMP) that provides data with regard to deconstruction or demolition measures, possible kinds of waste streams produced, and their treatment techniques, along with their treatment facility, mode of transport, and preventive steps to reduce CDW production. At this point, several reports in the literature do describe various theoretical blueprints to integrate the circular economy into CDW management via government policies and reforms (Baldassarre et al., 2019).

15.2.2 Plastic

The importance of the versatile polymer plastic cannot be denied, as its production is expected to double itself from the 300 million tons generated in 2015. By mass alone, the sector is responsible for the generation of more than 14% of the total solid waste produced across the globe. Unlike many other pollutants, plastics are known to exist in the environment for a long time, releasing microplastics that are heavily toxic and damaging to agriculture, biodiversity, and human health. The chronic harm to ocean and marine life due to the accumulation of plastic waste, leading to geological formations such as the Great Pacific Garbage patch, has led to the adoption of allocation of specific UN Sustainable Development Goals (SDG) with the sole purpose of targeting and enhancing sustainable plastic waste management resources in their member countries. In response, Europe has placed several initiatives on the table, including a ban on single-use disposable plastics, action plans for the CE, and other legislative agreements. But the necessity of plastic used in the form of medical supplies such as PPE, gloves, and masks, especially with regard to the COVID-19 pandemic, and the storage and supply of perishable and nonperishable foods cannot be avoided or reduced in the prospective future. The concept of waste hierarchy often indicates prevention is better than resource and recycle management. But overdependence on plastics throughout the global supply chains in all commercial sectors has a huge potential for reduction, while setting up an efficient framework based on CE is expected to provide a more viable solution to the plastic pandemic.

The CE action plan developed in 2015 addresses the recycling strategy required to tackle various forms of plastics across its supply chain, and approaches plastic waste management in two ways. The first strategy is to reduce reliance on the primary raw material for plastic production and replace it with recycled, secondary plastic products, whereas developing a closed loop economy based on the production of biodegradable bioplastics from the waste streams and energy recovery through incineration or combustion remains an economically feasible option for many countries (Hahladakis et al., 2020).

Currently four waste recycling processes in practice for plastic waste management can be differentiated into primary, secondary, tertiary, and quaternary recycling, as well as organic recycling. Primary recycling involves the conversion of industrial plastic waste that does not necessitate separate collection and segregation techniques. Secondary recycling is one of the most extensively used reprocessing techniques that utilizes mechanical methods, such as shredding, quenching, agglutination, and pelletization for decontamination and pretreatment of plastic wastes in the form of hard foams. The breakdown of different types of plastics, such as polyurethane, polyethylene, and polyvinyl chloride, into their monomeric units through depolymerization methods, such as thermolysis, hydrolysis, photodegradation, thermal cracking, pyrolysis, and gasification, is called tertiary recycling, otherwise known as chemical recycling. The resultant products from chemical recycling are supplied into another suitable and sustainable supply chain, and this is referred to as feedstock recycling.

One of the most widely used practices for plastic recycling is incineration, as open-space combustion of highly heterogeneous and poor-quality plastic wastes is often followed in underdeveloped or developing countries. This type of recycling is accompanied by air pollution due to the release of by-products formed from the burning of chlorines, bromines, and other halogens in the wastes, and the deposition of carbon pollutants into the residual soil at the dumpsites. This mode of combustion is also not in agreement with the concept of the circular economy. However, recovery of fuels such as refuse-driven fuel (RDF) from the organic portion of the waste is carried out by a process called mechanical biological treatment (MBT). This is a type of quaternary recycling, which also includes organic recycling, that is used for biological decomposition or composting of biodegradable plastic wastes. The organic recycling and energy recovery methodology in quaternary recycling seems to hold more potential for the circular economy due to the sustainable nature (Johansen et al., 2022).

15.2.3 Municipal solid waste (MSW)

Among all waste management systems or waste streams, the municipal solid waste stream is one of the most complicated to regulate and is multilayered in terms of governance but highly sustainable, even though it is not a major contributor to the global total waste generated. According to the CE action plan, this type of waste only constitutes around 10% of the total waste produced globally, but requires synergistic levels of management from various governing levels such as UN SDG for establishing policy frameworks, national governments to adopt them to their socioeconomic conditions, with the cooperation of the local governing authorities for setting up the collection, segregation, pretreatment, treatment, and transport of the wastes to the plants (Wielgosiński et al., 2021).

The principle of prevention of municipal solid waste can be complicated, as several studies show improvement of living standards and higher wages of the overall population increase consumption, thereby resulting in the production of additional municipal waste (Malinauskaite et al., 2017). Furthermore, higher living standards and wealth among the general public is also correlated with increased consumption of energy equivalent to more than 935 MTOE in 2019. This puts a heavy emphasis on the waste-to-energy (WtE) closed loop system that can readily benefit highly functional developed countries.

As countries of diverse socioeconomic scenarios endorse separate waste management operation for MSW, the EU has defined the concept for better understanding, which includes mixed wastes that are collected from markets, households, and other sources, which excludes sludge and sewage waste, construction and demolition waste, but does not mention food waste from any supply chain. In the allotted "waste hierarchy," despite the significance of WtE in the CE concept, the EU does not place emphasis on it as much as it encourages other MSW disposal systems, such as incineration and anaerobic digestion. The recommended technologies for MSW management by the EU are mainly production of solid-recovered fuels (SRFs) following incineration in combustion plants and cement kilns, using the consequently produced heat energy for temperature control, and for the production, distribution, and conversion of biomethane from biogas via anaerobic digestion (Velvizhi et al., 2020).

Gasification of SRF, pyrolysis, and anaerobic digestion are the important existing WtE technologies that help the completion of the recycling of the MSW waste steam, which is not possible by combustion or incineration alone. These WtE processes include biomass conversion processes along with thermochemical, biochemical, and physicochemical technologies that aid in the production of biogas, biomethane, bioethanol, heat, and electricity. The biomass waste recycling has been observed to positively impact greenhouse gas emissions, and is often referred to as the Organic Fraction of Municipal Solid Waste (OFMSW), which specifies the installation of biorefineries and biofuel generation for the appropriate recycling of biomass (Rolewicz-Kalińska et al., 2020). Further analysis and studies are required to put forth a strategy that can generate an efficient circular economy out of MSW.

15.2.4 Waste electric and electronic equipment (WEEE)

Nearly 50 million metric tons of waste from electrical and electronic equipment, otherwise referred to as e-waste, was generated across the globe in 2016, and is expected to reach a little less than 70 million metric tons in 2023, demonstrating an increase of 4%–5% every year. However, this industry maintains a high potential for CE application because of its composition of precious and rare earth elements (REE), which are critical raw material that holds high industrial demands for supply. The high demand for REE connects the need for recycling framework of the e-waste to the mining industry, as reprocessing e-waste for inexpensive but high-value secondary raw materials can be comparable to a profitable alternative to the urban mining industry. Unfortunately, as of 2017 estimates, only less than 20% of the total e-waste generated is properly segregated and guided into a recycling framework, which only amounts to a mere 8.9 million metric tons (Bressanelli et al., 2020). Moreover, similar to municipal waste, it has a direct correlation to the socioeconomic status of the population, since their expenditure ability leads to even more consumption of EEE.

Unlike other waste streams, the sustainability factor of e-waste management has been associated with the reduce and recycle strategies of 4Rs, rather than reuse and remanufacture. One of the main recommendations for reducing consumption of electrical and electronic equipment (EEE) is to design and produce equipment that is energy-efficient and lasts long or possesses longer lifetimes. As consumption is highly correlated to the high standards of living of the general public, recycling is considered a better alternative to pursue for long-term returns and sustainability. In contrast, mining of urban e-waste has several advantages, as it provides better yields of raw materials in comparison to conventional mining, and can help mitigate the environmental pollution and damage brought about by the latter. And the current narrative of the CE action plan requires the development of new, sustainable technologies that promote low-cost recovery and extraction of high-value raw materials and elements from WEEE.

15.2.5 Food waste biorefinery

Wasted food can be considered as the perfect example for emphasizing the need for the circular bioeconomy, because according to the Food and Agriculture Organization (FAO), more than one-third of the food harvested and produced is lost through its supply chain, which is directly proportional to the food scarcity, especially in view of the increasing population. Among the various categories of food industries, the beverage industry contributes around 26% of the total food waste generated, followed by the dairy, fruit and vegetable, and cereal industry, which generated 21%, 14.8%, and 12.9%, respectively. As per the United Nations Environment Programme (2021) report, copious amounts of foods being produced but wasted act as causative factors for the emissions of more than 10% of greenhouse gases. One of the targets of the UN Sustainable Development Goals (SDGs) is to track and reduce the kilograms per capita of global food waste generated at the country, retail, and consumer level by 50%. UNEP has also devised a framework to implement national strategies to measure the volume of food wastes generated at domestic as well as along the supply chains. Apart from the distinct wastes from each food industry and postharvest losses, food wastes are also generated to food spoilage, residual crops in the field, and leftovers from domestic and hospitality sectors across the supply chain.

Moreover, the impact of food waste is often reflected on a country's socioeconomic and environmental positions, as consumers of developed nations are associated with higher degrees of food wastage, whereas in developing nations, there are severe technological restrictions in the supply chain, resulting in food spoilage. According to the newly developed Emissions Database of Global Atmospheric Research (EDGAR-FOOD) database that tracks the global greenhouse gas (GHG) emissions of the global food supply chain, agrifood systems are major contributors, producing more than 16.5 billion tons of GHG emissions in 2019, with food supply chains overpowering deforestation, farming, and land use. While it is completely necessary to reduce the amount of food waste generated through policies, at the end of the day it is impossible to absolutely avoid food waste. Therefore, taking advantage of the homogeneous nature of the food wastes, it is possible to devise industrial biorefinery processes to produce bioproducts and biofuels. National and global food waste prevention campaigns and awareness via educational intervention can be promising to reduce food wastage; however, it is also essential to capitalize on the current food wastage to initiate a circular bioeconomy.

15.2.6 Wastewater management

Wastewater management can be considered a crucial sector, as most of the commercial industries depend on water supplies for their day-to-day functioning, and have been facing substantial stress due to water scarcity, affecting manufacturing volumes and financial gains. Industries also remain constant polluters of water due to the lack of proper wastewater management systems, causing disposal of effluents into marine and freshwater resources. Significant contributors of wastewater in 2019 were drainage from agricultural and other land use, along with industrial and urban sewage water. While more than 2000 km^3 of wastewater is being released back into the environment after industrial processes, a little less than 2000 km^3 is being taken up by agricultural uses every year. In the case of the European Union, the European Commission (EC) is responsible for bringing in wastewater management policies and reforms for the member states, and have taken up the concept of the circular economy (CE) to drive innovations to enhance water quality and reduce water scarcity.

Current strategies in wastewater management involve the 2R, 3R, and 4R methods, which can be considered as a progressive series of events that include reducing the volume of wastewater generated, engaging facilities for their reuse and recycling, and an additional step to close the loop in a linear economy model, recovery. The main two components of the wastewater generated in terms of management is the reclamation of the pollutants and recovery of the nutrients such as N, P, and K, as wastewater sludge is a major carrier of nutrients. Together with wastewater, the quantity of wastewater sludge has also been increasing consistently and remains an equally important population at the wastewater treatment plants (WWTPs). Sludge is the residual material that is generated during posttreatment in a WWTP, and is being generated at nearly 13 million tons per year in the European Union alone (Smol et al., 2020).

Even though there are several methods of wastewater treatment available, most WWTPs in the EU adopt a combination of physical, chemical, and biological wastewater treatment techniques in three different stages—primary, secondary, and tertiary. Treated water comes in two forms: potable, which is drinkable (domestic use), and nonpotable, which is specific for industrial use. Sludge and wastewater are rich sources of inorganic micronutrients such as nitrogen, phosphorus, and potassium, which are crucial components of organic fertilizers, and have been successfully recovered from WWTP leachates, sludge, and sewage ash. Taking into account the import dependency of Europe for fertilizers, the recovery of NPK from the domestic wastewater and sludge is highly advantageous to complete the loop in a circular economy. In the case of phosphorus recovery from wastewater or sludge, there are several physical and chemical methodologies, including crystallization, commercial processes such as PHOSPHAQ and AirPrex that extract phosphorus in the form of struvite, chemical precipitation that uses the concepts of adsorption and desorption, wet-chemical techniques such as Seaborne that involves addition of acids and alkali, and thermochemical treatments like AshDec that produces ashes. Often, application of treated sludge or wastewater has been deemed unfit for agricultural use due to the presence of concentrated heavy metals above the tolerable limits, even with a high yield of phosphorus. To remediate contamination of bioproducts due to heavy metals, further processing via ultrasonication, pyrolysis, microwave treatment, and gasification has also been evaluated. In addition, strategies that can enhance the bioavailability of the metals in bioproducts such as organic fertilizer are being assessed. Moreover, applying recovery techniques for wastewater treatment is expected to reduce the greenhouse gas emissions that contribute around 40% of the total emissions.

The recovery potential of wastewater and sludge is not limited to nutrients but is also applicable to heavy metals, biofuels like biogas, and biopolymers such as cellulose. Recovery of biological macromolecules such as carbohydrates, proteins, lipids, and other metabolic substrates is executed via processes such as protein precipitation, salt extraction, or a combination of one or two methods to improve the purity of the product. Similarly, one of the products that has been promising is the extraction and isolation of bacterial and fungal enzymes such as hydrolases, peroxidases, amylases, catalases, amylases, and proteases for industrial applications such as pharmaceuticals and detergents. Furthermore, WWTPs are a vast reservoir of bacterial species that are capable of producing biofilms, which is a rich source of polyhydroxyalkanoates (PHAs), a sustainable alternative to plastics. Production of bioplastics from wastewater and sludge is considered an efficient technique to implement and incorporate a circular economy (CE). Gaseous output from the WWTPs, such as biogas, hydrogen, and biofuels such as biodiesel, are also renewable energy alternatives produced through sludge treatment (Gherghel et al., 2019).

15.3 Role of microbial inoculants in waste stream treatment

As discussed earlier, biological treatments of various waste streams can be regarded as an appropriate strategic response to transfigure the linear economy to a circular or closed one effectively. Nevertheless, such technologies do not currently seem abundant. Despite the emergence of the circular economy as a key aspect for sustainability, the terminology and strategy around a "circular bioeconomy" is technically fresh and discussions regarding this are consistently few, and in the beginning stages. The significant chemical, physical, and biological techniques that are currently used to stratify waste management in diverse industrial processes have been discussed earlier, and therefore here we attempt to derive the role and function of microorganisms used for waste stream management so far, along with those that have been predicted to have potential for application.

15.3.1 Composting for organic waste management

In recent years, several studies have shown that the degradation of the biological or organic fraction of any waste stream to compost or humus using innate microbiome or additional microbial inoculants possesses an adequate advantage over other modes of treatment. The composting process is an age-old technique that employs conversion of organic wastes into humus by biological means, such as microorganisms with or without oxygen, which can be further used as a biofertilizer that improves the physicochemical parameters of the soil, including pH, porosity, and the capacity of the soil to suppress plant pathogens. It also increases the concentration of the organic carbon, which inadvertently helps the soil act as an improved carbon sink. While composting can be carried out in aerobic or anaerobic conditions, anaerobic digestion is more favorable due to the release of biogas, which is a suitable alternative for fossil fuels. On the other hand, the process of composting is complicated and complex due to the involvement of several accompanying parameters, such as the available levels of oxygen, temperature, C/N ratio, pH, porosity of soil, substrates, moisture content, and the microorganisms involved. Among the microbiological parameters, the microbial community is often changing, together with enzyme secretion, such as dehydrogenase activity. Moreover, the dynamics of the microbial populations and their synergy were studied by isolating and culturing them in specialized media, and several diverse species such as nonmycelial bacteria, obligate anaerobic bacteria, sulfur-reducing bacteria (SRB), fermentative bacteria, actinobacteria, and fungi, were identified for their functions in the composting process (Patchaye et al., 2018).

In any aerobic or anaerobic waste treatment, aerobic digestion, composting or anaerobic digestion could take place in four different phases that are microbially driven processes. In the primary phase, otherwise known as the mesophilic phase, primary decomposers made up of both heterogeneous fungi and bacteria degrade readily available, energy-rich macromolecules such as proteins and sugars, which generates a lot of heat and is thermodynamically favorable. The next phase is the thermophilic phase, where the temperature further increases up to 60°C, and thermophilic bacteria such as *Bacillus stearothermophilus*, *Thermus* sp., *Deinococcus* sp., and even nonsporing sulfur, as well as hydrogen-reducing bacteria commonly observed in hot springs such as *Hydrogenobacter* sp., have been identified. Similarly, it has also been noted that thermophilic actinobacteria and archaea could be involved in the process, following the destruction of normal mesophilic bacteria due to high temperatures. Since methanogenesis has also been detected in compost pits during this stage, the role of certain methanogenic archaea such as *Methanoculleus thermophilus*, *Methanosarcina thermophila*, *Methanobacterium formicicum*, and *Methanothermobacter* sp. have been studied after isolation and cultivation (Insam et al., 2010; Yamamoto and Nakai, 2019). Several studies have also reasserted the role of ammonia-oxidizing bacteria and archaea in the nitrogen transformation observed in composting of agricultural waste (Ren et al., 2018; Zhang et al., 2019).

Based on the decline of the availability of substrates, the activity of thermophilic microbiota also declines, and the temperature returns to the atmospheric temperature, which is marked by the recolonization of spore-forming mesophilic bacteria. The microbiota has a distinct community dominated by cellulolytic and ligninolytic bacteria such as *Clostridium*, *Cellulomonas*, *Nocardia*, and fungi such as *Fusarium*, *Paecilomyces*, and *Aspergillus* are also detected. These phases are referred to as the cooling phases, followed by the maturing and curing phases, the latter also being called the storage phase. There is a marked decline of the nitrifying bacteria, and arrival of species such as Mycobacteria and Verrucomicrobia (Insam et al., 2010).

However, as discussed in the Great Plate Count Anomaly, further information about the thermophilic actinobacteria and archaea is sorely lacking and requires considerable attention in research.

15.3.2 Anaerobic digestion reactors

As discussed previously, anaerobic digestion is favored due to the production of biogas, which is often undertaken for commercial purposes as green energy sources in specialized environments. Apart from biogas, the process also produces stabilized sludge as a by-product that is generally used as biofertilizer for agricultural lands, public parks, and range parks. The by-product is considered preferable to unprocessed manure due to the absence of malodor, decreased concentration of organic pollutants, and high bioavailability of nutrients. Extensive studies of the microbial community of the anaerobic reactors are frequently focused on the anaerobic bacterial phyla such as Proteobacteria, Actinobacteria, Firmicutes, Nitrospira, and Chloroflexi (Zhou et al., 2020; Pasalari et al., 2021). Most methanogens identified belong to Euryarchaeota, while microbiome studies using whole genome sequencing and loci-targeted sequencing have generated information about the presence of the Crenarchaeota and bacteria diversity that has not been actively studied via isolation and culturing. Enzyme-producing bacteria and fungi play a vital role in the anaerobic digestion process by decomposing complex organic molecules into simpler compounds that can then be further metabolized by other microorganisms to produce various bio-based products, such as biogas, a renewable energy source. Anaerobic bacteria and fungi that produce hydrolases, cellulases, esterases, and xylanases are important to the task of producing biogas in the presence of bulk polymers such as cellulase and lignin. Some of these bacteria and fungi that have been detected in different studies so far are *Clostridium*, *Ruminococcus*, *Bacteroides*, *Halocella*, *Acetivibrio*, *Neocallimastix*, *Piromyces*, *Orpimomyces*, and *Anaeromyces* (Insam et al., 2010; Zhou et al., 2020).

15.3.3 Microalgal biorefineries

Microalgae are photosynthetic prokaryotic or eukaryotic organisms that have high potential as third-generation feedstock for biofuels, as well as value-added products. However, the high cost associated with upstream and downstream processing of microalgae for the generation of biofuels such as biodiesel, biomethane, and biohydrogen or value-added products is not advantageous for economic feasibility. Apart from the value-added products such as biopolymers, polysaccharides, pigments, and biopeptides, microalgae have vast applications in the generation of biofertilizers, remediation of heavy metals, antibiotics, and nutrient recovery. The microalgal biorefinery approach applied for remediation of wastewaters aids the faster recovery of higher concentrations of nutrients such as nitrogen, phosphorus, potassium, and carbon, along with other micronutrients (Banu et al., 2020).

Several strains of microalgae have been identified through phycoprospecting for tolerance levels of different types of wastewaters such as agro-industrial, pharmaceutical, effluents, and domestic and municipal wastewater. A microalgal consortium consisting of indigenous species isolated from a wastewater treatment plant was compared with a mixture of *Neochloris oleoabundans* and *Chlorella vulgaris* for its efficiency in nutrient recovery from municipal wastewater, and the former was found to achieve more than 70% efficiency in phosphorus, nitrogen, and COD. Similar strains of high efficiency have been isolated and studied from sewage treatment plants, livestock and slaughterhouse wastewater, dairies, carpet industry, and rare earth element (REE) mining effluents. The microalgal strains included *Microspora*, *Oscillatoria*, *Tribonema*, *Stigeoclomium*, *Chlorella sorokinina*, *Scenedesmus*, *Chlamydomonas globosa*, and *Parachlorella*, which suggests that native microconsortia would reflect better on the environment in the long term (Nagarajan et al., 2020).

On a similar note, microalgal biorefineries of wastewater have also been studied for their capacity to remove heavy metals from tannery, electroplating, and mining industry effluents. Biosorption onto cellular surfaces, bioaccumulation, biotransformation, microprecipitation, complexation, and biosequestration into cellular organelles such as vacuoles, chloroplasts, or granular accumulation have been some common modes of action against an array of heavy metals such as cadmium (Cd), lead (Pb), chromium (Cr), mercury (Hg), arsenic (As), and other trace elements such as copper (Cu), nickel (Ni), and zinc (Zn) (Shen et al., 2020; Nagarajan et al., 2020).

15.4 Current nutrient recovery strategies

Review reports abound on the nutrient recovery technologies, and the challenges of nutrient recovery from different nutrient-rich wastewaters are available in the literature (Theregowda et al., 2019; Saliu and Oladoja, 2021; Gubernat et al., 2020; Bacelo et al., 2020; Rubaba et al., 2021; Vaneeckhaute et al., 2017; Campos et al., 2019). Recycling delivers multiple benefits by closing cycles and reducing waste, which translates into lower dependency on external resources, increasing the autonomy of producers and reducing their vulnerability to market and climate shocks. In addition, recycling organic materials and by-products offers great potential for agroecological innovations. This can be accomplished by sustainable use of nutrients and recycling the nutrients in organic side streams.

Agricultural and urban/municipal wastes are valuable nutrient recovery sources. However, they are mainly used as secondary sources of nutrients that are valorized into fertilizer components or directly to fertilizers after appropriate standardization (Jama-Rodzenska et al., 2021).

In general, four main classes can be distinguished (Fig. 1) among the available techniques for biowaste valorization and, at the same time, nutrient recycling. The first one is the mechanical methods that are the less advanced, which aim to lower the humidity and volume of the waste biomass and are rather considered as the preceding step before other techniques. Thermal techniques enable the utilization of higher temperatures and different levels of available oxygen during the process, thus resulting in different types of obtained ash. This method is sometimes used with simultaneously heat recovery. Obtained ashes are deprived of organic matter and are biologically safe. Because of their composition, they should not be used as soil amendments without further processing, as they can increase the pH of the soil environment and introduce heavy metals. At the same time they can be used as a component of fertilizer formulations because of their richness of micro- and macronutrients. The physical-chemical biowaste treatment methods engage the addition of some substrates that are chemically reactive with the

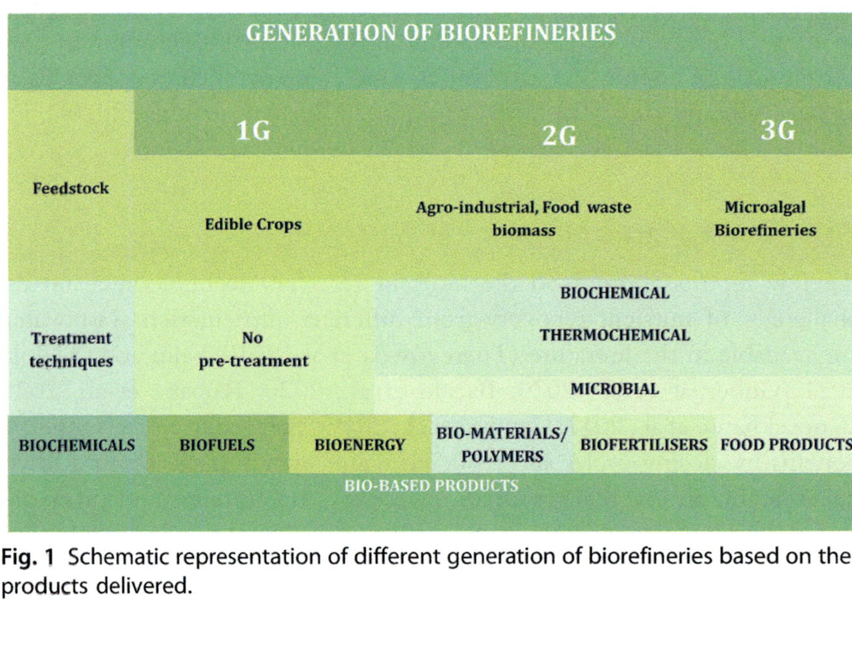

Fig. 1 Schematic representation of different generation of biorefineries based on their products delivered.

biomass's composition or change the biomass's physical properties, which enables the separation of nutrients available to plant form. Finally, the biological methods are discussed in more detail further in this chapter (Ranjani et al., 2019; Saliu and Oladoja, 2021; Gubernat et al., 2020; Bacelo et al., 2020; Rubaba et al., 2021; Vaneeckhaute et al., 2017; Campos et al., 2019).

In the scientific literature (Fig. 2), different methods are described, such as simple compaction/wrapping or compression, liquefaction, thermal processing (pyrolysis, combustion, gasification), and physiochemical processing. We can distinguish membrane filtration, adsorption, crystallization, freeze concentration, desalination, stripping, and acid/base digestion within the last one. Many scientific reviews are available in the literature describing in detail each of the mentioned methods and their different combinations (Lam et al., 2020; Lorick et al., 2020). Many of those processes are conducted on a technical scale. Their role is crucial in circulating nutrients passing through the wastewater treatment plant, such as precipitation of struvite, which is obligatory in a few countries in Europe (Cieślik and Konieczka, 2017; Macura et al., 2020). Therefore modifying those and combinations among them may result in significantly higher efficiency and lower environmental impact (Hollas et al., 2021; Macura et al., 2020). But, unfortunately, they are not lacking in weaknesses. The biggest disadvantage of the mentioned treatments of wastes is that the end product of a single process is not the final formulation ready to be applied on the field to serve as a source of nutrients.

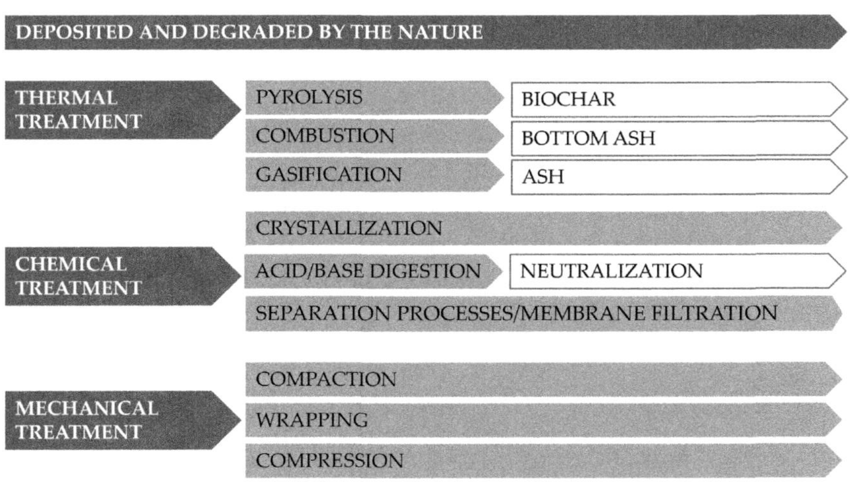

Fig. 2 Schematic representation of methods utilized for biowaste pretreatment and valorization.

In most cases, a combination of them is applied, resulting in the high recovery of nutrients and products ready to be used by the end-users. However, one very significant issue among the mentioned methods concerns the necessity of introducing additional agents (physical, mechanical, or chemical factors) necessary to perform the specific procedure of waste valorization. Moreover, those factors very often require resources or use energy (Fig. 3).

Different nutrient recovery technologies from wastewater have been proposed and assessed. However, there is no definitive nutrient recovery technology.

The biggest challenge of those techniques is producing a metal-free nutrient-rich product, with no biological risks, and delivering nutrients to agricultural production in an available form that enables the reuse of the nutrients. According to the words of Emilie Snauwaert (Belgium), an expert from the EIP-AGRI Focus Group, on nutrient recycling: "Promoting nutrient recycling is not only about technologies to produce biobased fertilizers, but also about practical tools to understand better the behavior of nutrients and their management on the farm." For obtained fertilizer (semi)products mentioned in this subsection, techniques should be designed in such a way that the nutrient will be delivered in a form that enables its uptake in the levels adopted to the needs of the plant, based on the nature of nutrient behavior in the soil environment. By this approach, besides the reuse of nutrients, we can increase the effectiveness of their use and mitigate the environmental impacts. This is possible, for example, by combining the formulations obtained by physiochemical methods with biological agents. Managing the nutrient properly guarantees efficiency and safety is based on understanding the soil system, which is a complex system of interactions between the three pillars that shape its properties. Plant roots and a set of soil microorganisms together with the soil content all affect the fertility, which ultimately translates into the growth, yield, composition, and nutritional properties of the obtained plant biomass (Saliu and Oladoja, 2021; Gubernat et al., 2020; Bacelo et al., 2020; Rubaba et al., 2021; Vaneeckhaute et al., 2017; Campos et al., 2019) (Fig. 4).

15.5 Conclusion

The current nutrient recovery technologies applied are predominantly chemical and physical in nature. Most of the studies concerning nutrient recovery from waste streams are restricted to microalgal applications to wastewater and other effluent treatments. Other microbial strains utilized

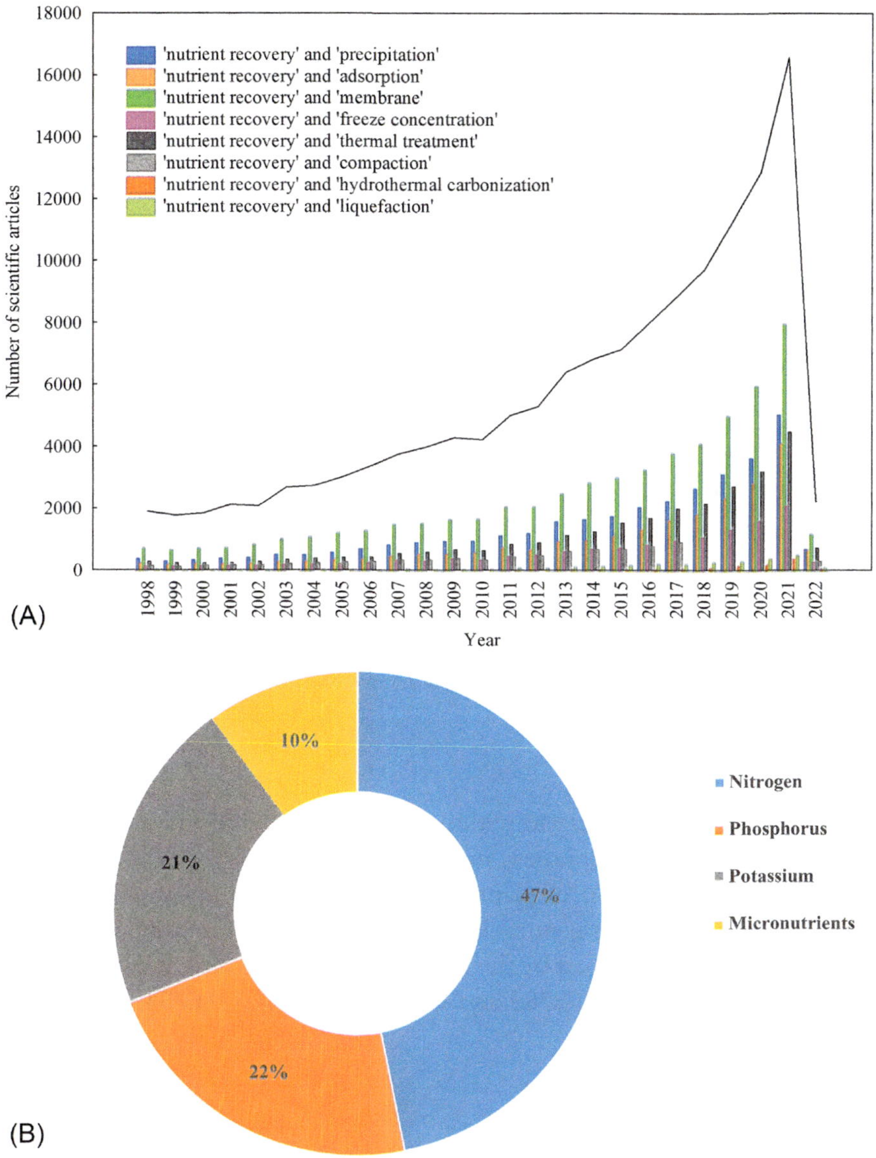

Fig. 3 (A) Number of articles published in the years >1998 and (B) share of articles on recycling by nutrients in published articles in 2021; according to ISI Web of Science database (December 2021).

in major waste streams such as e-waste and plastic waste streams do not seem abundant, even though bacterial as well as other microbiological metabolism is inherently available in the environment for integrating and enhancing the circular bioeconomy.

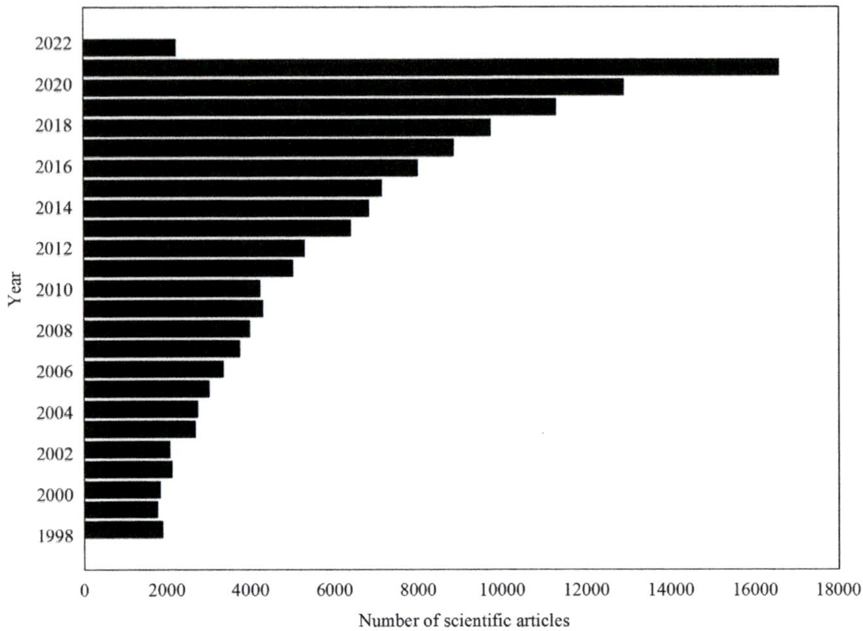

Fig. 4 Number of articles published in the years >1998 according to the ISI Web of Science database (search keyword "nutrient recovery," December 2021).

Funding

This project is financed in part within the framework of the National Centre for Research and Development in Poland within the framework of a grant entitled: Biofortification of plant biomass with selenium by utilization of biofertilizers obtained via microbiological solubilization—BioSeFert. This research was funded in part by the National Science Centre, Poland—Grant No. 2020/39/O/ST10/02783 and Grant No. 2021/42/E/ST10/00379.

References

Bacelo, H., Pintor, A.M.A., Santos, S.C.R., Boaventura, R.A.R., Botelho, C.M.S., 2020. Performance and prospects of different adsorbents for phosphorus uptake and recovery from water. Chem. Eng. J. 381, 122566.

Baldassarre, B., Schepers, M., Bocken, N., Cuppen, E., Korevaar, G., Calabretta, G., 2019. Industrial symbiosis: towards a design process for eco-industrial clusters by integrating circular economy and industrial ecology perspectives. J. Clean. Prod. 216, 446–460.

Banu, J.R., Kavitha, S., Gunasekaran, M., Kumar, G., 2020. Microalgae based biorefinery promoting circular bioeconomy-techno economic and life-cycle analysis. Bioresour. Technol. 302, 122822.

Bentsen, N.S., 2019. Biomass for biorefineries: availability and costs. In: Biorefinery. Springer, Cham, pp. 37–48.

Bourguignon, D., 2015. Understanding Waste Streams—Treatment of Specific Waste. European Parliamentary Research Service.

Bressanelli, G., Saccani, N., Pigosso, D.C., Perona, M., 2020. Circular economy in the WEEE industry: a systematic literature review and a research agenda. Sustain. Prod. Consum. 23, 174–188.

Campos, J.L., Crutchik, D., Franchi, O., Pavissich, J.P., Belmonte, M., Pedrouso, A., Mosquera-Corral, A., Val del Río, A., 2019. Nitrogen and phosphorus recovery from anaerobically pretreated agro-food wastes: a review. Front. Sustain. Food Syst., 2.

Castilla-Archilla, J., O'Flaherty, V., Lens, P.N., 2019. Biorefineries: industrial innovation and tendencies. In: Biorefinery. Springer, Cham, pp. 3–35.

Cieślik, B., Konieczka, P., 2017. A review of phosphorus recovery methods at various steps of wastewater treatment and sewage sludge management. The concept of "no solid waste generation" and analytical methods. J. Clean. Prod. 142 (4), 1728–1740.

Gherghel, A., Teodosiu, C., De Gisi, S., 2019. A review on wastewater sludge valorisation and its challenges in the context of circular economy. J. Clean. Prod. 228, 244–263.

Gubernat, S., Masłoń, A., Czarnota, J., Koszelnik, P., 2020. Reactive materials in the removal of phosphorus compounds from wastewater—a review. Materials 13 (15), 3377.

Hahladakis, J.N., Iacovidou, E., Gerassimidou, S., 2020. Plastic waste in a circular economy. In: Plastic Waste and Recycling. Academic Press, pp. 481–512.

Hollas, C.E., Bolsan, A.C., Venturin, B., Bonassa, G., Tápparo, D.C., Cândido, D., Antes, F. G., Vanotti, M.B., Szögi, A.A., Kunz, A., 2021. Second-generation phosphorus: recovery from wastes towards the sustainability of production chains. Fortschr. Mineral. 11 (9), 985.

Insam, H., Franke-Whittle, I., Goberna, M., 2010. Microbes in aerobic and anaerobic waste treatment. In: Microbes at Work. Springer, Berlin, Heidelberg, pp. 1–34.

Jama-Rodzenska, A., Sowinski, J., Koziel, J.A., Białowiec, A., 2021. Phosphorus recovery from sewage sludge ash based on cradle-to-cradle approach—mini-review. Fortschr. Mineral. 11, 985.

Johansen, M.R., Christensen, T.B., Ramos, T.M., Syberg, K., 2022. A review of the plastic value chain from a circular economy perspective. J. Environ. Manage. 302, 113975.

Lam, K.L., Zlatanović, L., van der Hoek, J.P., 2020. Life cycle assessment of nutrient recycling from wastewater: a critical review. Water Res. 173, 115519.

Leong, H.Y., Chang, C.K., Khoo, K.S., Chew, K.W., Chia, S.R., Lim, J.W., et al., 2021. Waste biorefinery towards a sustainable circular bioeconomy: a solution to global issues. Biotechnol. Biofuels 14 (1), 1–15.

Lorick, D., Macura, B., Ahlström, M., Grimvall, A., Harder, R., 2020. Effectiveness of struvite precipitation and ammonia stripping for recovery of phosphorus and nitrogen from anaerobic digestate: a systematic review. Environ. Evid. 9, 27.

Macura, B., Carolus, J., Barquet, K., Filippa, E.K., Järnberg, L., Lorick, D., Johannesdottir, S., Pedersen, S.M., Koskiaho, J., Haddaway, N.R., Okruszko, T., 2020. Circular nutrient solutions for agriculture and wastewater—a review of technologies and practices. Curr. Opin. Environ. Sustain. 45, 78–91.

Malinauskaite, J., Jouhara, H., Czajczyńska, D., Stanchev, P., Katsou, E., Rostkowski, P., Spencer, N., 2017. Municipal solid waste management and waste-to-energy in the context of a circular economy and energy recycling in Europe. Energy 141, 2013–2044.

Nagarajan, D., Lee, D.J., Chen, C.Y., Chang, J.S., 2020. Resource recovery from wastewaters using microalgae-based approaches: a circular bioeconomy perspective. Bioresour. Technol. 302, 122817.

Namlis, K.G., Komilis, D., 2019. Influence of four socioeconomic indices and the impact of economic crisis on solid waste generation in Europe. Waste Manag. 89, 190–200.

Pasalari, H., Gholami, M., Rezaee, A., Esrafili, A., Farzadkia, M., 2021. Perspectives on microbial community in anaerobic digestion with emphasis on environmental parameters: a systematic review. Chemosphere 270, 128618.

Patchaye, M., Sundarkrishnan, B., Tamilselvan, S., Sakthivel, N., 2018. Microbial management of organic waste in agroecosystem. In: Microorganisms for Green Revolution. Springer, Singapore, pp. 45–73.

Ren, L., Cai, C., Zhang, J., Yang, Y., Wu, G., Luo, L., et al., 2018. Key environmental factors to variation of ammonia-oxidizing archaea community and potential ammonia oxidation rate during agricultural waste composting. Bioresour. Technol. 270, 278–285.

Rolewicz-Kalińska, A., Lelicińska-Serafin, K., Manczarski, P., 2020. The circular economy and organic fraction of municipal solid waste recycling strategies. Energies 13 (17), 4366.

Rubaba, M., Walter, T., Mika, S., 2021. A systematic review and statistical analysis of nutrient recovery from municipal wastewater by electrodialysis. Desalination 498, 114626.

Ruiz, L.A.L., Ramón, X.R., Domingo, S.G., 2020. The circular economy in the construction and demolition waste sector—a review and an integrative model approach. J. Clean. Prod. 248, 119238.

Sáez, P.V., Osmani, M., 2019. A diagnosis of construction and demolition waste generation and recovery practice in the European Union. J. Clean. Prod. 241, 118400.

Saliu, T.D., Oladoja, N.A., 2021. Nutrient recovery from wastewater and reuse in agriculture: a review. Environ. Chem. Lett. 19, 2299–2316.

Shen, L., Wang, J., Li, Z., Fan, L., Chen, R., Wu, X., Li, J., Zeng, W., 2020. A high-efficiency Fe_2O_3@ Microalgae composite for heavy metal removal from aqueous solution. Journal of Water Process Engineering 33, 101026.

Smol, M., Adam, C., Preisner, M., 2020. Circular economy model framework in the European water and wastewater sector. J. Mater. Cycles Waste Manage. 22 (3), 682–697.

Tait, J., Wield, D., Bruce, A., Chataway, J., 2007. OECD International Futures Project on "the Bioeconomy to 2030: Designing a Policy Agenda". Health Biotechnology to 2030, OECD, Paris, France.

Theregowda, R.B., González-Mejía, A.M., Ma, X., Garland, J., 2019. Nutrient recovery from municipal wastewater for sustainable food production systems: an alternative to traditional fertilizers. Environmental Engineering Science 36 (7), 833–842.

United Nations Environment Programme, 2021. Food Waste Index Report 2021. UNEP, Nairobi.

Vaneeckhaute, C., Lebuf, V., Michels, E., et al., 2017. Nutrient recovery from digestate: systematic technology review and product classification. Waste Biomass Valorization 8, 21–40.

Velvizhi, G., Shanthakumar, S., Das, B., Pugazhendhi, A., Priya, T.S., Ashok, B., et al., 2020. Biodegradable and non-biodegradable fraction of municipal solid waste for multifaceted applications through a closed loop integrated refinery platform: paving a path towards circular economy. Sci. Total Environ. 731, 138049.

Wielgosiński, G., Czerwińska, J., Szufa, S., 2021. Municipal solid waste mass balance as a tool for calculation of the possibility of implementing the circular economy concept. Energies 14 (7), 1811.

Yamamoto, N., Nakai, Y., 2019. Microbial community dynamics during the composting process of animal manure as analyzed by molecular biological methods. In: Understanding Terrestrial Microbial Communities. Springer, Cham, pp. 151–172.

Zhang, L., Dong, H., Zhang, J., Chen, Y., Zeng, G., Yuan, Y., et al., 2019. Influence of FeONPs amendment on nitrogen conservation and microbial community succession during composting of agricultural waste: relative contributions of ammonia-oxidizing bacteria and archaea to nitrogen conservation. Bioresour. Technol. 287, 121463.

Zhou, L., Gao, Y., Yu, K., Zhou, H., De Costa, Y.G., Yi, S., Zhuang, W.Q., 2020. Microbial community in in-situ waste sludge anaerobic digestion with alkalization for enhancement of nutrient recovery and energy generation. Bioresour. Technol. 295, 122277.

Further reading

Crippa, M., Solazzo, E., Guizzardi, D., Monforti-Ferrario, F., Tubiello, F.N., Leip, A.J.N. F., 2021. Food systems are responsible for a third of global anthropogenic GHG emissions. Nat. Food 2 (3), 198–209.

FAO, 2013. Food Wastage Footprint: Impacts on Natural Resources: Summary Report. Food & Agriculture Organisation.

Oborne, M., 2010. The bioeconomy to 2030: designing a policy agenda. OECD Obs. 278, 35.

Tsegaye, B., Jaiswal, S., Jaiswal, A.K., 2021. Food waste biorefinery: pathway towards circular bioeconomy. Foods 10 (6), 1174.

CHAPTER 16

Exploring the potential of fungal endophytes: A quintessential source for novel secondary metabolites and beneficial aspects

H.N. Madhushree[a], Salman Siddeeqh[b], Shilpa Nayuni[c], R. Parthasarathy[a], D. Sruthi[a], C. Jayabaskaran[a], Ashwini Benjarwad[d], and H.C. Yashavantha Rao[a]

[a]Department of Biochemistry, Indian Institute of Science, Bengaluru, Karnataka, India
[b]Department of Maxillofacial Surgery and Diagnostic Sciences, College of Dentistry, King Saud Bin Abdulaziz University for Health Sciences, Riyadh, Kingdom of Saudi Arabia
[c]SPMVV-Women Biotech Incubation Facility, Sri Padmavati Mahila Visvavidyalayam, Tirupati, Andra Pradesh, India
[d]Department of Rasashastra and Bhaishajya Kalpana, SDM College of Ayurveda and Hospital, Hassan, Karnataka, India

16.1 Introduction

Nature is a potential source for the discovery of new therapeutic agents, with its diversity of millions of animals, plants, and microorganism species. Plant-associated biomolecules have played a significant role in the discovery and development of many clinically approved drug agents. These include camptothecin, vinaceuline, podophyllotoxin, vincristine, griseofulvin, vinblastine, and paclitaxel (Chandra, 2012; Noble, 2016). Endophytic fungi present in the tissues of asymptomatic plants have certainly been underexplored, but are now well known to produce new natural products, to be explored in the pharmaceutical, medical, and agricultural industries. They adopt some mechanisms to get inside the host tissues using mutualistic bonding and its biotransformation ability is being explored for synthesizing new bioactive compounds (Rao and Satish, 2015). Fungal endophytes produce secondary metabolites to contend with the epiphytes and plant pathogens to balance between plant defense and fungal virulence (Rao and Satish, 2016).

Microbial Inoculants
https://doi.org/10.1016/B978-0-323-99043-1.00002-5

16.1.1 Medicinal plants and their natural products

The field of pharmacognosy plays a pivotal role in the biodiscovery and development of novel drugs. Natural products are derived from diverse reservoirs, including microbial, medicinal plant, animal, and marine sources. Among these, medicinal plants are the most promising source for discovery of new bioactive metabolites (Newman et al., 2000; Baker et al., 2000). Even today, more than half of the world's population relies on medicinal herbs for their primary health care. The medicinal plants are a fundamental source for many pharmaceutical products to enter the market. Nearly one-fourth of the drugs approved by the Food and Drug Administration and the European Medical Agency are of plant origin. Long-familiar drugs that continue to be used today as therapeutics include the first plant molecular entity morphine (opioid analgesic) from *Papaver somniferum* Linn., aspirin from the bark of the willow tree, digitoxin (cardiac glycosides) from *Digitalis purpurea*, and the modern medicines paclitaxel (world's first multibillion-dollar anticancer drug) from *Taxomyces andreanae*, artemisinin (antimalarial drug) from *Artemisia annua*, camptothecin (anticancer drug) from *Camptotheca acuminate*, baccatin (anticancer drug) from *Taxus baccata*, are been derived from plants (Boucher et al., 2017; Veeresham, 2012).

Medicinal plant resources are inadequate to combat emerging new diseases due to slow growth, seasonal availability, and rare medicinal plant species (Chen et al., 2016). Indeed, whole-plant harvesting, indiscriminate collection, and overexploitation have generally resulted in diminution and scarcity of valuable medicinal plant diversity (Baker et al., 2007). For example, an expensive and very successful anticancer drug, paclitaxel (brand name Taxol) ($C_{47}H_{51}NO_{14}$), was derived from the plant genus *Taxus* (yew); consequently, for the chemotherapy of one patient, 2 g of paclitaxel is required, which consumes approximately four yew trees (Walsh and Goodman, 1999; Strobel et al., 1996; Gangadevi and Muthumary, 2007; Liu et al., 2016; Zhu and Chen, 2019). The bark of yews yields extremely low content of paclitaxel, which is neither economical nor eco-friendly. Thus an alternative source of procuring paclitaxel is from an endophytic fungus associated with the Pacific yew *Taxus brevifolia* that generates the same compound associated with the host, which is more economical and environmentally friendly (Stierle et al., 1993). This discovery paves the way to a new era of endophyte biotechnology (Liu et al., 2016).

16.1.2 Endophytic fungi and their host interactions

The term endophyte was first introduced by Heinrich Anton de Bary (1884), referring to 'any organism inhabiting the plant tissues.' The word originated from the Greek word meaning "in the plant" (endon Gr., within; phyton, plant). Endophytic fungi are a group of fungi having symbiotic association with the internal tissues of living plants, causing no apparent harm to the host (Sunitha et al., 2013; Verma et al., 2021). However, the existence of fungal association with asymptomatic plants has been known as of the mid-19th century. Endophytic fungi extensively colonize diverse parts of plants, such as leaves, stems, roots, and bark. The fungal endophytes are hyperdiverse, harboring all vascular plants globally. Each plant investigated harbors a minimum of one endophytic fungus and most plants, especially woody plants, contain several species (Faeth and Hammon, 1997). Hence, fungal endophytes are considered to be an important source of new secondary metabolites. In general, plant-inhabiting fungi are categorized into five groups: mycorrhizal, saprotrophic, epiphytic, endophytic, and pathogenic fungi. Among them, endophytic fungi are a portion of a mycobiome, with many functional roles in their life cycle or in relation to the plant or environmental factors (Alfaro and Bayman, 2011).

Endophytes express diverse symbiotic lifestyles such as antagonisms, obligate mutualisms, and commensalism, depending on the species. Thus different types of interactions between plants and endophytic organisms are expressed as a continuum (Rajamanikyam et al., 2017; Kumar et al., 2016). The interconnection between fungal endophytes and their host plant is due to their unique acclimatizing ability, which enables the endophytes to harmonize their growth with their host plant (Verma et al., 2012). Endophytes produce bioactive secondary metabolites that contribute to development of plant metabolites (Schulz et al., 2002; Rao et al., 2015a). Plants provide endophytes with nutrients, while the microbes in turn release active metabolites (Gao et al., 2015; Kumar et al., 2014). The released bioactive metabolites allow plants to have better access to nutrients for their root development (Taghavi et al., 2011). Endophytes are able to benefit plants by producing antimicrobial metabolites, phosphate solubilizing agents, iron chelators, nitrogen fixing, and insecticidal by-products (Knoth, 2014; Pathak et al., 2022). Endophyte biological activities have been documented, and play an important role in plant root elongation and growth promotion (Mangunwardoyo et al., 2012; Sheng et al., 2008). The various bioactive metabolites have been implicated in these systemic endophyte operations, including gibberellin, indole acetic acid, indole acetonitrile, and dcytokinin, which are involved in the development of host plants and other beneficial aspects. Endophytes enhance plant growth, including stomatal control, osmotic equilibrium change, increase of mineral absorption, modulation of root morphology, and alteration of metabolism (Kumar et al., 2014).

Fungal endophyte-host mutualism regulates the physiology of hosts and their symbionts to strengthen the capability to adjust to environmental stresses (Compant et al., 2010; Choudhary et al., 2014; Brader et al., 2014; Rao et al., 2015b). The interaction between host and endophytes in the dual culture technique indicates that chemotactic signaling is involved. Fungi isolated as endophytes do not have a set life history (Saikkonen et al., 2004; Rodriguez and Redman, 2008). The mutualistic relationship of endophytic fungi may lead to tolerance to drought, salts, and high temperatures (Malinowski and Belesky, 2000). The systemic and foliar endophytes are capable of decreasing herbivory by producing alkaloids that are poisonous to insects and vertebrates (Akello et al., 2007). Endophytes can defend the host by three key mechanisms: (a) rivalry between endophytes and pathogens (Lockwood, 1992), (b) biocidal or phytoalexin production (Rai et al., 2002), and (c) by inducing disease resistance in hosts (Gianinazzi et al., 1996).

16.1.3 Interactions between microbial endophytes and other symbionts

Microbial endophytes strongly establish relationships between the various community members of endophytic microorganisms inside the plant (fungi-fungi, fungi-bacteria, or bacteria-bacteria). There is a heterogeneous distribution of microorganisms within the plant tissue; a few of them clump in their secreted mucilage or appear to accumulate to particles (Fisher et al., 1992; Ezra et al., 2004). Various reports have documented the ability of bacterial and fungal endophytes for in vitro and in planta growth promotion and disease suppression (Mehmood et al., 2019; Soldan et al., 2019). Endophyte interspecies or intraspecies cross-talk mediate via quorum-sensing signals, and diffusible signaling molecules may trigger certain silent biosynthetic metabolic pathways (Kusari et al., 2012).

16.2 Potential role of endophytes in sustainable agriculture

Endophytes are a reliable source of bioactive compounds in the field of sustainable agriculture (Zhao et al., 2011). These microorganisms promote growth of plants by structural and morphological manipulations in the host via the production of phytohormones, auxins, ethylene, indole acetic acid, and gibberellic acid (Sturz et al., 2000). Endophytes produce siderospores and iron chelating agents, including pheonolate, catacholate, and α-hydroxycarboxylate functional groups that regulate internal iron concentrations, enhance pathogenicity, and assist in germination and asexual and sexual propagation (Renshaw et al., 2002; Rajkumar, 2010; Haas, 2014; Yadav and Yadav, 2017). Endophytes activate microbial processes that assist in acquisition of micronutrients like boron (B), copper (Cu), iron (Fe), manganese (Mn), zinc (Zn), and macronutrients like calcium (Ca), nickel (Ni), nitrogen (N), phosphorus (P), potassium (K), and sulfur (S) to stimulate growth of plant and crop productivity (Singh and Pandey, 2012; Lalitha, 2017). The macronutrient phosphorus is one of the most essential nutrients to promote plant growth. Phosphorus presents abundantly in soil as insoluble phosphate, which can be solubilized into orthophosphate by endophytic microorganisms by releasing organic acids like gluconic acid, salicylic acid, oxalic acid, citric acid, pyrroloquinoline, quinone, and benzene acetic acid to the soil (Chhabra et al., 2013; Oteino et al., 2015). Endophytes as biofertilizer in sustainable agriculture improve soil fertility, enhance crop yield, and reduce pollution and health risks. They thus act as an effective substitute for chemical fertilizers and pesticides that have adverse effects on the ecosystem. The plant-associated endophytes increase the production of bioactive metabolites that provide protection against biotic and abiotic stresses and improve crop yield (Wani et al., 2015; Verma et al., 2017).

16.2.1 Endophytic fungi and their role in natural product discovery

Endophytes are potent sources of natural products as they inhabit specialized niches and unique biotopes due to their potential to generate new and fascinating bioactive metabolites of biomedical, industrial, and agricultural significance (Schulz et al., 2002; Aly et al., 2011; Liu et al., 2012; Strobel and Daisy, 2003). The numbers of new natural products from fungal endophytes are continuously growing due to their proven ability to battle tumors, fungal infections, and antibiotic resistance developed by pathogenic bacteria and fungi (Wiyakrutta et al., 2004; Cota et al., 2008). Fungal endophytes produce various bioactive agents like antimicrobial, anticancer, antioxidant, antidiabetic, antimalarial, antiviral, immunomodulatory, insecticidal, and volatile antibiotics (Tan and Zou, 2001; Strobel et al., 2004; Gunatilaka, 2006; De Souza et al., 2011).

16.2.2 Antimicrobial compounds

The discovery of the first antibiotic compound penicillin from *Penicillium notatum* marked a new epoch in the scientific development of natural antibiotics. Since then, the quest for bioactive natural products has contributed to the discovery of a multitude of antimicrobial compounds, including cephalosporin from *Acremonium crysogenum*, chloramphenicol from *Streptomyces venezuela*, and macrolides from *Saccharopolyspora erythraea* (Zhu et al., 2014). Several new bioactive compounds have recently been isolated from fungal endophytes, including penicidones, xylarion, xylarosides, and benzoquinone derivatives (Kajula et al., 2010). Antimicrobial compounds derived from endophytes have a diverse functional and structural group, such as alkaloids, steroids, peptides, phenols, terpenoids, quinones, and flavonoids (Yu et al., 2010). Bioactive secondary metabolites produced by fungal endophytes serve as antimicrobial agents against human, animal, and phytopathogens (Tripathi et al., 2018). Antimicrobial compounds like phomenone, cryptocin, trichodermin, alterusin, ambuic acid, dihydroxycadalene, and nodulosporins have been derived from fungal endophytes to defend plants against phytopathogens. Endophytic fungi produce host plant secondary metabolites with therapeutic potential such as paclitaxel, azadira, camptothecin, hypericin, and podophyllotoxin. The antimicrobial compounds are also used as food preservatives to control food spoilage and food-borne diseases and increase the shelf life of food (Liu et al., 2008; Jagtap et al., 2009). Fungal genes encoding for antimicrobial compounds are clustered on chromosomes; since separate genera of fungi may generate the same metabolite, genetic clustering can encourage the production of secondary metabolites between fungal endophytes (Mousa and Raizada, 2013).

16.2.3 Anticancer compounds

Anticancer agents are the most crucial categories among cancer chemotherapeutic drugs. Today, cancer is the most common disease worldwide and is rapidly spreading. Cancer is a heterogeneous disease, arising from various types of tissues. There are more than 100 different types of cancer. Cancer is one of the major human life-threatening diseases, with a high rate of morbidity and mortality. Various reports indicate that natural bioactive compounds isolated from fungal endophytes serve as the best alternative source for the biodiscovery of anticancer drugs (Xie and Zhou, 2017; Parthasarathy et al., 2020). As previously mentioned, paclitaxel, which is an effective drug against ovarian and breast cancers, was isolated from *Taxus* species (Horwitz, 1994). Drugs like laulimalide, eleutherobin, isolaulimalide, discodermolide, and epothilones were found to be antimitotic to cells by inhibiting tubulin depolymerization taxanes (Salerno et al., 2010). A novel guanacastane diterpene compound Cercosporene F was extracted from *Cercospora* sp. inhabiting *Fallopia japonica*. Novel anticancer sesquiterpene compounds Phomoarcherins A-C were obtained from endophytic *Phomopsis archeri* inhabiting *Vanilla albidia* (Hemtasin et al., 2011). A novel 2,3-didehydro-19αhydroxy-14-epicochlioquinone B compound from *Nigrospora* sp. MA75 was isolated from the marine mangrove plant *Pongamia pinnata*.

16.2.4 Antioxidant compounds

Antioxidant compounds play a vital role in inhibiting cell and tissue damage. Biological reactions like normal cell metabolism and environmental conditions influence the production of reactive oxygen species (ROS) molecules like superoxide anion ($O_2^{\cdot-}$) and hydroxyl radical (OH−), in all life forms, from microbes to higher plants and animals (Nose, 2000; Birben et al., 2012). ROS and free radical-mediated reactions induce oxidative damage to biomolecules like proteins, lipids, and DNA (Shahidi and Ambigaipalan, 2015). Oxidative stress caused due to excess production of ROS and free radicals leads to the development of various disorders such as aging, malignancy, coronary arteries, cardiovascular disease, cystic fibrosis, hypovolemic shock, and other neurological disorders (Halliwell, 2009; Finkel and Holbrook, 2000).

Fungal endophytes, besides possessing strong potential, have become a potent source of natural antioxidants (Huang et al., 2007; Bérdy, 2012). The natural antioxidant compounds associated with endophytic fungi include flavipin, isopestacin, pestacin, isocoumarin derivatives, graphis lactone A, benzoquinones, phenol and phenolic acids, amines and amides, flavonoids, saponins, tanins, and many others. Among these phenolic compounds are key antioxidant components of secondary metabolites from fungal endophytes with antioxidative activity (de Carvalho et al., 2016; Goutam et al., 2018). The flavipin isolated from various species of endophytes including Chaetomium sp., *Aspergillus flavipes*, *Aspergillus terreus*, and *Epicoccum nigrum* exhibit antioxidant activity (Bamford et al., 1961; Ye et al., 2013). The pestacin and isopestacin naturally occurring as a racemic mixture isolated from *Pestalotiopsis microspora* exhibit potent antioxidant activity and antifungal activity against *Pythium ultimum* (Harper et al., 2003). The fungal endophyte *Chaetomium globosum* isolated from the medicinal plant *Adiantum capillusveneris* exhibited antioxidant activity (scavenging 99% at 100g/mL) (Selim et al., 2018).

16.2.5 Antidiabetic properties

Diabetes mellitus is the world's most prevalent endocrine condition, characterized by chronic hyperglycemia as a result of insulin secretion deficiencies (Lunze et al., 2013). Various reports indicate that bioactive metabolites from endophytic fungus can be a potent source for the management of diabetes through inhibition of alpha-amylase and alpha-glucosidase activity (Dhankhar et al., 2013). Several fungal endophytes derived from medicinal plants including *Aspergillus awamori* from *Acacia nilotica*, *Syncephalastrum* sp. from *Adhatoda beddomei*, *Dendryphion nanum* from *Ficus religios*, *Phomopsis* sp. from *Paeonia delavayila*, *Pseudomonas protegens* from *Piper auritum*, *Aspergillus* sp. and *Phoma* sp. from *Salvadora oleoides* Decne, *Alternaria* sp. from *Viscum album*, and many others showed antidiabetic potential by producing a peptide with alpha-amylase and glycosidase inhibitory activity. The endophytic fungi *Nigrospora oryzae* from the *Combretum dolichopetalum* plant yielded three biosynthetic compounds (*S*)-(+)-2-*cis*-4-*trans*-abscisic acid, 4-des-hydroxyl altersolanol A, and 7′-hydroxy-abscisic acid that exhibit strong antidiabetic activity (Uzor et al., 2017).

16.2.6 Antimalarial compounds

Malaria is a disease caused by single-celled obligate intracellular parasites like Plasmodium, viz. *P. falciparum*, *P. malariae*, *P. ovale*, *P. vivax*, and *Pheidole knowlesi*. Among these, the infection caused by *Plasmodium falciparum* in humans, which is transmitted via the bites of infected female anopheles mosquitoes, is a major public health concern and the leading cause of malaria deaths globally (Kavunga-Membo et al., 2018). Though various drugs act as antiplasmodial, the biometabolites isolated from endophytic fungi are also a prominent natural source for malaria prophylaxis (Ateba et al., 2018). The culture extract of the endophytic fungus *Fusarium* sp. segregated from the roots of *Mentha longifolia* (Lamiaceae) yielded a novel metabolite cyclodepsipeptide, namely Fusaripeptide A, including three known compounds, cyclopentanol, adenosine, and 2[(2-hydroxypropionyl) amino] benzamide; it showed a significant antimalarial potency towards *P. falciparum* (D6 clone) with an IC_{50} value of 0.34 μM compared to the artemisinin IC_{50} value of 0.57 μM (Ibrahim et al., 2017). The bioactive metabolites derived from *Preussia* sp. associated with leaves and bark of *Enantia chlorantha* comprise compounds such as preussiafurans A, asterric acid, and cissetin, which expressed their activity against chloroquine resistant (CQR) *P. falciparum* (NF54) erythrocytic stages, using a 3*H*-hypoxanthine incorporation assay (Talontsi et al., 2014).

16.2.7 Antiviral compounds

The biodiscovery of endophytic fungal metabolites with antiviral activity is an emerging area of research, due to the limited number of vaccines to combat viral diseases. The key obstacle to the detection of antiviral compounds is most likely due to the lack of antiviral screening methods. No endophytic fungal metabolites have been licensed for antiviral therapy so far, although several of these have demonstrated a high propensity for different viral diseases (Roy, 2017). A novel secondary metabolite 14-nordrimane-type sesquiterpenoid called phomanolide and a known compound (−)-6-methoxymellein obtained from the culture extract of fungal endophytes *Phoma* sp. associated with the roots of *Aconitum vilmorinianum* plants exhibited potent antiviral activities against influenza A virus (A/Puerto Rico/8/34, H1N1) with IC_{50} values of 2.96 ± 0.64 and $20.98 \pm 2.66\,\mu g/mL$, respectively. The culture extracts from three endolichenic fungus strains *Nigrospora sphaerica*, *Alternaria alternata*, and *Phialophora* sp. produced heptaketides like alternariol with an IC_{50} value of $13.5\,\mu M$ and alternariol-9-methyl ether with an IC_{50} value of $21.3\,\mu M$ showed antiviral activity against herpes simplex virus (He et al., 2012). Endophytic fungi such as *Penicillium* sp., *Phomopsis* sp., *Aspergillus* sp., *Fusicoccum* sp., *Muscodor* sp., *Guignardia* sp., *Pestalotiopsis* sp., and *Curvularia* sp., isolated from *Garcinia gummigutta* plants, exhibited weak to moderate antiviral activity against herpes simplex virus type 1 (ATCC VR-260).

16.2.8 Immunomodulatory compounds

Immunomodulatory drugs have been used as a preventive or therapeutic approach to enhance or inhibit host defense response. Endophytic fungal-derived immunomodulatory metabolites are used to treat numerous diseases, like autoimmune disease, inflammatory disorders, and multiple myeloma (Lee et al., 1995). The culture extract of endophytic fungus *Fusarium solani* isolated from the medicinal plant orchid *Dendrobium officinale* in a solid-state fermentation method generated two polysaccharides: DGS1 includes arabinose, glucose, mannose, and galactose with a molar ratio of 2.9:13.4:3.0:1, respectively, and DGS2 contains arabinose, glucose, mannose, and galactose with a molar ratio of 3.5:8.1:2.1:1, respectively. These could significantly enhance the levels of TNF-α, IL-6, and nitric oxide (NO) by activating TNF-α, IL-6, and inducible nitric oxide synthase (iNOs) gene expression, respectively, and they displayed strong immune-suppressive activity (Zeng et al., 2019). The secondary metabolites include phenol, 2,4-bis

(1,1-dimethyl ethyl), benzene propanoic acid, and 3,5-bis(1,1-dimethyl) produced from the endophytic fungus *Fusarium solani* associated with the root segments of *Withania somnifera* showed potent immunomodulatory activity. The biometabolites obtained from the fungal endophyte *A. alternata* associated with the medicinal plant *Acalypha indica* showed immune-suppressive activity against the insect pest *Spodoptera litura* (Kaur et al., 2015). The bioactive metabolites hydroxybenzoic acid, xanthorrhizol acid, and orsellinic acid obtained from the crude extract of endophytic fungus associated with the leaves of *Ageratum conyzoides* expressed potential immunomodulatory activity using the cyclophosphamide-induced immunosuppression method (Ujam et al., 2019).

16.2.9 Insecticidal compounds

The need for biological insecticides is rising tremendously to overcome the use of synthetic chemical pesticides exhibiting a massive impact on human, animal, and agricultural crops. Fungal endophyte-derived bioactive compounds also serve as an ideal eco-friendly alternative for replacing the use of chemical derivatives for insect management. The secondary metabolite oxazolidine-2-one moiety extracted from a marine algae-derived fungal endophyte *Acremonium vitellinum* displayed potential insecticidal activity against the cotton bollworm, *Helicoverpa armigera*, with an IC_{50} value of 0.56–0.03 mg/mL in contrast to matrine with an LC_{50} value of 0.24 0.01 mg/mL (Chen et al., 2018). The endophytic fungal extract of *Aspergillus sojae* from the medicinal plant *Plectranthus amboinicus* displayed strong insecticidal activity against cotton leaf worm *S. litura* (Elango et al., 2020). The endophytic entomopathogenic fungal species like *Beauveria bassiana* and *Metarhizium anisopliae* have been recorded to provide effective biological protection against plant pests (Pava-Ripoll et al., 2011; Akutse et al., 2013; Resquín-Romero et al., 2016; Russo et al., 2019).

16.3 Conclusion

Natural sources are an important tool to discover bioactive constituents with unique structures and mechanisms of action. Plant-associated fungal endophytes are well known for the discovery of new bioactive secondary metabolites. Researchers should focus more on their molecular studies and optimum fermentation conditions to scale up production. Endophytic fungi are a promising and interesting source to produce biologically active metabolites and needs to explore more. Indeed, novel potential drugs can be discovered from these fungal endophytes and provide a path to develop new and effective chemotherapeutic drugs.

Acknowledgments

Dr. H.C. Yashavantha Rao acknowledges the Young Scientist award and a research grant from DHR-HRD Indian Council of Medical Research (YSS/2020/0000054/PRCYSS) and University Grants Commission (DSKPDF Scheme: No. F.4-2/2006 (BSR)/BL/17-18/0234) Government of India, New Delhi. Dr. R. Parthasarathy extends thanks to DST-SERB, Govt. of India, New Delhi for NPDF Grant (Ref. No. PDF/2017/001184), and ICMR-Research Associate Fellowship (Ref. No. 45/1/2020-DDI-BMS). Dr. D. Sruthi acknowledges DHR (Young Scientist-HRD scheme; YSS/2019/000035/PRCYSS) and DST-SERB (NPDF/2017/000339), Govt. of India, New Delhi.

References

Akello, J., Dubois, T., Gold, C.S., Coyne, D., Nakavuma, J., Paparu, P., 2007. *Beauveria bassiana* (Balsamo) vuillemin as an endophyte in tissue culture banana (Musa spp.). J. Invertebr. Pathol. 96, 34–42.

Akutse, K.S., Maniania, N.K., Fiaboe, K.K.M., Berg, V.D.J., Ekesi, S., 2013. Endophytic colonization of *Vicia faba* and *Phaseolus vulgaris* (Fabaceae) by fungal pathogens and their effects on the life-history parameters of *Liriomyza huidobrensis* (Diptera: Agromyzidae). Fungal Ecol. 6 (4), 293–301.

Alfaro, A.P., Bayman, P., 2011. Hidden fungi, emergent properties: endophytes and microbiomes. Annu. Rev. Phytopathol. 49, 291–315.

Aly, A.H., Debbab, A., Proksch, P., 2011. Fungal endophytes: unique plant inhabitants with great promises. Appl. Microbiol. Biotechnol. 90, 1829–1845.

Ateba, J., Toghueo, R., Awantu, A., Mbaning, B., Gohlke, S., Sahal, D., Lenta, B., 2018. Antiplasmodial properties and cytotoxicity of endophytic fungi from *Symphonia globulifera* (Clusiaceae). J. Fungi 4 (2), 70.

Baker, W., van den Broek, A., Camon, E., Hingamp, P., Sterk, P., Stoesser, G., Tuli, M.A., 2000. The EMBL nucleotide sequence database. Nucleic Acids Res. 28, 19–23.

Baker, D.D., Chu, M., Oza, U., Rajgarhia, V., 2007. The value of natural products to future pharmaceutical discovery. Nat. Prod. Rep. 24, 1225–1244.

Bamford, P.C., Norris, G.L.F., Ward, G., 1961. Flavipin production by *Epicoccum* species. Trans. Br. Mycol. Soc. 44, 354–356.

Bérdy, J., 2012. Thoughts and facts about antibiotics: where we are now and where we are heading. J. Antibiot. 65, 385–395.

Birben, E., Sahiner, U.M., Sackesen, C., Erzurum, S., Kalayci, O., 2012. Oxidative stress and antioxidant defense. World Allergy Organ. J. 5 (1), 9–19.

Boucher, H.W., Ambrose, P.G., Chambers, H.F., Ebright, R.H., Jezek, A., Murray, B.E., Newland, J.G., Ostrowsky, B., Rex, J.H., 2017. White paper: developing antimicrobial drugs for resistant pathogens, narrow-spectrum indications, and unmet needs. J. Infect. Dis. 216, 228–236.

Brader, G., Compant, S., Mitter, B., Trognitz, F., Sessitsch, A., 2014. Metabolic potential of endophytic bacteria. Curr. Opin. Biotechnol. 27, 30–37.

Chandra, S., 2012. Endophytic fungi: novel sources of anticancer lead molecules. Appl. Microbiol. Biotechnol. 95 (1), 47–59.

Chen, S.L., Yu, H., Luo, H.M., Wu, Q., Li, C.F., Steinmetz, A., 2016. Conservation and sustainable use of medicinal plants: problems, progress, and prospects. Chin. Med. 11 (1), 1–10.

Chen, D., Zhang, P., Liu, T., Wang, X.-F., Li, Z.-X., Li, W., Wang, F.-L., 2018. Insecticidal activities of chloramphenicol derivatives isolated from a marine alga-derived endophytic fungus, *Acremonium vitellinum*, against the cotton bollworm, *Helicoverpa armigera* (Hübner) (Lepidoptera: Noctuidae). Molecules 23 (11), 2995.

Chhabra, S., Brazil, D., Morrissey, J., Burke, J.I., O'gara F, Dowling, D.N., 2013. Characterization of mineral phosphate solubilization traits from a barley rhizosphere soil functional metagenome. Microbiol. Open 2, 717–724.

Choudhary, B., Nagpure, A., Gupta, R.K., 2014. Fungal cell-wall lytic enzymes, antifungal metabolite(s) production, and characterization from *Streptomyces exfoliates* MT9 for controlling fruit-rotting fungi. J. Basic Microbiol. 54 (12), 1295–1309.

Compant, S., Clement, C., Sessitsch, A., 2010. Plant growth-promoting bacteria in the rhizo- and endosphere of plants: their role, colonization, mechanisms involved and prospects for utilization. Soil Biol. Biochem. 42, 669–678.

Cota, B.B., Rosa, L.H., Caligiorne, R.B., Teles-Rabello, A.L., Almeida-Alves, T.M., et al., 2008. Altenusin, a biphenyl isolated from the endophytic fungus *Alternaria* sp., inhibits trypanothione reductase from *Trypanosoma cruzi*. FEMS Microbiol. Lett. 285, 177–182.

de Carvalho, P., de Oliveira, S.E., Chagas-Paula, D., Luiz, J., Ikegaki, M., 2016. Importance and implications of the production of phenolic secondary metabolites by endophytic fungi: a mini-review. Mini-Rev. Med. Chem. 16, 259–271.

De Souza, J.J., Vieira, I.J., Rodrigues-Filho, E., Braz-Filho, R., 2011. Terpenoids from endophytic fungi. Mol. Ther. 16, 10604–10618.

Dhankhar, S., Dhankhar, S., Yadav, J.P., 2013. Investigations towards new antidiabetic drugs from fungal endophytes associated with Salvadora oleoides Decne. Med. Chem. 9, 624–632.

Elango, D., Manikandan, V., Jayanthi, P., Velmurugan, P., Balamuralikrishnan, B., Ravi, A. V., Shivakumar, M.S., 2020. Selection and characterization of extracellular enzyme production by an endophytic fungi *Aspergillus sojae* and its bio-efficacy analysis against cotton leaf worm, *Spodoptera litura*. Curr. Plant Biol., 100153.

Ezra, D., Hess, W.M., Strobel, G.A., 2004. New endophytic isolates of *Muscodor albus*, a volatile-antibiotic-producing fungus. Microbiology 150 (12), 4023–4031.

Faeth, S.H., Hammon, K.E., 1997. Fungal endophytes in oak trees. I. Long-term patterns of abundance and associations with leaf miners. Ecology 78, 810–819.

Finkel, T., Holbrook, N.J., 2000. Oxidants, oxidative stress and the biology of ageing. Nature 408 (6809), 239–247.

Fisher, P.J., Petrini, O., Lappin-Scott, H.M., 1992. The distribution of some fungal and bacterial endophytes in maize (Zea mays L.). New Phytol. 122, 299–305.

Gangadevi, V., Muthumary, J., 2007. Endophytic fungal diversity from young, mature and senescent leaves of *Ocimum basilicum* L. with special reference to Taxol production. Indian J. Sci. Technol. 1, 1–15.

Gao, Y., Liu, Q., Zang, P., Li, X., Ji, Q., et al., 2015. An endophytic bacterium isolated from *Panax ginseng* CA Meyer enhances growth, reduces morbidity and stimulates ginsenoside biosynthesis. Phytochem. Lett. 11, 132–138.

Gianinazzi, P.V., Dumas-Gaudot, E., Golotte, A., Tahiri-Alaoui, A., Gianinazzi, S., 1996. Cellular and molecular defence-related root responses to invasion by arbuscular mycorrhizal fungi. New Phytol. 133, 45–47.

Goutam, J., Singh, R., Vijayaraman, R.S., Meena, M., 2018. Endophytic fungi: carrier of potential antioxidants. In: Fungi and Their Role in Sustainable Development: Current Perspectives, pp. 539–551.

Gunatilaka, A.A.L., 2006. Natural products from plant-associated microorganisms: distribution, structural diversity, bioactivity, and implications of their occurrence. J. Nat. Prod. 69, 509–526.

Haas, H., 2014. Fungal siderophore metabolism with a focus on *Aspergillus fumigatus*. Nat. Prod. Rep. 31 (10), 1266–1276.

Halliwell, B., 2009. Free radicals and antioxidants: a personal view. Nutr. Rev. 52 (8), 253–265.

Harper, J.K., Arif, A.M., Ford, E.J., Strobel, G.A., Porco, J.A., Tomer, D.P., Grant, D.M., 2003. Pestacin: a 1,3-dihydro isobenzofuran from *Pestalotiopsis microspora* possessing antioxidant and antimycotic activities. Tetrahedron 59 (14), 2471–2476.

He, J.W., Chen, G.D., Gao, H., Yang, F., Li, X.X., Peng, T., Yao, X.S., 2012. Heptaketides with antiviral activity from three endolichenic fungal strains *Nigrospora* sp., *Alternaria* sp. and *Phialophora* sp. Fitoterapia 83 (6), 1087–1091.

Hemtasin, C., Kanokmedhakul, S., Kanokmedhakul, K., Hahnvajanawong, C., Soytong, K., Prabpai, S., Kongsaeree, P., 2011. Cytotoxic pentacyclic and tetracyclic aromatic sesquiterpenes from *Phomopsis archeri*. J. Nat. Prod. 74 (4), 609–613.

Horwitz, K.B., 1994. How do breast cancers become hormone resistant? J. Steroid Biochem. Mol. Biol. 49 (4–6), 295–302.

Huang, W.Y., Cai, Y.Z., Xing, J., Corke, H., Sun, M., 2007. A potential antioxidant resource: endophytic fungi from medicinal plants. Econ. Bot. 61, 14–30.

Ibrahim, S.R.M., Abdallah, H.M., Elkhayat, E.S., Al Musayeib, N.M., Asfour, H.Z., Zayed, M.F., Mohamed, G.A., 2017. Fusaripeptide a: new antifungal and anti-malarial cyclodepsipeptide from the endophytic fungus *fusarium* sp. J. Asian Nat. Prod. Res. 20 (1), 75–85.

Jagtap, N.S., Khadabadi, S.S., Ghorpade, D.S., Banarase, N.B., Naphade, S.S., 2009. Antimicrobial and antifungal activity of *Centella asiatica* (L.) urban, umbeliferae. Res. J. Pharm. Technol. 2, 328–330.

Kajula, M., Tejesvi, M.V., Kolehmainen, S., Makinen, A., Hokkanen, J., Mattila, S., 2010. The siderophore ferricrocin produced by specific foliar endophytic fungi *in vitro*. Fungal. Biol. 114, 248–254.

Kaur, H.P., Singh, B., Thakur, A., Kaur, A., Kaur, S., 2015. Studies on immunomodulatory effect of endophytic fungus *Alternaria alternata* on *Spodoptera litura*. J. Asia Pac. Entomol. 18 (1), 67–75.

Kavunga-Membo, H., Ilombe, G., Masumu, J., Matangila, J., Imponge, J., Manzambi, E., Muyembe, J.J., 2018. Molecular identification of plasmodium species in symptomatic children of Democratic Republic of Congo. Malar. J. 17 (1).

Knoth, J.L., 2014. Biological nitrogen fixation and biomass accumulation within poplar clones as a result of inoculations with diazotrophic endophyte consortia. New Phytol. 201 (2), 599–609.

Kumar, U., Aggarwal, C., Paul, S., Annapurna, K., 2014. Endophytes as biocontrol agents of plant pathogens and insects. Kavaka 41, 92–95.

Kumar, A., Singh, R., Yadav, A., Giri, D.D., Singh, P.K., Pandey, K.D., 2016. Isolation and characterization of bacterial endophytes of *Curcuma longa* L. 3Biotech 6, 60.

Kusari, S., Herweck, C., Spiteller, M., 2012. Chemical ecology of endophytic fungi: origins of secondary metabolites. Chem. Biol. 19, 792–798.

Lalitha, S., 2017. Plant growth-promoting microbes: a boon for sustainable agriculture. In: Dhanarajan, A. (Ed.), Sustainable Agriculture Towards Food Security. Springer Singapore, Singapore, pp. 125–158.

Lee, J., Lobkovsky, E., Pilam, N.B., Strobel, G.A., Clardy, J., 1995. Subglutinols A and B: immunosuppressive compounds from the endophytic fungus *Fusarium subglutinans*. J. Org. Chem. 60, 7076–7077.

Liu, X., Dong, M., Chen, X., Jiang, M., Lv, X., Zhou, J., 2008. Antimicrobial activity of an endophytic *Xylaria* sp. YX-28 and identification of its antimicrobial compound 7-amino-4-methylocoumarin. Appl. Microbiol. Biotechnol. 78, 241–247.

Liu, X., Bolla, K., Ashforth, E.J., Zhuo, Y., Gao, H., Huang, P., Stanley, S.A., Hung, D.T., Zhang, L., 2012. Systematic-guided bioprospecting for bioactive microbial natural products. Antonie van Leewuwenhoek 101, 55–66.

Liu, W.C., Gong, T., Zhu, P., 2016. Advances in exploring alternative Taxol sources. RSC Adv. 6 (54), 48800–48809.

Lockwood, J.L., 1992. Exploitation competition. In: Carroll, G.C., Wicklow, D. (Eds.), The Fungal Community—Its Organization and Role in the Ecosystem. Marcel Dekker, New York, pp. 243–263.

Lunze, K., Singh, T., Walter, M., Brendel, M.D., Leonhardt, S., 2013. Blood glucose control algorithms for type 1 diabetic patients: a methodological review. Biomed. Signal Proc. Cont. 8 (2), 107–119.

Malinowski, D.P., Belesky, D.P., 2000. Adaptations of endophyte-infected cool-season grasses to environmental stresses: mechanisms of drought and mineral stress tolerance. Crop Sci. 40, 923–940.

Mangunwardoyo, W., Suciatmih, S., Gandjar, I., 2012. Frequency of endophytic fungi isolated from *Dendrobium crumenatum* Sw. (pigeon orchid) and antimicrobial activity. Biodiversitas J. Biol. Div. 13 (1).

Mehmood, A., Hussain, A., Irshad, M., Hamayun, M., Iqbal, A., Khan, N., 2019. *In vitro* production of IAA by endophytic fungus *aspergillus awamori* and its growth promoting activities in Zea mays. Symbiosis 77, 225–235.

Mousa, W.K., Raizada, M.N., 2013. The diversity of anti-microbial secondary metabolites produced by fungal endophytes: an interdisciplinary perspective. Front. Microbiol. 4 (65), 1–18.

Newman, D.J., Cragg, G.M., Snader, K.M., 2000. The influence of natural products upon drug discovery. Nat. Prod. Rep. 17, 215–234.

Noble, R., 2016. Anti-cancer alkaloids of *Vinca rosea*. Pharmacol. Orient. Plants 7, 61–78.

Nose, K., 2000. Role of reactive oxygen species in regulation of physiological functions. Biol. Pharma. Bul. 23, 897–903.

Oteino, N., Lally, R.D., Kiwanuka, S., Lloyd, A., Ryan, D., Germaine, K.J., 2015. Plant growth promotion induced by phosphate solubilizing endophytic *Pseudomonas* isolates. Front. Microbiol. 6, 745.

Parthasarathy, R., Chandrika, M., Rao, H.C.Y., Kamalraj, S., Jayabaskaran, C., Pugazhendhi, A., 2020. Molecular profiling of marine endophytic fungi from green algae: assessment of antibacterial and anticancer activities. Process Biochem. 96, 11–20.

Pathak, P., Rai, V.K., Can, H., Singh, S.K., Kumar, D., Bhardwaj, N., Roychowdhury, R., de Azevedo, L.C.B., Kaushalendra, Verma, H., Kumar, A., 2022. Plant-endophyte interaction during biotic stress management. Plan. Theory 11, 2203. https://doi.org/10.3390/plants11172203.

Pava-Ripoll, M., Angelini, C., Fang, W., Wang, S., Posada, F.J., St Leger, R., 2011. The rhizosphere-competent entomopathogen Metarhiziumanisopliae expresses a specific subset of genes in plant root exudate. Microbiology 157, 47–55.

Rai, M.K., Varma, A., Pandey, A.K., 2002. Antifungal potential of *Spilanthes calva* after inoculation of *Piriformospora indica*. Mycoses 47, 479–481.

Rajamanikyam, M., Vadlapudi, V., Amanchy, R., Upadhyayula, S.M., 2017. Endophytic fungi as novel resources of natural therapeutics. Braz. Arch. Biol. Technol. 60 (0).

Rajkumar, M., 2010. Potential of siderophore-producing bacteria for improving heavy metal phytoextraction. Trends Biotechnol. 28 (3), 142–149.

Rao, H.C.Y., Baker, S., Rakshith, D., Satish, S., 2015a. Molecular profiling and antimicrobial potential of endophytic *Gliomastix polychrome* CLB32 inhabiting *Combretum latifolium* blume. Mycology 6, 176181.

Rao, H.C.Y., Santosh, P., Rakshith, D., Satish, S., 2015b. Molecular characterization of an endophytic *Phomopsis liquidambaris* CBR-15 from *Cryptolepis buchanani* Roem and impact of culture media on biosynthesis of antimicrobial metabolites. 3Biotech 5, 165–173.

Rao, H.C.Y., Satish, S., 2015. Genomic and chromatographic approach for the discovery of polyketide antimicrobial metabolites from an endophytic *Phomopsis liquidambaris* CBR-18. Front. Life. Sci. 8 (2), 200–207.

Rao, H.C.Y., Satish, S., 2016. Intra-specific differentiation of fungal endosymbiont *Alternaria longissima* CLB44 using RNA secondary structure analysis and their anti-infective potential. Naturwissenschaften 103 (7–8), 69.

Renshaw, J.C., Robson, G.D., Trinci, A.P.J., Wiebe, M.G., Livens, F.R., Collison, D., Taylor, R.J., 2002. Fungal siderophores: structures, functions and applications. Mycol. Res. 106 (10), 1123–1142.

Resquín-Romero, G., Garrido-Jurado, I., Delso, C., Ríos-Moreno, A., Quesada-Moraga, E., 2016. Transient endophytic colonizations of plants improve the outcome of foliar applications of mycoinsecticides against chewing insects. J. Invertebr. Pathol. 136, 23–31.

Rodriguez, R., Redman, R., 2008. More than 400 million years of evolution and some plants still can't make it on their own: plant stress tolerance via fungal symbiosis. J. Exp. Bot. 59 (5), 1109–1114.

Roy, B.G., 2017. Potential of small-molecule fungal metabolites in antiviral chemotherapy. Antiviral Chem. Chem. 25 (2), 20–52.

Russo, M.L., Scorsetti, A.C., Vianna, M.F., Cabello, M., Ferreri, N., Pelizza, S., 2019. Endophytic effects of *Beauveria bassiana* on corn (Zea mays) and its herbivore, Rachiplusia nu (Lepidoptera: Noctuidae). Insects 10 (4), 110.

Saikkonen, K., Wali, P., Helander, M., Faeth, S.H., 2004. Evolution of endophyte-plant symbioses. Trends Plant Sci. 9 (6), 275–280.

Salerno, J.M., Najdowski, C.J., Stevenson, M.C., Wiley, T.R., Bottoms, B.L., Vaca Jr., R., Pimentel, P.S., 2010. Psychological mechanisms underlying support for juvenile sex offender registry laws: prototypes, moral outrage, and perceived threat. Behav. Sci. Law 28 (1), 58–83.

Schulz, B., Boyle, C., Draeger, S., Rommert, A.K., Krohn, K., 2002. Endophytic fungi: a source of novel biologically active secondary metabolites. Mycol. Res. 106 (9), 996–1004.

Selim, K., Elkhateeb, W., Tawila, A., El-Beih, A., Abdel-Rahman, T., El-Diwany, A., Ahmed, E., 2018. Antiviral and antioxidant potential of fungal endophytes of Egyptian medicinal plants. Fermentation 4 (3), 49.

Shahidi, F., Ambigaipalan, P., 2015. Phenolics and polyphenolics in foods, beverages and spices: antioxidant activity and health effects—a review. J. Funct. Foods 18, 820–897.

Sheng, X.F., Xia, J.J., Jiang, C.Y., He, L.Y., Qian, M., 2008. Characterization of heavy metal-resistant endophytic bacteria from rape (*Brassica napus*) roots and their potential in promoting the growth and lead accumulation of rape. Environ. Pollut. 156, 1164–1170.

Singh, A., Pandey, J., 2012. Metal contamination and health risk from consumption of organically grown vegetables influenced by atmospheric deposition in a seasonally dry tropical region of India. Bull. Environ. Contamin. Toxicol. 89, 384–389.

Soldan, R., Mapelli, F., Crotti, E., Schnell, S., Daffonchio, D., Marasco, R., Fusi, M., Borin, S., Cardinale, M., 2019. Bacterial endophytes of mangrove propagules elicit early establishment of the natural host and promote growth of cereal crops under salt stress. Microbiol. Res. 223, 33–43.

Stierle, G., Strobel, S., Stierle, D., 1993. Taxol and taxane production by *Taxomyces andreanae*, an endophytic fungus of Pacific yew. Science 260 (5105), 214–216.

Strobel, G., Daisy, B., 2003. Bioprospecting for microbial endophytes and their natural products. Microbiol. Mol. Biol. Rev. 67, 491–502.

Strobel, G., Hess, W., Ford, E., 1996. Taxol from fungal endophytes and the issue of biodiversity. J. Ind. Microbiol. Biotechnol. 17, 417–423.

Strobel, G., Daisy, B., Castillo, U., Harper, J., 2004. Natural products from endophytic microorganisms. J. Nat. Prod. 67, 257–268.

Sturz, A.V., Christie, B.R., Nowak, J., 2000. Bacterial endophytes: potential role in developing sustainable systems of crop production. Crit. Rev. Plant Sci. 19 (1), 1–30.

Sunitha, V., Nirmala, D., Srinivas, C., 2013. Extracellular enzymatic activity of endophytic fungal strains isolated from medicinal plants. World J. Agric. Res. 9 (1), 01–09.

Taghavi, S.N., Weyens, J., Vangronsveld, J., Lelie, D., Pirttilä, A.M., Frank, A.C., 2011. Improved phytoremediation of organic contaminants through engineering of bacterial endophytes of trees. In: Pirttila, A., Frank, A. (Eds.), Endophytes of Forest Trees. Forestry Sciences. vol. 80. Springer, Dordrecht, pp. 205–216.

Talontsi, F.M., Lamshöft, M., Douanla-Meli, C., Kouam, S.F., Spiteller, M., 2014. Antiplasmodial and cytotoxic dibenzofurans from *Preussia* sp. harbored in *Enantia chlorantha* Oliv. Fitoterapia 93, 233–238.

Tan, R.X., Zou, W.X., 2001. Endophytes: a rich source of functional metabolites (1987–2000). Nat. Prod. Rep. 18 (4), 448–459.

Tripathi, V.C., Satish, S., Horam, S., Raj, S., Arockiaraj, J., Pasupuleti, M., et al., 2018. Natural products from polar organisms: structural diversity, bioactivities and potential pharmaceutical applications. Polar Sci. 18, 147–166.

Ujam, N.T., Eze, P.M., Chukwunwejim, C.R., Okoye, F.B., Esimone, C.O., 2019. Antimicrobial and immunomodulatory activities of secondary metabolites of an endophytic fungus isolated from *A. conyzoides*. Curr. Life Sci. 5 (1), 19–27.

Uzor, P.F., Osadebe, P.O., Nwodo, N.J., 2017. Antidiabetic activity of extract and compounds from an endophytic fungus *Nigrospora oryzae*. Drug Res. 67 (05), 308–311.

Veeresham, C., 2012. Natural products derived from plants as a source of drugs. J. Adv. Pharm. Technol. Res. 3, 200–201.

Verma, V.C., Singh, S.K., Kharwar, R.N., 2012. Histological investigation of fungal endophytes in healthy tissues of *Azadirachta indica* A. Juss. Kasetsart J. (Nat. Sci.) 46, 229–237.

Verma, P., Yadav, A.N., Kumar, V., Singh, D.P., Saxena, A.K., 2017. Beneficial plant-microbes interactions: biodiversity of microbes from diverse extreme environments and its impact for crop improvement. In: Plant-Microbe Interactions in Agro-Ecological Perspectives. Springer, Singapore, pp. 543–580.

Verma, H., Kumar, D., Kumar, V., Kumari, M., Singh, S.K., Sharma, V.K., Droby, S., Santoyo, G., White, J.F., Kumar, A., 2021. The potential application of endophytes in management of stress from drought and salinity in crop plants. Microorganisms 9, 1729. https://doi.org/10.3390/microorganisms9081729.

Walsh, V., Goodman, J., 1999. Cancer chemotherapy, biodiversity, public and private property: the case of the anti-cancer drug taxol. Soc. Sci. Med. 49 (9), 1215–1225.

Wani, Z.A., Ashraf, N., Mohiuddin, T., Riyaz-Ul-Hassan, S., 2015. Plant-endophyte symbiosis, an ecological perspective. Appl. Microbiol. Biotechnol. 99, 2955–2965.

Wiyakrutta, S., Sriubolmas, N., Panphut, W., Thongon, N., Danwisetkanjana, K., et al., 2004. Endophytic fungi with anti-microbial, anti-cancer and anti-malarial activities isolated from Thai medicinal plants. World J. Microbiol. Biotechnol. 20, 265–272.

Xie, S., Zhou, J., 2017. Harnessing plant biodiversity for the discovery of novel anticancer drugs targeting microtubules. Front. Plant Sci. 8, 720.

Yadav, A., Yadav, K., 2017. Exploring the potential of endophytes in agriculture: a mini review. Adv. Plants Agric. Res. 6 (4), 00221.

Ye, Y., Xiao, Y., Ma, L., Li, H., Xie, Z., Wang, M., Liu, J., 2013. Flavipin in *Chaetomium globosum* CDW7, an endophytic fungus from *Ginkgo biloba*, contributes to antioxidant activity. Appl. Microbiol. Biotechnol. 97 (16), 7131–7139.

Yu, H., Zhang, L., Li, L., Zheng, C., Guo, L., Li, W., Qin, L., 2010. Recent developments and future prospects of antimicrobial metabolites produced by endophytes. Microbiol. Res. 165 (6), 437–449.

Zeng, Y.J., Yang, H.R., Wu, X.L., Peng, F., Huang, Z., Pu, L., Lou, W.Y., 2019. Structure and immunomodulatory activity of polysaccharides from *Fusarium solani* DO7 by solid-state fermentation. Int. J. Biol. Macromol. 137, 568–575.

Zhao, J., Shan, T., Mou, Y., Zhou, L., 2011. Plant-derived bioactive compounds produced by endophytic fungi. Mini-Rev. Med. Chem. 11 (2), 159–168.

Zhu, L., Chen, L., 2019. Progress in research on paclitaxel and tumor immunotherapy. Cell Mol. Biol. Lett. 24 (1), 1–11.

Zhu, H., Swierstra, J., Wu, C., Girard, G., Choi, Y.H., van Wamel, W., Sandiford, S.K., van Wezel, G.P., 2014. Eliciting antibiotics active against the ESKAPE pathogens in a collection of actinomycetes isolated from mountain soils. Microbiology 160, 1714–1725.

Further reading

Amna, T., Puri, S.C., Verma, V., Sharma, J.P., Khajuria, R.K., Musarrat, J., Spiteller, M., Qazi, G.N., 2006. Bioreactor studies on the endophytic fungus *Entrophospora infrequens* for the production of an anti-cancer alkaloid camptothecin. Can. J. Microbiol. 52, 189–196.

Arnold, A.E., Herre, E.A., 2003. Canopy cover and leaf age affect colonization by tropical fungal endophytes: ecological pattern and process in *Theobroma cacao* (Malvaceae). Mycologia 95, 388–398.

Batta, Y.A., 2013. Efficacy of endophytic and applied *Metarhizium anisopliae* (Metch.) Sorokin (Ascomycota: Hypocreales) against larvae of *Plutella xylostella* L. (Yponomeutidae: Lepidoptera) infesting *Brassica napus* plants. J. Crop Prot. 44, 128–134.

Clark, R.B., Zeto, S.K., 2000. Mineral acquisition by arbuscular mycorrhizal plants. J. Plant Nutr. 23 (7), 867–902.

Deshmukh, S.K., Verekar, S.A., Bhave, S.V., 2015. Endophytic fungi: a reservoir of antibacterials. Front. Microbiol. 5, 1–43.

Fei, G.E., Yao, T., Qian, G., Qi-sen, M.A., Liangjun, Y., 2016. Comparison on antimicrobial activity of 40 endophytic fungi strains from Ginkgo biloba and preliminary analysis on ingredients with antimicrobial activity. Chin. Tradit. Herb. Drug. 47, 1554–1559.

Kour, D., Rana, K.L., Yadav, A.N., Yadav, N., Kumar, M., Kumar, V., Saxena, A.K., 2019. Microbial biofertilizers: bioresources and eco-friendly technologies for agricultural and environmental sustainability. Biocatal. Agric. Biotechnol., 101487.

Kuldau, G.C.B., 2008. Clavicipitaceous endophytes: their ability to enhance resistance of grasses to multiple stresses. Biol. Control 46, 57–71.

Kumar, A., 2022. Microbial Biocontrol: Sustainable Agriculture and Phytopathogen Management. vol. 1 Springer Nature Chem, Cham, Switzerland, pp. 1–369.

Mukundi, M.J., 2015. Antidiabetic effects of aqueous leaf extracts of *Acacia nilotica* in alloxan induced diabetic mice. J. Diabetes Metabol. 6 (7).

Patridge, E., Gareiss, P., Kinch, M.S., Hoyer, D., 2016. An analysis of FDA-approved drugs: natural products and their derivatives. Drug Discov. Today 21, 204–207.

Phongpaichit, S., Nikom, J., Rungjindamai, N., Sakayaroj, J., Hutadilok-Towatana, N., Rukachaisirikul, V., Kirtikara, K., 2007. Biological activities of extracts from endophytic fungi isolated from Garcinia plants. FEMS Immunol. Med. Microbiol. 51 (3), 517–525.

Rao, H.C.Y., Rakshith, D., Harini, B.P., Satish, S., 2017a. Antimicrobial profiling and molecular identification of *Alternaria arborescens* CLB12, a myco-endosymbiont inhabiting *Combretum latifolium* blume. J. Biol. Act. Prod. Nat. 7 (1), 1–9.

Rao, H.C.Y., Rakshith, D., Harini, B.P., Gurudatt, D.M., Satish, S., 2017b. Chemogenomics driven discovery of endogenous polyketide anti-infective compounds from endosymbiotic *Emericella variecolor* CLB38 and their RNA secondary structure analysis. PLoS One 12 (2), e0172848.

Schulz, B., Boyle, C., 2005. The endophytic continuum. Review. Mycol. Res. 109 (6), 661–686.

Stierle, A., Strobel, G., Stierle, D., Grothaus, P., Bignami, G., 1995. The search for a taxol-producing microorganism among the endophytic fungi of the Pacific yew, *T. brevifolia*. J. Nat. Prod. 58, 1315–1324.

Tejesvi, M.V., Pirttilä, A.M., 2011. Potential of tree endophytes as sources for new drug compounds. In: Pirttilä, A.M., Frank, A.C. (Eds.), Endophytes of Forest Trees: Biology and Applications. Springer, New York, pp. 295–312.

Varga, T., Hixson, K.K., Ahkami, A.H., Sher, A.W., Barnes, M.E., Chu, R.K., Doty, S.L., 2020. Endophyte-promoted phosphorus solubilization in populus. Front. Plant Sci., 1585.

Yashavantha Rao, H.C., Satish, S., 2015. Genomic and chromatographic approach for the discovery of polyketide antimicrobial metabolites from an endophytic *Phomopsis liquidambaris* CBR-18. Front. Life Sci. 8 (2), 200–207.

Index

Note: Page numbers followed by *f* indicate figures and *t* indicate tables.

K

L

M

R

S

T

9780323990431